# Science and Philosophy in the Indian Buddhist Classics

## COMPENDIUM COMPILATION COMMITTEE

*Chair*
Tromthok Rinpoche, Abbot of Namgyal Monastery

*General Series Editor*
Thupten Jinpa, PhD

*Advisory Members*
Geshe Yangteng Rinpoche, Sera Me Monastic College
Geshe Thupten Palsang, Drepung Loseling College
Gelong Thupten Yarphel, Namgyal Monastery

*Editors*
Geshe Jangchup Sangye, Ganden Shartse College
Geshe Ngawang Sangye, Drepung Loseling College
Geshe Chisa Drungchen Rinpoche, Ganden Jangtse College
Geshe Lobsang Khechok, Drepung Gomang College

# SCIENCE AND PHILOSOPHY

IN THE

# INDIAN BUDDHIST CLASSICS

VOLUME 1

## *The Physical World*

CONCEIVED AND INTRODUCED BY

His Holiness the Dalai Lama

Developed by the
Compendium Compilation Committee

Translated by Ian James Coghlan

Edited with contextual essays
by Thupten Jinpa

Wisdom Publications
199 Elm Street
Somerville MA 02144 USA
wisdompubs.org



A translation of *Nang pa'i tshan rig dang lta grub kun btus, vol. 1*. Dharamsala, India: Ganden Phodrang Trust (Office of His Holiness the Dalai Lama), 2014.



*Library of Congress Cataloging-in-Publication Data* is available.

ISBN 978-1-61429-472-6    ebook ISBN 978-1-61429-492-4

21 20 19 18 17
5 4 3 2 1

Cover and interior design by Gopa & Ted2, Inc. Cover art by Getty Images.
Set in Diacritical Garamond Premier Pro 11/14.

Wisdom Publications' books are printed on acid-free paper and meet the guidelines for permanence and durability of the Production Guidelines for Book Longevity of the Council on Library Resources.

♻ This book was produced with environmental mindfulness.
For more information, please visit wisdompubs.org/wisdom-environment.

Printed in the United States of America.

# Contents

# Preface

## General Editor's Note

This is volume 1 of a four-volume series that brings together classical Buddhist scientific and philosophical explorations on the nature of reality within a framework accessible to the contemporary reader. This ambitious series was conceived by His Holiness the Dalai Lama himself and compiled under his visionary supervision.

As the Dalai Lama explains in his introduction, the creation of this compilation is grounded in an understanding that it is possible to distinguish three domains in the contents of the great Buddhist treatises of classical India. First is the *scientific dimension*, which relates to empirical claims about not only the outer physical world but also the inner world of our experience, including the underlying principles that govern their functions and relationships. Second is the *philosophical dimension*, primarily statements presenting the ultimate truth or truths about reality. Finally, there is what might be called the *religious dimension*, which relates to Buddhist practice and the path to enlightenment. The Dalai Lama believes that, as the exchange of knowledge among the world's cultures and languages becomes increasingly common, the insights contained in the works of the great Indian Buddhist thinkers, especially the more scientific and philosophical aspects, should be made accessible to contemporary readers.

The first two volumes in the series cover the scientific domain: volume 1 presents the physical world and volume 2 presents the mind sciences. Volumes 3 and 4 will focus specifically on the philosophical dimension of the Buddhist heritage. The remarkable accomplishment of the first two volumes is the gathering together in one place of insights of scientific interest from the great Indian Buddhist thinkers. Thanks to the Dalai Lama's vision, for the first time the contemporary reader has the opportunity to directly engage with ideas of these key Buddhist thinkers from a scientific perspective, read their own words, and follow the line of their arguments.

In their original context, the presentations compiled in this series are embedded within a larger framework that includes philosophical reflections as well as the soteriological goal of awakening. The extraction and organization of these views within the framework of scientific inquiry is in itself a revolutionary achievement in the history of Buddhist thought. That all the classical sources gathered in these two volumes on science are drawn from the Tengyur—the Tibetan translations of the Indian Buddhist treatises—also makes this compilation an important gift to the world from the Tibetan tradition.

The Buddhist science that arose in India is as ancient as Greek science, and its methods and insights speak to us from a long bygone era. For many it is a reality that now lives only in the pages of ancient treatises, whose sounds and smells have long dissipated, and whose logic, art, and wisdom remain obscure. But equally it may be said that such impressions are misleading, for Buddhist science and its insights have relevance to us in our time. And this is not just within traditional Eastern cultures where Buddhism remains a living tradition. Buddhist science has two components: the external science of the material universe plus the internal science on the nature of the mind itself. Each relies on the other, but it is internal Buddhist science that has achieved unique and profound insights into the nature of mind, and such insights are specifically relevant to the modern world.

Thanks to influential works like A. L. Basham's *The Wonder That Was India*, consciousness of classical India as the cradle of a highly advanced civilization rich in art, culture, and religion has long been mainstream. Scholarly research during last few decades has demonstrated the sophistication of India's rich philosophical heritage as well. In addition, archeological excavations of the cities of Mohenjo Daro and Harappa of the Indus Valley civilization have brought to light the sophistication of ancient India's technological knowledge and skills. Now these two volumes on Buddhist science have the potential to fill in one important gap in our knowledge of classical India, namely the achievements in scientific thinking that occurred within the Indian Buddhist traditions.

Perhaps the most exciting feature of these two volumes on science in Indian Buddhist classics is their contribution to the history of ideas. The current discipline of the history of ideas, especially the history of science, is undeniably Eurocentric, with little attention paid to civilizations outside the Western world. This volume clearly documents a sophisticated tradition of scientific thinking in India, with investigations of atomic theory, the relativity of time, the concept of multiple world systems, embryonic development, the function of brain, and microorganisms within the human body. As a resource for the history of ideas, these two volumes have the potential to bring focused attention to the intellectual achievements of great thinkers like Nāgārjuna, Asaṅga, Vasubandhu, Dignāga, and Dharmakīrti. These volumes also open up the possibility to engage in a more comprehensive cross-cultural comparison between the scientific thinking of classical India and the West, thus offering a basis for developing a truly inclusive global narrative of the history of ideas.

To assist the contemporary reader, the six parts of each volume are all introduced with brief essays, from myself for volume 1 and from my colleague John Dunne for volume 2. These essays aim to (1) provide a larger context to the topics in each section, (2) offer helpful signposts to the contemporary reader so that he or she can ably navigate terrain covered, and (3) draw attention to possible parallels in Western scientific and philosophical thinking.

Personally, it has been both a profound joy and an honor to be part of this ambitious project. First and foremost, I would like to offer my deepest gratitude to His Holiness the Dalai Lama for his vision as well as his leadership of this valuable initiative. As stated in his introduction, in creating this compilation, His Holiness is sharing with the world the wisdom and insights of classical India that have been such an important resource for the Tibetan people for over a millennium. Never have I met anyone who cares so much about the world and thinks so much about the welfare of humanity with such constancy.

I thank the Tibetan editors who have worked so diligently, over a period of several years, in creating this compilation and patiently putting up with the frequent critical feedback and editorial changes I offered as the general editor. I would like to thank Ian Coghlan for taking on the most challenging task of translating this important volume into English, and patiently incorporating the extensive suggestions offered for fine-tuning the language of the English text. I would like to thank our editor at Wisdom Publications, David Kittelstrom, and his assistant Mary Petrusewicz for their careful copyediting of the English translation for publication.

Last but not least, I would like to offer my deep gratitude to the Ing

Foundation for its generous patronage of the Institute of Tibetan Classics, and to the Scully Peretsman Foundation for its support of my own work, which made it possible for me to dedicate so much time to the creation of the Tibetan volumes, help edit and refine this English translation, and prepare the contextual essays.

Through the publication of this English volume may the wisdom of the great Buddhist masters of classical India touch people everywhere, across boundaries of geography, language, culture, and religion.

*Thupten Jinpa*

## Translator's Note

The main sources of Buddhist material science are the early sūtras, Abhidharma works, and the treatises of Nālandā University scholars, treatises often composed in Sanskrit and tested through exacting academic debate. Centuries later these works were translated into Tibetan by Tibetan lotsāwas, many of whom travelled to India, enduring the heat and dangers of the Indian plains. Contemporary translators of Tibetan works may often draw therefore on two source languages, Sanskrit and Tibetan, each acting as an authoritative normative reference. But in terms of a living tradition, only the Tibetan system survives, and it alone remains a genuine living resource, one that can be consulted to unravel the many difficult terms and concepts encountered.

Having two source languages aides in identifying the intention of the early authors. The extant Tibetan tradition provides insight into how early Tibetan translators dealt with issues such as ambiguity and etymology involved in translating Sanskrit works, and how they located viable indigenous Tibetan terms that spoke to a Tibetan audience. There are times also when the terms remain obscure in both languages, and the translator must persevere until a word or phrase becomes clear and some light is cast upon the meaning.

In brief, translation implies compromise, for in general a translator cannot reproduce every element of the original work. Translation also implies interpretation, for the translator must identify and present what he or she considers the central elements of any passage, with the understanding that other elements may become secondary, lose importance, or even be discarded. Balancing the demands of accuracy and readability, the translator's job seems to be more an art than a science. Since the reader who lacks access to the original work in Tibetan must rely on the translator, it is my hope that the energy expended by the reader in tackling this work will be rewarded.

I wish to thank His Holiness the Dalai Lama, whose vision has inspired this project, whose guidance has ensured its completion, and whose admonition to maintain a good motivation remains a constant goal. I wish also to thank the Compendium Compilation Committee, and in particular the committee chair Tromthok Rinpoche, abbot of Namgyal Monastery, for his valued and generous assistance. I had the privilege of first meeting Rinpoche in Sera in 1980, when he, at the request of Geshe Ngawang Dhargyey, kindly agreed to teach me Tibetan grammar when I first began my debate studies.

In terms of the translation project itself, I wish to thank Thupten Jinpa, the general series editor, for his suggestion to join the project and for his continued encouragement and assistance. I also express my gratitude to the four Tibetan edition editors with whom I had the privilege of meeting frequently to consult on issues concerning this volume: Geshe Jangchup Sangye of Ganden Shartse, Geshe Ngawang Sangye of Drepung Loseling, Geshe Chisa Drungchen Rinpoche of Ganden Jangtse, and Geshe Lobsang Khechok of Drepung Gomang. They proved to be a delight to work with, a wealth of knowledge and information regarding the Tibetan and Sanskrit sources, and insightful commentators on difficult and obscure points. Also I wish to thank the committee advisory members—Geshe Yangteng Rinpoche of Sera Me, Geshe Thupten Palsang of Drepung Loseling, and Gelong Thupten Yarphel of Namgyal Monastery—for their valued and generous assistance.

The editing of the English translation was supervised by Thupten Jinpa, who undertook a thorough edit of my initial draft translation, and I am indebted to his counsel and advice on a wide range of issues related to rendering this work in English. So too I wish to thank senior editor David Kittelstrom for his assistance in navigating the complexities of readability, grammar, and syntax, and copyeditor Mary Petrusewicz for her valued input in many areas.

Also I wish to thank Dechen Rochard, the translator of volume 2 in this series, for her input on the many and varied issues related to translation, to Tenzin Tsepak, translator of His Holiness, for our conversations on terminology and related issues, and especially to Jampel Lhundrup for his valued assistance on the ground in Dharamsala for the duration of the project.

May the translation of this work provide ready access to these materials for the Western reader.

*Ian Coghlan*

# Introduction

## My Encounter with Science[1]

In my childhood I had a keen interest in playing with mechanical toys. After reaching India in 1959, I developed a strong wish to engage with scientists to help expand my own knowledge of science as well as to explore the question of the relationship between science and religion. The main reason for my confidence in engaging with scientists rested in the Buddha's following statement:

> Monks and scholars, just as you test gold
> by burning, cutting, and polishing it,
> so too well examine my speech.
> Do not accept it merely out of respect.

The Buddha advises his disciples to carefully analyze when they engage with the meaning of his words, just as a goldsmith tests the purity of gold through burning, cutting, and rubbing. Only after we have gained conviction through such inquiry, the Buddha explains, is it appropriate to accept the validity of his words. It is not appropriate to believe something simply because one's teacher has taught it. Even with regard to what he himself taught, the Buddha says, we must test its validity for ourselves through experimentation and the use of reason. The testimony of scriptures alone is not sufficient. This profound advice demonstrates the centrality of sound reasoning when it comes to exploring the question of reality.

In Buddhism in general, and for the Nālandā masters of classical India in particular, when it comes to examining the nature of reality, the evidence of direct perception is accorded greater authority than both reason-based inference and scripture. For if one takes a scripture to be an authority in describing the nature of reality, then that scripture too must first be verified as authoritative by relying on another scriptural testimony, which in turn must be verified by another scripture, and so on, leading to an infinite regress. Furthermore, a scripture-based approach can offer no proof or rebuttals against alternative standpoints proposed by opponents who do not accept the validity of that scripture. Even among scriptures, some can be accepted as literal while some cannot, giving us no reliable standpoints on the nature of reality. It is said that to cite scripture as an authority in the context of inquiring into the nature of reality indicates a misguided intelligence. To do so precludes us from the ranks of those who uphold reason.

In science we find a similar approach. Scientists take experimentation and the logic of mathematics as arbiters of truth when it comes to evaluating the conclusions of their research; they do not ground validity in the authority of some other person. This method of critical inquiry, one that draws inferences about the unobservable, such as atomic particles, based on observed facts that are evident to our direct perception, is shared by both Buddhism and contemporary science. Once I saw this shared commitment, it greatly increased my confidence in engaging with modern scientists.

With instruments like microscopes and telescopes and with mathematical calculations, scientists have been able to carefully analyze phenomena from atomic particles to distant planets. What can be observed by the senses is enhanced by means of these instruments, allowing scientists to gain new inferences about various facts. Whatever hypothesis science puts forth must be verified by observation-based experiments, and similarly Buddhism asserts that the evidence of direct perception must ultimately underpin critical inquiry. Thus with respect to the way conclusions are drawn from evidence and reasoning, Buddhism and science share an important similarity. In Buddhism, however, empirical observation is not confined to the five senses alone; it has a wider meaning, since it includes observations derived from meditation. This meditation-based empirical observation grounded in study and contemplation is also recognized as part of the means of investigating reality, akin to the role scientific method plays in scientific inquiry.

Since my first visit to the West, a trip to Europe in 1973, I have had the opportunity to engage in conversations with great scientists, including the noted twentieth-century philosopher of science Sir Karl Popper, the quantum physicist Carl Friedrich von Weizsäcker, who was the brother of the last

West German president and also a colleague of the famed quantum physicists Werner Heisenberg and David Bohm.[2] Over many years I have had the chance to engage in dialogues with scientists on a range of topics, such as cosmology, neurobiology, evolution, and physics, especially subatomic-particle physics. This latter discipline of particle physics shares methods strikingly similar to those found in Buddhism, such as the Mind Only school's critique of the external material world that reveals that nothing can be found when matter is deconstructed into its constitutive elements, and similarly the statements found in the Middle Way school treatises that nothing can be found when one searches for the real referents behind our concepts and their associated terms. I have also on numerous occasions had dialogues with scientists from the fields of psychology and the science of mind, sharing the perspectives of the Indian tradition in general, which contains techniques of cultivating tranquility and insight, and the Buddhist sources in particular, with its detailed presentations on mind science.

Today we live in an age when the power of science is so pervasive that no culture or society can escape its impact. In a way, there was no choice but for me to learn about science and embrace it with a sense of urgency. I also saw the potential for an emerging discourse on the science of mind. Recognizing this, and wishing to explore how science and its fruits can become a constructive force in the world and serve the basic human drive for happiness, I have engaged in dialogue with scientists for many years. My sincere hope is that these dialogues across cultures and disciplines will inspire new ways to promote both physical and mental well-being and thus serve humanity through a unique interface of contemporary science and mind science. Thus, when I engage in conversations with scientists, such as in the ongoing Mind and Life dialogues, I have the following two aims.[3]

The first concerns expanding the scope of science. Not only is the breadth of the world's knowledge vast, advances are being made year by year that expand human knowledge. Science, however, right from its inception and especially once it began to develop quickly, has been concerned primarily with the world of matter. Unsurprisingly, then, contemporary science focuses on the physical world. Because of this, not much inquiry in science has been made into the nature of the person—the inquirer—as well as into how memory arises, the nature of happiness and suffering, and the workings of emotion. Science's advances in the domain of the physical world have been truly impressive. From the perspective of human experience, however, there are dimensions of reality that undoubtedly lie outside the current domain of scientific knowledge. It is of vital importance that the science of mind takes its place among the current fields of human investigation. The brain-based explanations in contemporary science about the different classes of sensory experience will be enriched by incorporating a more expanded and detailed understanding of the mind. So my first goal in my dialogues with scientists is to help make the current field of psychology or mind science more complete.

Not only do Buddhism and science have much to learn from each other, but there is also a great need for a way of knowing that encompasses both body and mind. For as human beings we experience happiness and suffering not only physically but mentally as well. If our goal is to promote human happiness, we have a real opportunity to pursue a new kind of science that explores methods to enhance happiness through the interface of contemporary science with contemplative mind science. It is my belief that, while acknowledging the great contribution that science has made in advancing human knowledge, our ultimate aim should always be to help create a comprehensive approach to understanding our world.

This takes us to the second goal behind my dialogues with scientists—how best to ensure that science serves humanity. As humans, we face two kinds of problems, those that are essentially our own creation and those owing to natural forces. Since the first kind is created by we humans ourselves, its solution must also be within our human capacity. In contemporary human society, we do not lack knowledge, but the persistence of problems that are our own creation clearly demonstrates that we lack effective solutions to these problems. The obstacle to solving these problems is the presence in the human mind of excessive self-centeredness, attachment, anger, greed, discrimination, envy, competitiveness, and so on. Such problems also stem from deficits in our consideration of others, compassion, tolerance, conscientiousness, insight, and so on. Since many of the world's great religions carry extensive teachings on these values, I have no doubt that such teachings can serve humanity through helping to overcome the human-made problems we face.

The primary purpose of science is also to benefit and serve humanity. Discoveries in science have brought concrete benefits in medicine, the environment, commerce, travel, working conditions, and human relationships. There is no doubt that science has brought great benefits when it comes to alleviating suffering at the physical level. However, since mental suffering is connected with our perception and attitude, material progress is not enough. Even in countries where science has flourished greatly, problems like theft and violent disputes persist. As long as the mind remains filled with greed, anger, conceit, envy, and so on, no matter however perfect our material facilities, a life of genuine happiness is not possible. In contrast, if we possess qualities like contentment and loving-kindness, we can enjoy a life of happiness even without great material facilities. Happiness in life is primarily a function of the state of the mind.

If contemporary society were to pay more attention to the science of mind, and more importantly, if science were to engage more with societal concerns, including fundamental human values, I believe that this could lead to great advancement and novel outcomes. Although science has not concerned itself with the enhancement of ethics and the cultivation of basic human values such as kindness, since science has emerged as a means to serve humanity, it should never be completely divorced from the values that are of great importance to the flourishing of human society.

In Indian philosophical traditions in general, and in Buddhism in particular, one finds many techniques for training the mind, such as the cultivation of tranquility (*śamatha*) and insight (*vipaśyanā*). These definitely have the potential to make important contributions to contemporary psychology as well as to the field of education. The mental-training techniques developed in these traditions are uniquely potent for alleviating mental suffering and promoting greater inner peace. So my second goal for my dialogues with contemporary scientists is to see how these techniques, as well as their underlying insights, can be best harnessed to the task of transforming our contemporary education system so that our society does not suffer from a deficit in basic ethics.

Today no aspect of human life is not impacted by science and technology. Science occupies a central place in both our personal and our professional lives. It is critically important that we reflect on the ultimate purpose of science, on what larger consequences and impact science can have in our world. In the early part of the twentieth century, many believed that the spread of science would erode faith in religion. Yet today, in the beginning of this twenty-first century, there seems to be a renewed interest in ethics in general and, in particular, the insights of those ancient traditions that contain systematic presentations of mind science and philosophy.

## Three Domains in the Subject Matter of Buddhist Texts

In our society, all sorts of immoral acts are committed on a regular basis. We observe murder, theft, cheating, violence against others, exploitation of the weak, misuse of public goods, abuse of alcohol and other addictive substances, and disregard for societal responsibility. We also see people suffer from social isolation, from vengefulness, envy, extreme competitiveness, and anxiety. I see all these as consequences primarily of our neglect of ethics and basic human qualities such as kindness. It is essential for us to pay attention to the means that would help promote basic ethics. The profound interdependence of today's world calls us to create a society permeated by kindness.

What kind of foundation is necessary for this? Since religion-based ethical teachings are grounded in the philosophical views of their respective faith traditions, an ethics contingent on religion alone will exclude those who are not religious. If ethics is contingent on religion, it will be ignored by those who adhere to no religious faith. We do not need to be religious to see the value of kindness; we can discern it by observing our everyday life. Even animals survive by relying on the care of others.

Furthermore, impulses for empathy, kindness, helpfulness, and tolerance seem naturally present in small infants, well before the influence of religious faith begins. Looking to these innate qualities and their associated behaviors as a foundation, I have striven to promote an approach to ethics and basic human values that does not rely on the perspectives of a specific religious tradition. My reason is simply this: We can enjoy a life of peace and happiness without religion. In contrast, if we are divorced from human love and kindness, our very survival is at risk; and even if we do survive, our life becomes devoid of joy and trapped in loneliness.

We can promote ethics on the basis of a specific religion, but prioritizing the perspective of one religion over others is problematic in today's deeply interconnected and global society, which is characterized by a multiplicity of religions and cultures. For an approach to the promotion of ethics to be universal, it must appeal to the fundamental values we share as human beings. If we neglect these basic human values, who can we blame for the negative consequences? Thus, when I speak of secular ethics, I am speaking of these fundamental values that are inherent to human nature, and that are in fact the very foundation of the ethical teachings of the world's religious traditions.[4]

Historically, there have been societies where respect was accorded to the perspectives of both believers and nonbelievers. For example, although the materialist Cārvāka school was the object of vehement critiques from other schools in ancient India, it was a custom to refer to the upholders of that viewpoint in honorific terms. Consonant with this ancient tradition, when India gained its independence in the twentieth century, the country adopted a secular constitution independent of any specific religious faith. This establishment of a secular constitution was not to show disrespect for religion; it was to promote peaceful coexistence among all religious faiths. One of the major forces behind the adoption of this secular constitution was Mahatma Gandhi, himself a deeply religious person. Conscious of this important historical precedent, I feel no apprehension in promoting a secular universal approach to ethics.

My own personal view is that, in general, people should remain within their own traditional religions. Changing faith can lead to difficulties for oneself, and it can also undermine the basis of interreligious harmony. With this belief I have never harbored any intention to make converts or convince followers of other religions to become Buddhists. What is appropriate for believers is to contribute to the common good by practicing those aspects of the teachings that can serve humanity as a whole. Such teachings are definitely present in all the world's main religions.

Within Buddhism, for example, I see two things with the greatest potential to serve everyone, regardless of their faith. One is the presentation on the nature of reality, or "science," as found in the Buddhist treatises, and the second encompasses the methods or techniques for training the mind to alleviate mental suffering and promote greater inner peace. In this regard it is important to differentiate among three distinct domains within the subject matter of the Buddhist sources: the presentations (1) on the natural world, or science, (2) on philosophy, and (3) on religious beliefs and practice. In general, when one speaks of religion or religious practice, it is linked with faith in a source of refuge. In this religious sense, Buddhism, too, is relevant only to Buddhists and has no particular connection to those who follow other religions and those who have no religious faith. Clearly presentations rooted in religious faith are not universally applicable, especially when we recall that among today's world population, as many as a billion human beings identify themselves nonbelievers.

Buddhist philosophy contains aspects, such as the principle of dependent origination, which can be relevant and beneficial even to those outside the Buddhist faith. This philosophy of dependent origination can of course conflict with standpoints that espouse a belief in a self-arisen absolute being or an eternal soul, but for others, this philosophy can help expand their outlook and enable them to see things in life from multiple angles, which prevents the narrow fixation that blames everything on a single cause or condition. I see great benefit in extracting the scientific and philosophical explorations found in Buddhist texts and presenting them independently of the strictly religious teachings. This allows someone who is not Buddhist to learn about the Buddhist scientific explorations of reality as well as Buddhist philosophical insights. It also gives many people the opportunity to learn how Buddhist traditions have developed their worldview and their philosophical outlook on the ultimate nature of reality.

Take, for example, the Buddha's first teaching, the four noble truths, which is common to all Buddhist traditions. In this teaching, we can observe a clear differentiation among the "ground" (the nature of reality), the path, and the result. The statements on the nature of the four truths, e.g., "This is the noble truth of suffering," present the ground; the statements on the function of the truths, e.g., "Suffering is to be known," present the path; and the statements pertaining to the agent and the fruits of the path, e.g., "Suffering is to be known, yet there is nothing to be known," explain how the result of the path comes to be actualized. My point is that, whether the presentation is of philosophy or of ethical precepts, the fundamental approach in the Buddhist texts is to ground them in an understanding of the nature of reality.

In what is called the Mahāyāna, or Great Vehicle, too, the presentation on the two truths (conventional and ultimate) is the ground, the presentation of the two aspects (method and wisdom) is the path, and the presentation of the two buddha bodies (the form and truth bodies) is the result. All of these are grounded in an understanding of the nature of reality. Even in the case of the highest aim in Buddhism—the attainment of the two buddha bodies, or the buddhahood that is the embodiment of the four buddha bodies[5]—the potency to actualize these aims can be found in the innate mind of clear light that resides naturally within us. The presentations found in the Buddhist sources are developed on the basis of an understanding of the nature of reality. If we look at the way the words of the Buddha were interpreted in the treatises composed by the great Buddhist thinkers of the past, such as the masters of Nālandā University, there too the subject matter of the entire corpus of Buddhist texts, including those that were translated into Tibetan running into more than three hundred volumes, fall into the threefold classification of the ground, the path, and the result.

As stated, the content of the Buddhist texts can be grouped within the three domains of (1) the nature of reality, or science, (2) philosophical tenets or views, and (3) religious practice, namely the presentation of the path and the way in which the results of the path are actualized. I see great benefits if we engage with the works in the Kangyur (the scriptures) and Tengyur (the treatises) on the basis of critically examining whether their contents present science, philosophy, or religious practice.

## Buddhist Presentations of Reality, or Buddhist Science

### *Philosophical Outlook and Methods of Inquiry*

In general, the word *science* refers to a body of knowledge about the world that is obtained through a particular method and that is verifiable by anyone repeating the same experiment. The term can be applied to both the acquired body of knowledge and to the method by which such knowledge is obtained. More broadly, *science* can refer also to a systematic method of investigation. When a scientist explores a particular question, first he or she develops a hypothesis. Then, through experiments, certain results are found, and these findings are then subjected to the scrutiny of one's colleagues. When the findings of different scientists come to converge, these findings come to be accepted as part of the larger body of scientific knowledge. The means by which these discoveries are made is characterized as scientific. This basic feature of the scientific method seems to accord with the two criteria of existence proposed in Buddhist Madhyamaka texts: for something to exist, (1) it must be known by a conventionally valid cognition, and (2) it must not be contravened by some other conventionally valid cognition.[6]

In the Buddhist tradition, the principle of "four reliances" sets out the basic discernments necessary when inquiring into the nature of reality: (1) When analyzing a particular claim, we must not draw conclusions based on the renown of the person who is making the statement; rather we should draw conclusions on the basis of critically examining what the person has said. (2) With respect to what has been stated, too, we should not deduce the truth or falsity of a statement from its literary merits or the quality of the writing; the content of the statement is more important than its form. (3) With respect to the contents of the statement, too, we should not trust those that may have been stated provisionally for an expedient purpose; rather we should accord greater importance to the definitive meaning that pertains to the actual nature of reality. (4) With respect to the definitive meaning, too, we should accord greater importance to the observations of direct perception and not be content with mere conjectures or word-based understandings. These four conditions should be applied when we engage in analysis, irrespective of the topic.

As for the actual analysis, the texts speak of the "four principles of reason." Not only do these four principles demonstrate the Buddhist outlook on the natural world, if we examine carefully, we can also say that they embody the entire Buddhist presentation about reality. The first, the *principle of nature*, explains how in the Buddhist understanding there is no absolute beginning or end to the universe when viewed from the perspective of its constitutive elements. For example, even though we can speak of a beginning to the macroscopic world, whether the structure of the world or its manifest aspects, we cannot posit an absolute beginning at the level of ultimate constitutive elements. For if there is such a beginning, that first source will have to be either devoid of cause or be created by some transcendent being. Similarly, if one posits an absolute beginning to consciousness, whose essential character is that of subjectivity, this would mean that consciousness emerges from the aggregation of matter, which is contrary to the nature of subjective experience. Were this the case, how would we explain the subsequent arising, on the basis of that alleged first moment of consciousness, of multiple distinct streams of cognition? So, in the Buddhist sources, the fact that material things maintain their own integrity and propagate their continuity and the fact that consciousness exists as characterized by subjective experience are stated simply as part of the principle of nature. Things give rise to their own continuum—*that is the way it is.*

Śāntideva makes an important distinction with respect to this principle of nature. In his *Engaging in the Bodhisattva's Deeds*, Śāntideva explains how things arise in the world through a natural process of evolution. However, when this process reaches a point where the basis for the experience of happiness and suffering of a sentient being comes into play, at that point the process has gone beyond the simple principle of nature. At that point, he states, the principle of nature is still the basis, but adventitious factors, such as the intentional activity of a person, have also intervened and become part of the causal process.[7]

The second principle, the *principle of dependence*, relates to the way the features an effect displays are contingent on the characteristics of its cause. In other words, all the diverse attributes that exist in an effect are byproducts of the aggregation of the various characteristics present in the cause. To give an easy example, say there is to be an excellent crop with healthy stalks, unaffected by pests, and crops ripening at the right time, and so on. The presence of seeds is not enough; other important conditions must converge as well, such as soil, warmth, moisture, fertilizer, and dedicated human effort. One cannot say excellent crops are due merely to nature.

Within this principle of dependence, we can distinguish three types of dependence: (1) causal dependence, (2) the way the integrity of a whole depends on its constitutive parts—even the perception of a whole is dependent upon the perception of its parts—and (3) dependence in terms of conceptual designation, which is to say that given everything exists as mere conventional designation, their very identity as distinct phenomena is a function of the mind labeling them. This last sense of dependence is extremely profound.

The *principle of function*, the third principle, refers to specific functions individual entities perform because of their distinct natures and the functions they support. For example, the seed is the primary cause of a sprout, but conditions such as moisture and fertilizer each perform their own unique functions, and it is through the combination of these causes and conditions that a specific effect arises. In general, the earth element functions to ground and support, water functions to cohere, fire to mature, and the wind to enhance. It is functions such as these that are referred to when speaking of the principle of function.

The final principle, the *principle of evidence*, refers to what allows us to draw inferences, such as "If such and such is the current condition, then such and such will be the future state," "Given these diverse features of a given effect, there must have been such and such diverse characteristics in the cause too," and "Such and such is impermanent because it comes into being at certain times but not at others."

### *The Nature of the Objective World*

Briefly, the presentation of the nature of reality in classical Buddhist texts is fourfold: (1) the nature of the objective world, (2) the presentation of the mind, the subject, (3) how the mind engages its object, and (4) the means, such as the science of logical reasoning, by which the mind engages its object. This overarching framework has been adopted for the presentations in *Science and Philosophy in the Indian Buddhist Classics*.

With respect to the first, the objective world, we can distinguish three categories. First is the category of evident observable facts, those things for whose cognition we do not need to rely on either logical reasoning or someone else's verbal testimony. Second are those facts not directly observable that can be known on the basis of logical reasoning. Finally, some facts we simply cannot discern through either direct perception or evidence-based logical reasoning. This last category must be known on the basis of the words of a reliable person.

Take our own health as an example. Symptoms of ill health such as the loss of body luster, shortness of breath, and inflammation belong to the first category of facts, *observable evident facts*. On the basis of these symptoms, we can diagnose the specific illness and perhaps the cause of that illness as well. These facts that are inferred are known as the *slightly concealed facts*, the second level of facts. Some facts in this second category are opaque to humans in general; some are categorized as such contextually in relation to a particular place or time. Among the third level of facts, *extremely concealed facts*, is, for example, our date of birth. This is something we can know only from the testimony of others; we have no other means of knowing it. This third category includes also the truth about the subtle workings of cause and effects, as well as the answers to what are the causes and conditions for the existence of such diversity of species of sentient beings in this world.

When dealing with the third category, it becomes essential in Buddhist inquiry to rely on the authority of scriptures. However, such scriptural testimony must fulfill certain criteria. To begin with, the content of such a scripture must not be undermined by direct perception and valid logical reason; also the words of that scripture must be free of internal contradictions, and the utterer of that scripture must have no exceptional circumstantial reason for teaching that particular scripture. In our everyday life we use this third type of inference relying on others' testimony. For example, we come to believe certain facts about the world based on what we read in the papers, and we accept numerous facts about the past based on historical works. Even in science too, scientists subscribe to the conclusions of other researchers shared through scientific publications without themselves undertaking the same experiments.

This mode of inquiry into the nature of reality through the use of direct observation, reason-based inference, and scriptural testimony, consonant with the threefold classification of the facts of the world, existed in the Buddhist tradition from its inception. Nonetheless, it is Dignāga—appearing in the fifth century—and his commentator Dharmakīrti in the seventh century who were responsible for developing a comprehensive science of logical reasoning within the Buddhist tradition. The Madhyamaka philosophy based primarily on critical reasoning arose during the time of the Buddha himself and was refined especially around the second century of the Common Era by Nāgārjuna. However, I think the emergence of a distinct and complete system of logic and epistemology (*pramāṇa*) within the Buddhist tradition must be attributed to Dignāga and Dharmakīrti.

In general, the science of logic and epistemology was well established through the treatises of the non-Buddhist Naiyāyika school, well before Dignāga and Dharmakīrti. Dignāga further advanced the science of logic and epistemology through his innovations, such as the invention of the triple criteria of a logical proof as well as his *apoha*, or "differentiation," theory of meaning in the philosophy of language. In his writings on logic and epistemology, Dignāga presented detailed critiques of the standpoints of other Indian schools such as the Naiyāyika, Vaiśeṣika, and Sāṃkhya. In response, the Naiyāyika thinker Uddyotakara and the Mīmāṃsā philosopher Kumārila critiqued Dignāga. And if we analyze how the views of these thinkers were, in turn, chosen as objects of sustained critique by Dharmakīrti, what we find is the development of a highly advanced tradition of logic and epistemology in classical India.

There are other alternative systems of classification of reality presented in the Buddhist sources. For example, there is the fivefold classification of (1) material form, the visible, (2) primary mind, (3) concomitant mental factors, (4) nonassociated conditioning factors, and (5) unconditioned phenomena. There are also the classifications with specific purposes, such as that of twelve bases and eighteen elements.[8] Thus, taking into account their specific natures, characteristics, purposes, and types, there are various ways in which reality is parsed and subsumed within each other in Buddhist sources.

The first category of reality in the fivefold classification is material form. In presenting this category, Buddhist sources examine the nature of matter and differentiate it into eleven material forms, which include both obstructive and nonobstructive forms, mental-object forms, as well as the great elements and their derivatives (see part 2 of this volume). Explorations of material form include also the topic of the formation of the natural world (part 5), as well as the topic of atomic particles, the ultimate constituents of the physical world (part 3). Also addressed is the question of whether the nature of material objects can be explained on the basis of aggregation of atoms or whether, ultimately, what is perceived as external matter is nothing but perceptions of the mind. Within the theory of atoms, there is also the view of *space particles* that act as the basis for the other natural elements. Even with respect to a single particle, its characteristics can differ vastly depending on the distinct perspectives of two beings interacting with that individual particle at the same locus, differences as great as between the sky and the earth.

When speaking of natural elements, Buddhist sources refer also to a fifth element, known as the *space element*. By "natural elements" we should not understand only composite things; it also encompasses some things that remain in the form of potencies. Furthermore, since atomic particles are said to derive from the great elements, this would indicate that there are natural elements that are the sources of these particles. What we perceive in our everyday experience as earth, water, fire, and air, although labeled as "elements," are in fact quite coarse. Furthermore, in the Buddhist sources, there are detailed presentations of physiology, such as the gross, intermediate, and subtle levels of the body; the primary and secondary wind energies; the 21,600 cycles of breath that flow within the body in a single day, and so on (part 6).

With respect to the origin and evolution of the natural world and its inhabitants (part 5), Asaṅga, for example, explains this on the basis of three conditions—an *absence of prior intelligence*, the condition of *impermanence*, and the condition of *potentiality*. Based on these three conditions, Asaṅga explains how the cosmos and its inhabitants evolve purely through the function of mere conditionedness and how they do not come into being through the design of a creator. This view is the standard Buddhist position on the question. This standpoint is highly compatible with the basic viewpoint of science.

This compatibility aside, there are explanations found in the Buddhist sources, such as in the Abhidharma texts, describing the shape and size of the planets, the passage of sun and moon, how solar and lunar eclipses occur, and so on that are based on the existence of Mount Meru at the center, with earth being flat and the heavenly bodies like the sun, moon, and the stars circling the earth. I believe these cosmological explanations are simply the received views of the time, based on sources such as the ancient Vedas. These specific claims about cosmology are in direct conflict with confirmed findings of contemporary science. I have not held these as part of my worldview for quite a long time.

Buddhist sources also describe how the world first comes to form, then abide, disintegrate, and become empty, and how during the empty stage there remains space or empty particles. It is from these particles that another new world comes to form that abides, disintegrates, and becomes empty. In this way, in a repeated cycle, the universe is said to retain its endless continuity. These space particles, described in the texts of Kālacakra tantra,[9] are not conceived of as something observable with a physical mass. Nevertheless, they persist as the sources for the emergence of the entire material world. Through the forces of these space particles, the four natural elements evolve from the subtle to coarser levels, and these later dissolve from the grosser to the subtler levels, ultimately reverting back into space particles. Thus these space particles are the foundation for both the emergence and the dissolution of the world systems. The Kālacakra texts explain how and at what stage these space particles act as the basis for the emergence of a particular world system. Before forming, a world system remains empty, since its entire material basis exists in the form of empty space particles. When the potencies of the karma of the sentient beings that will later come to inhabit that particular world ripen and begin to exert influence, the wind particles coalesce and set in motion the formation of that world.

Although contemporary science provides a profound explanation for the emergence of the universe out of a big bang, important questions remain. What existed before the big bang? Where did the big bang come from? What conditions gave rise to this big bang? What led to the process that created the conditions for the emergence of life on Earth? What is the relationship between the natural world and sentient beings that came to evolve within it?

According to the explanations in the Buddhist sources, there is a connection between the way the outer natural world is formed and the formation of the sentient beings that come to inhabit it. Furthermore, there is a connection between the corporeal body we possess and a subtle body at a deeper level that, in turn, evolves through a process traceable finally to the subtlest state, where there exists an indivisible union of wind and mind. So we find in Buddhist sources the view of how the human body represents in a subtle way the entire universe.

In biology, according to the principles of Darwinian theory, and in particular the process of natural selection, questions of how the diversity of species came to evolve and what might be the origin of life are explored. Evolutionary science explains how life came to emerge through increasingly complex aggregations of molecules and how living organisms propagate their own kinds through reproduction and so on. Thus science explains how five characteristics must be present in something that is alive.[10] Since biology's account of the origin of life is based on the notion of cells, it explains how all living things are composed of cells. When asked what are the first cells, the explanation is that they evolved through an extremely long natural process.

Now in evolutionary science's account of how life first came to emerge, we find the idea of how a series of random selections leads to the emergence of distinct species. When examined from the Buddhist perspective, this view resembles those who speak of the emergence of the universe through pure chance or without any cause. I find this particular aspect of evolutionary theory to be problematic.

Speaking of biology, I have asked scientists on a number of occasions the following questions: Why is it that modern evolutionary view does not accept fundamental human qualities such as compassion to be part of basic human nature? And how is it that the impulse for helping and kindness are not recognized as drivers for human behavior and the basis of flourishing?

In the Buddhist sources, time (part 4) is not identified as something independent of matter and mind. Time is defined as a construct on the basis of matter or consciousness. The past, that which has ceased, functions not only by opening the way for a present arising, it also helps to make that which is yet to be a reality. The shortest moment of time is identified as the time it takes for a single atom to turn. There are also coarser measurements of the shortest moment of time, such as a sixtieth-fifth—or a three hundred and sixty-fifth—of a finger snap. That countless submoments can be differentiated even within such a short moment of time is clear from the citations from the *Flower Ornament Sūtra*.[11] There it is explained through the descriptions of the qualities of bodhisattvas at different levels that even though there is an immense difference between an eon and an instance, the two need not be contradictory if judged from the perspectives of two distinct persons.

### *The Mind and Reasoning*

I explained above that Buddhist science, or its presentation on reality, can be grouped under four main topics: (1) the nature of the objective world, (2) the presentation of the mind, the subject, (3) how the mind engages its object, and (4) the means, such as the science of logical reasoning, by which the mind engages its object. Of these, I have commented on the first topic. The remaining three topics are part of the sciences of mind, and I explain these in my introduction to volume 2 in this series.

## BUDDHIST PHILOSOPHY

Philosophy represents the summation of the conclusions about the nature of reality developed through critical inquiry. In *Science and Philosophy in the Indian Buddhist Classics*, philosophy will be treated in volumes 3 and 4, but I will touch on it briefly here. In Buddhism, works explicitly presenting philosophical views evolved early. We see this with the appearance of the *Questions of King Menander* before the Common Era, the Abhidharma treatises starting around the first century CE, the six philosophical treatises of Nāgārjuna shortly thereafter,[12] and so on. There also appeared, in Buddhism's classical era, treatises in which the principal views of both Buddhist and non-Buddhist Indian schools were presented together in a single work and critically examined. For example, Bhāvaviveka composed his *Blaze of Reasoning* in the fifth century, and in the eighth century Śāntarakṣita authored his *Compendium on Reality*.

Buddhism's basic philosophy is encapsulated in what are known as the view of the four seals, or axioms: all conditioned things are impermanent, all contaminated things are characterized by suffering, all phenomena are empty and devoid of selfhood, and nirvāṇa is peace. *Impermanence* refers to the fact that things, right from their birth, do not remain static even for a single moment. This is because things do not depend on some third condition for their disintegration; the very causes that produce them also make them susceptible to disintegration. We can see this truth of impermanence for ourselves if we contemplate deeply the gross changes we observe in things. The statement that "all contaminated things are characterized by suffering" indicates how our existence is bound to a causal nexus of undisciplined states of mind that keeps it under their power. As for the statement "nirvāṇa is peace," Dharmakīrti identifies this with the possibility of eliminating pollutants from the mind. He establishes the existence of such a state of freedom through reasoning so that we do not need to rely on faith alone to explain it. The teaching on no-self relates principally to the ultimate nature of reality, namely that things do not exist the way they appear to.

All Buddhist schools reject the existence of a self that is eternal, unitary, and autonomous. Yet many Buddhist schools assume that what we call "self" or "person" must nonetheless exist in some form. We find the assertion that the self exists on the basis of the aggregates, with some proposing that all five aggregates constitute that person and others positing the mind alone to be the person. Some, recognizing that the six types of consciousness are unstable like bubbles in water, assert eight classes of consciousness and posit foundational consciousness (*ālayavijñāna*) to be the real person. Others, seeing faults in identifying the person with the aggregates, assert a self (or person) that is neither identical to nor different from the aggregates.

As we can see, there is a divergence of interpretations and subtleties among the various Buddhist schools with respect to the meaning of no-self. The Tibetan tradition relies chiefly on the interpretation of the Perfection of Wisdom sūtras by Nāgārjuna and his disciples. In this view, the meaning of no-self is understood by way of dependent origination. Two types of selflessness are differentiated from the perspective of their bases (persons and phenomena), but there is no difference in subtlety in what is negated; in both contexts, it is independent existence.[13] The very fact that things are dependently originated establishes that they are devoid of self-existence. When we think, for example, in terms of the designator and the designated, the knower and the known, the agent and the act, and so on, we can see the utter mutuality and contingency of these things. If the table in front of us, for example, were to exist objectively without depending on conceptual designation, the table itself could provide the criteria of what constitutes a table from its own side. This is not the case. We have no choice but to accept that what we call "table" is posited by the mind.

What we see is a mutual dependence. The objective world exerts constraints on the mind, and the mind in turn exerts constraints on the objective world. Take the simple example of a handwritten letter *a*. So many factors converge that are part of its dependent reality. There is, for example, the shape of the letter, the pen that wrote it, the ink used to write it, the paper on which it is written, the person who wrote it, the intention of the writer, the convention that established this letter, those who accept this as a letter, and the cultural environment in which this letter has a meaningful usage. Without these, its existence as a letter is simply impossible. The nature of all things is exactly like this. Therefore things are explained as having a nature of dependence requiring so many other factors for their existence. This is why Madhyamaka thinkers such as Candrakīrti[14] speak of how things are unfindable when subjected to ultimate analysis and of how their existence can only be posited as designated by the mind. This view is strikingly similar to explanations found in contemporary physics about how nothing can be found to possess reality when analyzed at the subatomic level.

Another important philosophical view in Buddhist texts is that of the two truths. We find the language of "two truths" in the non-Buddhist Indian philosophical schools as well. In Buddhism, all four schools of thought equally accept the notion of two truths, but what constitutes these two varies from school to school.[15] Between the two Mahāyāna schools, for example, there is not much difference in the way the Cittamātra (Mind Only) and Madhyamaka (Middle Way) schools define the two truths. Nonetheless there is a substantial difference in the specific examples they give for those two truths.

In brief, *ultimate truth* pertains to the ultimate nature of things while *conventional truth* relates to perspectives rooted in the apparent world. Both the Madhyamaka and Cittamātra schools explain ultimate truth in terms of emptiness. Cittamātra speaks of the emptiness of external reality or the emptiness of subject-object duality, while Madhyamaka speaks of the emptiness of real existence of everything, even the minutest particles of matter. Conventional truth encompasses the entirety of the everyday reality we perceive—the natural world, the beings who inhabit it, arising and disintegration, progress and decline, cause and effect, happiness and suffering, good and bad, and so on. In short, the clay pot of flowers we see in front of us is conventional truth, while its absence of objective existence—that this pot cannot be found when sought through ultimate analysis—is its ultimate truth.

That pot is empty at the very moment it is perceived, and it can be perceived while simultaneously being empty. Madhyamaka thinkers explain this by saying that the two truths have the same nature but are conceptually distinct. When the Buddhists speak of the way things exist, they maintain that we need to transcend both extremes—the extreme of reification and the extreme of denial—and view things simply as they are.

## Buddhist Religion

Generally speaking, although aspects of the Buddhist tradition that fall under religion are connected with faith, the basic framework of Buddhist religious practice is grounded in the principle of causality, which is part of the laws of nature. For example, the impulse to shun pain is part of our natural disposition, and our existence as conditioned beings is the basis for the arising of suffering. Therefore, Buddha taught the *reality of suffering* as the first truth of our existence. Since suffering necessarily arises from a cause, he identified the second truth as the *origin of suffering*. These two truths pertain to the cause and effect of suffering. What is the cause of suffering? Its ultimate source is explained as ignorance, and since this ignorance can be brought to an end, the Buddha taught the third truth, the *cessation* of suffering and its origin. Since such a cessation must also have a cause, the Buddha taught the truth of *the path*, the means of attaining such a cessation. There is thus a cause-and-effect pair of truths pertaining to the attainment of freedom. Clearly the foundation of Buddhist practice described in the four noble truths is the natural law of cause and effect.

When Dharmakīrti introduces the truth of cessation, he demonstrates the possibility of bringing an end to ignorance, the cause of suffering. Nowhere does he speak of the need to demonstrate the truth of cessation by relying on scriptural authority. Furthermore, Dharmakīrti offers a profound explanation of suffering and its origin in terms of the sequence of the twelve links of dependent origination, and cessation and the path in terms of the reverse order of the twelve links.[16] Since happiness and suffering are characteristics of sentient experience, no account of them can be divorced from sentient experience. Therefore Dharmakīrti also offers an extensive account of cause and effect as it relates to the inner world of experience. Furthermore, when one speaks of Dharma (religion) in Buddhism, its true meaning must be understood in terms of the attainment of nirvāṇa. The term Dharma refers to the means and the path that lead to nirvāṇa as well as the scriptures taught by the teacher, the Buddha, that present this path.

Having now distinguished three domains of subject matter in the Buddhist sources—(1) scientific presentations about the natural world, (2) philosophy, and (3) religious practice—we might ask from what sources the presentations in this series on the first two dimensions, science and philosophy, are developed. Among the Buddhist classics available in Tibetan, we have the two canonical collections introduced above. The precious collection of the Kangyur contains translations of the Buddha's words as embodied in the "three baskets" (Tripiṭaka), containing both sūtra and tantra teachings. The precious collection of the Tengyur contains the treatises of great masters such as the seventeen Nālandā masters that include Nāgārjuna and Asaṅga, the two trailblazers prophesized by the Buddha.[17]

The Tibetan translations that comprise the Kangyur and the Tengyur are the largest body of Indian Buddhist texts extant today anywhere. Today modern scholars who engage in objective studies of Indian Buddhist sources state that these Tibetan collections not only contain the largest number of texts, they represent the best translations and most comprehensive Buddhist canon. Many of these works composed in classical Indian languages, especially Sanskrit, were entirely lost in their original language through changes of history and environmental conditions. Only a few of the great works remain in original Sanskrit. In the Pali canon, we find scriptures associated with the Theravāda tradition but not of other

Buddhist schools, such as the Mahāyāna sūtras and tantras. Although a great number of Buddhist texts were translated into Chinese, modern researchers say that because of the character of the Chinese language, those translations tend to be looser and do not match the rigorous correspondence, both in terms and meaning, found in the Tibetan translations. Today, therefore, the Tibetan language is the storehouse for the Buddha's scriptural teachings in their entirety. By offering access to the complete system of the Indian Buddhist tradition encompassing all three vehicles—the shared teachings, the Mahāyāna, and the Vajrayāna—there is simply no alternative to the literary heritage of the Tibetan language.

## THE THREE BUDDHIST COUNCILS

It might be helpful for the readers of this series to have some understanding of the ultimate sources of the classical texts in the Tibetan canon. I offer below a brief account.

According to both the Mahāyāna and non-Mahāyāna traditions, from his first public sermon on the four noble truths in Vārāṇasī seven days after his enlightenment until his final nirvāṇa at Kuśinagara, the Buddha traveled for forty years across central India and taught in numerous places to countless disciples in accordance with their needs and dispositions. These discourses were later compiled within the Tripiṭaka. For example, in the summer of the Buddha's final nirvāṇa, convened by senior monk Mahākāśyapa, the first Buddhist council took place at Rājagṛha (modern-day Rajgir) in the Saptaparṇa cave. At this council, each of the three compilers opened with the phrase "Thus I once heard" and ended with the statement "Everyone gathered praised what the Blessed One had taught." Mahākāśyapa recited and compiled the Abhidharma basket, the arhat Upāli compiled the Vinaya (discipline) basket, and the arhat Ānanda compiled the Sūtra basket. Thus began the process of upholding the scriptural baskets that are part of the common teachings of the Buddha.

About a hundred years later, a second Buddhist council took place. Buddhist schools generally concur on how this second council took place, and an account of it is found in the Vinaya texts.

As for the third Buddhist council and any additional councils, the scriptures do not explicitly mention them, and there is some divergence of opinion among the Buddhist schools on this question. According to the Sarvāstivāda school, for example, a Buddhist council took place during the reign of King Kaniṣka in Kashmir at the monastery of Kuṇḍalavana. This council was said to have been attended by five hundred arhats including the elder Pūrṇika, five hundred bodhisattvas including masters Vasumitra and Aśvaghoṣa, and fifteen hundred paṇḍitas of great learning. In this account, formal recognition was accorded at this council to all eighteen schools as presenting authentic teachings of the Buddha, and the disputes among these schools were settled in accordance with the Dharma. Some later scholars do not accept that elders from all eighteen schools gathered at this council. In any case, it is said that at this council the discourses of the Tripiṭaka that had not been committed to writing were written on copper plates.[18] The *Great Treatise on Differentiation* (*Mahāvibhāṣā*) was also said to have been composed or compiled as a result of this council. Some modern scholars suggest that this Kuṇḍalavana Monastery could possibly have been located in the modern-day Kangra district in northern India.

According to the Theravāda tradition, the third Buddhist council took place 218 years following the Buddha's final nirvāṇa during the seventeenth year of the reign of Aśoka (circa 250 BCE). It took place in Pāṭaliputra (Patna) under the leadership of Tissa Moggaliputta and was attended by one thousand elders. It is stated that over a period of nine months, the discourses of the Tripiṭaka were compiled. Their account seems to indicate it was only the scriptures of the Theravāda school that were compiled at this council. The Theravāda tradition also holds the view that the Pali Tripiṭaka as well as their commentaries were first written down in 27 BCE during the reign of the Sinhalese king Vaṭṭagāmani.

As for the scriptures that are unique to the Mahāyāna tradition, they were not compiled at the above-mentioned Buddhist councils. They are understood to have been compiled by Mañjuśrī, Avalokiteśvara, Vajrapāṇi, and so on in various locations perceptible only to the minds of disciples with pure vision. On how the tantric scriptures were compiled, the tantras themselves contain diverse explanations. The non-Mahāyāna Śrāvaka schools accept the Buddha's sermon on the four noble truths to be the only turning of the wheel of Dharma and recognize all other discourses of the Buddha to be further elaborations of this sermon. Mahāyāna schools, on the other hand, accept three turnings of the wheel of Dharma. The second turning pertaining to the absence of characteristics occurred on Vulture

Peak at Rājagṛha, while the third, pertaining to clear differentiation, took place in Vaiśālī. After the Buddha's death, for around four hundred years before the appearance of the glorious Nāgārjuna, the Mahāyāna sūtras remained mostly outside the purview of common human perception. Nāgārjuna made many of these Mahāyāna sūtras flourish widely and, in this way, blazed the trail of the Mahāyāna tradition and the Madhyamaka school of philosophy. Composing numerous treatises, Nāgārjuna elucidated the stages of the truth of emptiness, which is the explicit subject matter of the Perfection of Wisdom sūtras.

At that time, most Śrāvaka schools had not accepted the Mahāyāna sūtras to be authentic scriptures taught by the Buddha, and so these sūtras remained an object of debate for some time. Therefore Nāgārjuna himself, and later Maitreya and Bhāvaviveka, presented numerous logical arguments to establish the Mahāyāna sūtras as authentic teachings of the Buddha. Even in Śāntideva's time (eighth century CE), there still seemed to have been a need to defend the authenticity of the Mahāyāna sūtras.[19]

Asaṅga initiated the Cittamātra (Mind Only) school and made known the Five Treatises of Maitreya. He also composed a vast number of texts himself, such as the collection on the levels (*bhūmi*) in which he elucidated the stages of the path. These represent the implicit subject matter of the Perfection of Wisdom sūtras.

Over time, the Mahāyāna sūtras came to be established as authentic teachings of the Buddha himself, and the Mahāyāna tradition, including the Vajrayāna teachings, greatly flourished in central India as well as in various other regions of the Indian continent. Not only that, the teachings of all three vehicles spread to numerous places in Central Asia as well as to China. Even in the lands where today the Theravāda tradition flourishes, such as Burma, Siam, and Sri Lanka, there appear to be numerous signs that indicate the Mahāyāna teachings flourished there in its early stages.

## Translations of Sūtras and Treatises into Tibetan

Because Tibet, a land of snows, is high in altitude and has a terrain that is difficult for travelers, it remained isolated from other lands for a long time. In the fifth century, however, during the reign of King Lha Thothori Nyentsen, the first Buddhist scriptures reached Tibet. Then in the seventh century, Emperor Songtsen Gampo took for his brides two princesses, one a daughter of the Chinese emperor and the other a daughter of the Nepalese king, rulers of two lands where the Dharma flourished before arriving in Tibet. These two princesses brought with them very special icons of Buddha Śākyamuni and constructed the Jokhang and Ramoché temples in Lhasa, initiating the tradition of paying homage to the images and making offerings to them. The emperor dispatched many intellectually gifted Tibetan youth, including Thönmi Sambhoṭa, to India to learn Sanskrit and Buddhism. Thönmi then invented the characters for writing Tibetan and also developed the Tibetan grammar, taking Sanskrit grammar as his model. In this way, he helped develop the Tibetan language to a point where it could match the perfection of Sanskrit itself. This towering achievement made it possible for rigorous and faithful translations of Buddhist sūtras and treatises to emerge. This achievement is one of the great wonders in the history of the Tibetan people. Over time Tibet, or Purgyal as it is known, came to acquire the title of the Land of Dharma (*chos ldan gyi zhing*). Today, that Tibet continues to remain a resource capable of sharing the teachings of the Buddha, both scriptural as well as actual realizations, is due to the kindness of Thönmi Sambhoṭa in inventing the Tibetan writing system and developing the grammar, as well as that of the translators and the Indian paṇḍitas.

Having invented Tibetan writing and grammar, Thönmi himself translated several Buddhist sūtras and tantras, but there appears to have been few other very early translators. Only some time later did more systematic and large-scale translation activity take place in Tibet. The scriptures Thönmi himself translated include *Calling Witness with a Hundred Prostrations*, the *Jewel Casket Sūtra*, and the *Dependent Origination Sūtra*.[20] There is also a reference to Thönmi having translated twenty-one scriptures associated with Avalokiteśvara at the command of the emperor, but it is difficult to say which scriptures these would be within the present Kangyur collection. Thus the earliest Tibetan translations of Buddhist scriptures include those translated by Thönmi himself and the early translations done by Dharmakośa and Lhalung Dorjé of the *Jewel Clouds Sūtra* (*Ratnamega-sūtra*), the *Descent into Laṅka Sūtra* (*Laṅkāvatāra-sūtra*), and so on; similarly, some early medical treatises are said to have been translated into Tibetan in collaboration with Chinese monks.

In the eighth century, during the reign of Emperor Tridé Tsukten, the temples of Chimphu, Drakmar Drinsang, and Phangthang were constructed. There, Drenka Mūlakośa, Nyak Jñānakumāra, and others translated sūtras such as the *One Hundred on Karma* (*Karmaśataka*), the *Golden Light Sūtra* (*Suvarṇaprabhāsottama-sūtra*), and so on. Then, during the reign of the Dharma king Trisong Detsen, seeing that for Buddhism to become firmly established in Tibet would require the invitation of incomparable scholars of the Dharma as well as monastics from India, he invited two great masters. One was the grand abbot Śāntarakṣita, an indisputable Nālandā scholar who was born into Sahor royalty but had chosen to become a monk. He was an upholder of the Mūlasarvāstivāda Vinaya tradition and a highly accomplished philosopher who had mastered the great ocean of Buddhist as well as non-Buddhist philosophical systems of thought. The second master the emperor invited was Padmasambhava, to whom Tibet owes deep gratitude. The precious Guru Rinpoché (Padmasambhava) helped dispel external as well as internal obstacles, and with this the great monastery of Samyé was successfully constructed.

The grand abbot Śāntarakṣita kindly advised the Tibetan emperor that, if he wished to have the Buddhadharma firmly established in the land of Tibet, the scriptures taught by the Buddha as well as their commentarial treatises must be translated into Tibetan. He also advised that bright young Tibetans become monks so that the monastic order could be established in the country. In accordance with this advice, the emperor established a center for translation at Samyé Monastery. There, hundreds of Tibetan translators gathered and began the systematic process of translating the scriptures and the treatises. This continued until the king Langdarma initiated various activities to suppress Buddhism in Tibet.

In any case, many great masters and translators emerged in Tibet in the eighth and ninth centuries. These included not only Śāntarakṣita and Padmasambhava but other great Indian paṇḍitas like Vimalamitra and nearly a hundred Tibetan translators, such as the "seven who are awake,"[21] Kawa Peltsek, and Chokro Lui Gyaltsen. During this long period of translation activity, an important need was felt for a timely reform of the Tibetan language, especially with respect to a specific set of Tibetan terms. Thus, during the reign of Emperor Tridé Songtsen, Indian masters, including Surendrabodhi, Śrīlendrabodhi, Dhānaśīla, and Bodhimitra, as well as Tibetan translators such as Ratṇarakṣita, Dharmaśīla, Jñānasena, Jayarakṣita, Mañjuśrīvarma, Ratendraśīla, and so on, convened a meeting.

They established the norms for translating key Dharma terms from Sanskrit into Tibetan and compiled the bilingual glossaries the *Mahāyutpatti* and the *Nighaṇṭu*, the latter known also as the middle-length *Vyutpatti*.[22] Composing these glossaries, the Indian paṇḍitas and the Tibetan translators undertook a reform of the Tibetan written language and established a sound tradition of translation from Sanskrit sources into Tibetan. Since then, thanks to this standardization, there remained a unified consistent system of translation from Sanskrit and other Indian languages into Tibetan. This kind of systematic approach and standardization of translation rarely occurred elsewhere in history, and it is another unique achievement of the Tibetan people. It unquestionably constitutes a great contribution to the discipline of translation as well.

Consonant with the advances being achieved in the work of translation, all the translations of sūtras and treatises that were housed at the Denkar Palace were cataloged in the *Denkarma Catalog* under the supervision of great translators like Kawa Peltsek and Lui Wangpo. In this catalog, the scriptures and treatises were organized into twenty-seven categories with 725 individual entries. This catalog was the first attempt to create a classificatory and cataloging system for Buddhist scripture in Tibet. This too shows the remarkable intellect of the ancient Tibetans. We also read about how, again in the ninth century, the translators Kawa Peltsek and Chökyi Nyingpo and others cataloged all the scriptures and treatises found at the Chimphu Palace and compiled the *Chimphu Catalog*. This work is no longer available. Not long afterward, what is today known as the *Phangthangma Catalog* was created, cataloging the scriptures and treatises found at the pillarless Phangthang Palace. In these two catalogs, the titles of the individual texts as well as their size in *bampo*[23] and page numbers were recorded. As for classification, it appears that the texts that have closely related themes were grouped together. Somewhere around the mid-ninth century, beginning with Emperor Langdarma's activities of suppression, Buddhism in the Tibetan kingdom became splintered, and with it the discipline of formal translation too came undone, and the scholars from India and Nepal were scattered. Though even during this long period, it appears that isolated translation activities persisted in border regions such as in Ngari in the west.

In the tenth century, the Ngari rulers Lha Lama Yeshe Ö and his nephew Jangchup Ö sent many young Tibetans to Kashmir to study

Buddhism. Among these, the most accomplished ones, like the great translator Rinchen Sangpo (958–1055) and Ngok Lekpai Sherap, returned to Tibet, having mastered Sanskrit and become learned in the sūtras and the tantras. Inviting Indian paṇḍitas like Śraddhākaravarma to participate, they translated numerous sūtras and tantras. Especially following the arrival in Tibet of the glorious incomparable Atiśa (982–1054), many treatises, including Haribhadra's *Light on Ornament of Clear Realizations* (*Abhisamayālaṃkārāloka*) and Bhāvaviveka's *Blaze of Reasoning*, were translated into Tibetan. During the period of the degeneration of Buddhism in Tibet translations had appeared, some accurate and others less so. Questions were raised about the authenticity of some of these texts. One remarkable contribution the great translator Rinchen Sangpo is said to have made was to search for the Indian originals of all existing Tibetan translations and separating out those for which no Indian equivalents were found. It is from the time of Rinchen Sangpo that the demarcation is made between the old translations and new translations.

Famous translators that appeared in the eleventh and twelfth centuries after Rinchen Sangpo include Drokmi Śākya Yeshé, Taktsang Shönu Tsöndrü, Khyungpo Naljor, Naktso Tsultrim Gyalwa, Rongzom Chösang, Gö Khukpa Lhetsé, Lha Lama Shiwa Ö, Lokya Sherap Tsek, Patsap Nyima Drak, and so on. Around sixty well-known translators flourished during this period, as is evident from the historical sources. Most importantly, the great translator Ngok Loden Sherap (1059–1109) undertook new translations of many important philosophical works, especially the works associated with Maitreya as well as the treatises on logic and epistemology. He undertook new translation work, revised and edited existing translations, and wrote summary introductions to the important Indian Buddhist treatises. In these ways, Loden Sherap made it possible for the scholastic study of these Indian Buddhist texts to become firmly established in Tibet. Well-known Indian paṇḍitas who came to Tibet during this period include, in addition to most importantly Atiśa, the paṇḍita Gayadhara, the paṇḍita Smṛtijñānakīrti, and others.

From the thirteenth to the seventeenth centuries, still numerous Tibetan translators appeared who rendered into Tibetan texts on sūtra and tantra, as well as Sanskrit grammar, medical science, and poetics. These include, among others, Pang Lotsāwa Lodrö Tenpa, Shongtön Dorjé Gyaltsen, Butön Rinchen Drup, Dratsepa Rinchen Namgyal, Gö Shönu Pal, Taktsang Lotsāwa Sherap Rinchen, Shalu Lotsāwa Chökyong Sangpo, and Jonang Tarānātha. Thus, over a period of more than a thousand years, Tibetan translators rendered into Tibetan most of the important works of the Indian Buddhist tradition—scriptures representing the teachings of all three vehicles and the four classes of tantra, most of the treatises that expound the meanings of these scriptures, as well as those that are part of the common fields of knowledge, such as Sanskrit grammar, logic and epistemology, art, and medicine. This service of the Tibetan translators is something we will never be able to fully repay, not just us Tibetans but everyone who shares an interest in the classical Indian tradition.

On a final note, within the Tibetan translations of scriptures and treatises as contained in the twin collections of Kangyur and Tengyur, there appear to be a few works that were translated from Indian languages other than Sanskrit.[24] In the Kangyur, around twenty-four scriptures, including the *Great Nirvāṇa Sūtra*, were translated from the Chinese version, while thirteen commentarial treatises, such as Wonch'uk's great commentary on the *Unraveling the Intention Sūtra*, were also translated from Chinese. In the thirteenth century, Butön's teacher, the translator Nyima Gyaltsen, translated under the supervision of the Sri Lankan master Ānandaśrī thirteen different sūtras from Pali, such as the *Preamble to the Jātaka* (*Jātakanidāna*) and the *Brahma's Net* (*Brahmajāla Sutta*). Otherwise, although there exist in the Kangyur numerous sūtras with titles similar to entries found in the Pali canon, given the differences in their contents, these do not seem to have been translated from the Pali. Similarly, according to Chomden Rikral's catalog, eighteen sūtras, including the *Questions of the Goddess Vimalaprabhā* (*Vimalaprabhāparipṛcchā*) and the *Prophecy of Khotan* (*Li yul lung bstan pa*), and in the Tengyur Vasubandhu's commentary on the *Dependent Origination Sūtra* as well as the commentary on the *Descent into Laṅka Sūtra*, appear to have been translated from Khotanese. Apart from these texts, most of the collection of 5,892 texts contained in the Kangyur and Tengyur were translated directly from Sanskrit.

## On This Compendium of Buddhist Science and Philosophy

The teachings of the Buddha are so vast that the Buddhist tradition speaks of 84,000 heaps of Dharma. Based on their subject matter as well as their forms, they are classified into twelve or nine branches of excellent teaching.[25] When condensed, these classes are subsumed within the three baskets of the canon. Unlike the world's other major religions, the Buddhist tradition's canons contain an extremely large number of texts. Even in the case of the part translated into Tibetan, there are more than five thousand individual texts in over 320 large volumes. The size of the collection means that it would be difficult for a person to read the entire collection even once. So a tradition emerged, from the period of the trailblazers, to extract the essential points from this vast body of scripture and present them in accord with the interests and capacities of the aspirants, in accessible formats such as compendiums and manuals. For example, Nāgārjuna composed the *Compendium of Sūtras*; Śāntideva too composed a compendium of sūtras[26] as well as the *Compendium of Training*; the glorious Atiśa too composed the *Extensive Compendium of Sūtras* (*Mahāsūtrasamuccaya*); and the trailblazer Asaṅga composed his *Compendium of Abhidharma*. Similarly, Dignāga wrote his *Compendium on Logic and Epistemology* (*Pramāṇasamuccaya*), bringing together the essential points of numerous works he had authored previously, such as his *Analyses*[27] and his short verse texts. All of these various compendiums proved to be of tremendous benefit to subsequent students of Buddhism.

Taking these precedents as our inspiration, I recognize that in today's time, too, presentations based on the words of the excellent teacher, the Buddha, pertaining to the basic nature of reality as well as associated philosophical concepts can be a source of benefit to humanity, irrespective of whether one is Buddhist or non-Buddhist, religious or not religious. My aspiration has been to see the creation of these compendiums in a format consistent with the approach of contemporary academic scholarship. This way these presentations can benefit many people. So several years ago I discussed this vision with others and tasked a group of scholars to initiate the project. Today, this group has completed the work of creating compendiums on the presentations on the nature of reality and on philosophy, the first two domains within the threefold division of the subject matter of Buddhist texts. With great efforts the compilers have gathered a vast number of citations from authoritative sources relevant to these two domains. Thus my wish to see such compendiums on science and philosophy from the Buddhist classics—wherein the presentations on these two domains are explained separately in their own rights—has today become a reality. I offer my appreciation to the compendium editors as well as to those senior scholars who have advised the editors. I also thank the translators who have rendered these volumes into other languages.

Given that the scriptures and their commentarial treatises in the Buddhist classics are so vast and profound in their meaning, it is conceivable that there are shortcomings in these volumes in the form of omission, over-reading, or even error. At the very least, what the editors have achieved is a series that demonstrates with clarity that there exists within the subject matter of the Buddhist texts three distinct domains of science, philosophy, and religious practice. If, in the future, there should be a need for additional material or deletion of some elements, the structure is now in place so that such modifications can be made easily.

In conclusion, I would like to share my hope that these volumes on the presentations on the nature of reality and their associated philosophical concepts from the Buddhist sources will make an important contribution to our collective human knowledge by offering the gift of a new set of insights. I pray that these volumes become a source of great benefit to many people.

*The Buddhist monk Tenzin Gyatso, The Dalai Lama*
*Introduction translated into English by Thupten Jinpa*

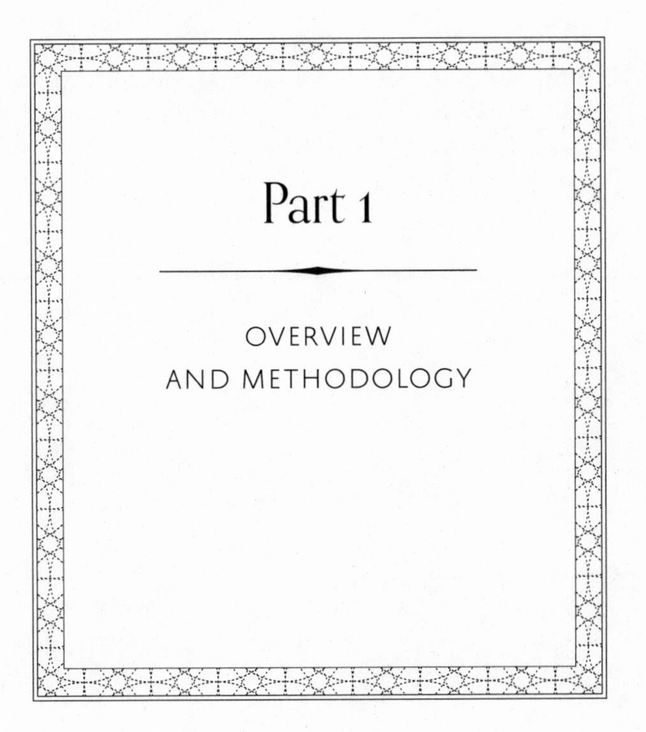

# Part 1

## OVERVIEW AND METHODOLOGY

## Science in What Sense?

Approaching a volume such as this one, the first question a contemporary reader might ask is: Can the presentation of the empirical world in this special compilation be justifiably characterized as science? Today we think of science primarily in terms of its unique method involving testing hypotheses with experiments and control of variables, but this is a fairly recent development. "Science" in its original sense refers to knowledge about the natural world and the laws or principles that govern their behavior. What distinguished this type of knowledge from other forms, even in this classical sense, is its grounding in empirical observations and its intersubjective nature—such knowledge must be findable by others so long as they engage in the same method of inquiry.

With this definition, the views in this series about the facts of our empirical world can be characterized as scientific in a broad sense. Classical Buddhist thinkers were not content simply to make claims about our physical and mental world; they also strove to understand the natural laws governing life. They were interested in the building blocks of reality, such as the atomic constituents of the physical world, how the macroscopic world is formed through aggregation of smaller units, how the cosmos evolved, what defines life and how it emerged, and so on. It's no wonder archeological evidence demonstrates that at the ancient Buddhist university of Nālandā, in northern India, there once existed even an astronomical observatory, which was used by the fourth-century astronomer Āryabhaṭa. This said, just as in the history of science in the West, Buddhist explorations about our natural world were not uniform; they were characterized by a multiplicity of views, arguments, debates, and revisions.

## The Abhidharma Background

It is a truism to say that the ultimate source of Buddhist thought must lie in the teachings of the Buddha Gautama, the founder of Buddhism who appeared in India sometime between the sixth and fifth centuries BCE.

Although decisively determining which of the scriptures attributed to the Buddha—and there are many—were actually taught by him during his teaching career of some forty years is difficult, most early sources agree that what the Buddha taught is captured within the framework of the four noble truths: *suffering exists, there is a cause to suffering, suffering can be extinguished*, and *there is a path that leads to such freedom*. A central message of the Buddha was that since the source of much suffering lies within our own mind, rooted especially in a false belief in an enduring self, it's through transforming our mind that we can alleviate our suffering. The insight into "no self" (*anātman*) came to be closely associated with the Buddha, with its classic analysis of the person in terms of five constituents that make up its existence. Referred to as the *five aggregates*, they are form, feelings, discrimination, formation, and consciousness. Over time this schema of five aggregates evolved into a sophisticated framework within which the entire world of conditioned phenomena came to be subsumed and analyzed.

Evidence suggests, however, that a systematic approach to organizing the Buddha's discourses and drawing out their larger philosophical implications, such as the development of a comprehensive worldview, began sometime in the second or first century BCE. Contemporary scholars refer to this time as the dawn of the Abhidharma period. Literally "highest Dharma (teaching)" or "directed toward the Dharma," the term *abhidharma* contrasts with the word *sūtra*, the latter referring to the discourses given by the Buddha. While the sūtras are narrative accounts of the Buddha's sermons as he traversed the Indian subcontinent, the treatises on Abhidharma are far more scholastic, with almost no narrative elements. They mine the expressions of doctrine in the sūtras and try to distill them in a comprehensive and systematic fashion. What was anecdotal in the sūtras becomes in the Abhidharma a complete system of thought. The influential fourth-century Indian thinker Vasubandhu defines *abhidharma* as "immaculate wisdom together with its attendant factors." As a body of thought and its associated texts, however, Abhidharma is a system of knowledge that enables one to discriminate and analyze existents according to their generic or particular characteristics, establish the true meaning of the Buddha's discourses, and help eliminate defilements of the mind and progress along the path to enlightenment. Over time there emerged what are known as the Seven Books of Abhidharma, which provided the basis for the composition of the *Great Treatise on Differentiation* (*Mahāvibhāṣā*), where what were deemed to be the received positions of the influential Abhidharma school the Sarvāstivāda came to be codified. The Theravāda Abhidharma, which developed parallel to the Sarvāstivāda, also canonized its own seven Abhidharma treatises.

A characteristic feature of Abhidharma texts is their tendency toward taxonomic organization. Matrices group factors of existence together based on shared attributes. The texts also present discursive explanations aimed at clarifying complex points. At the heart of Abhidharma thought is the notion of a *dharma*, a "factor of existence." This use of the word *dharma* is distinct from the same word when it is used to refer to the Buddha's teachings. A *dharma* here is an irreducible unit of reality with its own intrinsic nature, and the view of the world as being comprised of irreducible dharmas came to be referred to as the "dharma theory." The Abhidharma project thus entails a reductive analysis of the composite objects of our everyday experience into their constitutive elements, so that we come to know our own existence and the world as "they really are," not as they appear to our naïve perspective.

By the second century of the Common Era at the latest in the influential Sarvāstivāda school, all existent factors came to be classified within a comprehensive fivefold taxonomy: (1) form or matter, (2) the primary mind, (3) mental concomitants, (4) nonassociated conditioning factors that are neither physical nor mental, and (5) unconditioned phenomena. The first four, characterized by causal interaction and subject to production and destruction, subsume the category of conditioned phenomena. In contrast, those in the fifth category are devoid of arising or cessation. Through their reductive analysis the Sarvāstivāda Abhidharma enumerates seventy-five dharmas and classifies them as follows: matter (eleven), the mind (one), mental concomitants, or mental factors (forty-six), conditioning factors (fourteen), and unconditioned factors (three).

Although the specifics of the Abhidharma taxonomy as well as the underlying dharma theory came to be disputed by other Buddhist schools, the basic classification of reality into distinct categories of conditioned and unconditioned, as well as the differentiation of the conditioned world into the physical, the mental, and abstract conditioning factors (time, action, and so on), became established as part of the foundational norms of critical inquiry in Buddhist thought. However, with the growing influence of the Sautrāntika school (possibly around the third century) and especially after Dignāga's fifth-century philosophical contributions, there was a shift from ontology to epistemology. With this epistemological turn, questions such as "What is real?" "What defines existence?" and "Do universals exist?" came to dominate the Indian Buddhist world.

## Classifying Phenomena in Early Buddhist Texts

Part 1 of this volume begins with a brief comment on how different systems of classification of phenomena are utilized by Buddhist texts in their investigation of the natural world. We already noted the framework of the five aggregates, which though originally serving as a basis for analyzing the nature of the person later emerged as an overarching system for classifying the conditioned world. There are, however, two further classification systems pervasive in the Buddha's discourses—those of the twelve bases (*āyatana*) and the eighteen elements (*dhātu*). The term *āyatana* literally connotes "gateway for arising," and it is through these twelve that consciousness and the various mental factors come to arise and proliferate. The twelve bases consist of the six sense faculties—eye, ear, nose, tongue, body, and the mind—and their corresponding objects—forms, sounds, smells, tastes, textures, and mental objects. This twelvefold taxonomy encompasses all the factors necessary for the arising of mental processes. They are also thought to encompass everything that exists, with the category of "mental object" subsuming even the unconditioned.

In the classification of eighteen elements, the category of mind is further differentiated into six types of consciousness—visual, auditory, olfactory, gustatory, tactile, and mental. In this model, we now have three sets of six: the six sense faculties, their corresponding objects, and the six classes of consciousness that arise from the combination of the senses and their objects. In this system, the eighteen are referred to as *elements*, a translation of the Sanskrit *dhātu*, which has a broad range of meanings even within Buddhism. Vasubandhu defines the term in this context as a "species" or "type" as well as an "origin." Connecting these two senses of the term, he explains how each of the eighteen elements is the origin for a continuum of its own type. In its original context in the Buddha's discourses, the eighteen element schema was probably a pragmatic list to explore more deeply the teaching on no-self by analyzing the person in terms of sense faculties, objects, and consciousnesses. Later these classification systems became integral to the "discrimination of dharmas" (*dharmapravicaya*), recognized as the principal method of insight into reality.

## The Structure and Sources of the Present Compilation

The first two volumes of *Science and Philosophy in the Indian Buddhist Sources* are organized around five topics: (1) the nature of the world, the object of knowledge, (2) mental phenomena, the cognizing subject, (3) how the mind engages its object, (4) the means by which the mind engages its object, and (5) the nature of the person that knows. In this compilation, the first topic is most extensive, occupying the entire first volume. The remaining four themes are treated in volume 2. This fivefold framework takes its inspiration from the highly influential epistemological works of Dignāga and his primary interpreter Dharmakīrti (seventh century), whose writings profoundly shaped philosophical thinking in Indian Buddhism as well as the Tibetan tradition that inherited it.

The Indian Buddhist classics to which the editors of our compilation are referring constitute a vast number of texts grouped into two canonical collections in Tibetan translation. Known as the Kangyur, literally "translated sacred words," and Tengyur, "translated treatises," the first contains scriptures and the second over three thousand treatises. These texts were translated, sometimes multiple times, beginning in the seventh century, and the canon we have today was largely closed by the fourteenth century, after Buddhism had disappeared from India. In the present series, the Tibetan editors tell us which classical Indian Buddhist sources they relied on for each of the five broad themes. That said, this usage in no way suggests that the authors of these classical treatises saw their texts as relevant only in narrow domains of inquiry. Many of these works were written with a comprehensive aim of presenting all the knowledge essential for awakening—from regulating difficult emotions to attaining the highest wisdom of the buddha.

## Four Schools of Classical Buddhist Philosophy

One thing helpful to keep in mind when reading this series is the framework of the four classical Buddhist philosophical schools. As much as the Tibetan editors of these volumes have striven to present views of individual Indian Buddhist thinkers rather than established standpoints of the philosophical schools they are thought to belong to, the reading of these Indian thinkers remains inextricably colored by the received views on the positions of these schools. This differentiation of Buddhist philosophy into the perspectives of four main schools and critical comparison of their views began in India, at the latest by the sixth century, and became a major focus in the Tibetan tradition as well. These four schools are Vaibhāṣika, Sautrāntika, Cittamātra (Mind Only), and Madhyamaka (Middle Way). Broadly defined, the first school includes all the Abhidharma schools, such as the Sarvāstivada and the Theravāda school. Sautrāntikas, literally meaning the "proponents of sūtra" (as opposed to the Abhidharma), reject dharma theory with its monadic taxonomy of reality and propose a simpler, more epistemologically oriented, ontology. The core positions presented in Vasubandhu's autocommentary to his *Treasury of Knowledge* as well as large parts of the views presented in the epistemological works of Dignāga and Dharmakīrti are recognized as standpoints of this school. Key historical figures of the Cittamātra or Mind Only school, known also as Yogācāra, are Asaṅga (fourth century) and his brother Vasubandhu; moreover, the more-evolved standpoints in the writings of Dignāga and Dharmakīrti are understood as representing the Mind Only school, which denies external material reality. The final school, the Madhyamaka, was founded by the influential Buddhist thinker Nāgārjuna. The central tenet of this school is the concept of emptiness, a radical relativistic standpoint that maintains that nothing exists outside the world of dependent origination; nothing, in other words, exists independently or possesses an intrinsic nature. In contemporary parlance, this is a nonreductionist and antirealist school that questions the very project of seeking any metaphysically complete description of reality. Key figures of this school also include Āryadeva, Bhāvaviveka, Buddhapālita, Candrakīrti, and Śāntideva.

## Principles and Methods Underlying Buddhist "Scientific" Inquiry

An important topic in part 1 relates to what in contemporary parlance we might call "methodological issues." These include (1) basic philosophical views on regulative principles or laws in nature, (2) epistemological theories pertaining to sources of knowledge and their scopes, and (3) logical principles—the law of identity, the law of contradiction, and so on—essential for gaining knowledge about the world.

Though not found in the Pali sources, some early Sanskrit Buddhist texts exhort the adoption of the principle of "four reliances": rely on the teaching, not on the person; rely on the meaning, not the words; rely on the definitive meaning, not the provisional meaning; and rely on wisdom knowledge, not ordinary intellectual understanding. Similarly, in an early Buddhist scripture, the Buddha admonishes his monk disciples not to take his words at face value, out of faith and reverence to him, but to accept their validity based on one's own investigation, just as a goldsmith tests gold by burning, cutting, and polishing it. Although for Buddhists, historically, the Buddha's words did carry enormous weight, for the more scientifically and philosophically minded schools, there emerged a hierarchy for evaluating sources of knowledge, with direct perception the most reliable, followed by inference, and finally scriptural testimony, which is valid only within an extremely narrow scope of inquiry, when dealing with topics or facts inaccessible to ordinary cognition.

This prioritization of direct perception over inference, reason over scripture, has an ontological corollary whereby the world, the object of knowledge, is characterized by three distinct types of facts. There is, first and foremost, the world of *evident facts*, which are directly perceptible to our senses; second, there are *slightly obscure facts*, which though not accessible to our senses can be inferred based on observed facts that are logically related to those hidden dimensions. Finally, there are *extremely obscure facts*, such as the specific details of the subtle workings of the law of karma, which for ordinary humans are knowable only by relying on the scriptures.

Now, it is in knowledge about the second domain, slightly obscure facts, as well as interpretation of the first domain, evident facts, where reason comes to play a critical role in Buddhist inquiry. Just as in ancient Greek thought as well as its descendent, contemporary Western scientific thinking, reason in Buddhist thought is premised on acceptance of a basic set of logical principles. These include the laws of identity, contradiction, and the excluded middle, and also the laws connected with logical relationships, such as between universals and their instantiations and between causes and their effects. It's only by taking into account these logical principles that reason can connect what is observed to something unobserved but logically related and use that to gain reliable inferential knowledge. We can see that even for reason-derived knowledge, validity depends upon a process of verifiability traced to empirical knowledge of an observed fact.

Another important methodological principle identified in this section is what is known as the "four principles of reason." These are the principles of nature, function, dependence, and logical evidence. The idea here is that the laws of evidence or proof can only operate on the basis of the preceding three principles. One can use the logical reasoning "if *a* then *b*" because there is a causal relationship between the two that supports the principle of dependence. And this kind of dependence is possible by virtue of the fact that the two relata individually possess distinct natures of their own endowed with specific functions they perform. This entire chain is supported by the fact that things are the way they are in reality, the principle of nature. Essentially, when the line of analysis comes to the principle of nature, the process of inquiry ends, for here the only explanation that can be offered is simply this: "That's the way it is."

Can Buddhism, at least as characterized above, be viewed as a form of empiricism? There is also the related question, raised earlier, of whether what is in this compilation can be justifiably characterized as "science." My own sense is that the answer to the first question is both yes and no, depending on the context. Yes, the Buddhist standpoint, at least shaped by Dignāga and Dharmakīrti's epistemology, can be broadly viewed as some form of empiricism with respect to specific domains of knowledge—that is, when it comes to knowledge of the first category of facts, empirically evident facts, as well as the second category of facts, those that though empirically opaque can be logically inferred. As observed before, in these two domains, it is direct perception that remains the ultimate source of our knowledge. The answer is no if we are to speak of our knowledge of all the facts. Clearly, Buddhism does not subscribe to the view that all facts are accessible by our senses or by knowledge derived from our senses.

How strict are the boundaries among these three domains of facts? Are these objective features of reality or are they relative to the capacity of the mind observing them? In other words, do these distinctions collapse when it comes to the Buddha's wisdom?

These are important questions that may delimit the interface between Buddhism and science. Even so, it seems clear that in both the physical world and the inner mental world, Buddhist method, like science, appears to entail, at least in expectation, that all theories should be judged against empirical observation rather than being prejudged based on some *a priori* reasoning or scriptural statements. Judging from the competing views found in the texts as well as the multitude of arguments supporting or critiquing specific positions, there does seem to be a shared consensus among Buddhist theorists that veracity of one's propositions should be judged against the criteria of empirical evidence, internal coherence, and reason. It's this kind of methodological commitment that is behind the Dalai Lama's famous statement that if science were to ever prove the absence of rebirth, the Buddhists will have to accept such a conclusion.

Part 1 concludes with a brief introduction to the Tibetan art of reasoning known as the Collected Topics (*bsdus grva*, pronounced *düra*), a systematic approach applying the fundamental principles of logic in dialectics. This training in Collected Topics ("collected" or extracted from the great Indian Buddhist classics on logic and epistemology) is, in fact, the first stage in Tibetan monastic education, preparing the student in the art of debate and reasoning so that he or she can critically engage in the study of the great philosophical classics. The reason for its inclusion here, at the beginning of the compilation, is also help sensitize the reader to the role reasoning plays in formulating and further refining our views about reality.

## Further Reading in English

For an accessible introduction to Abhidharma philosophy, see Jan Westerhoff, "Abhidharma Philosophy," in *The Oxford Handbook of World Philosophy*, edited by William Edelglass and Jay L. Garfield (New York: Oxford University Press, 2011).

For a succinct historical introduction to Abhidharma, see part 1 of Collet Cox, *Disputed Dharmas: Early Buddhist Theories of Existence* (Tokyo: The International Institute of Buddhist Studies, 1995).

For an engaging and personal discussion on the methodological issues related to Buddhist exploration of the nature of reality and their parallels with the scientific method, see chapter 2 of the Dalai Lama, *The Universe in a Single Atom* (New York: Morgan Road Books, 2005).

For a collection of informative essays on the relationship between Buddhism and science, see Alan Wallace, ed., *Buddhism and Science: Breaking New Ground* (New York: Columbia University Press, 2003).

For a detailed yet accessible introduction to the Tibetan system of debate and reason as developed in the Collected Topics, see Daniel Perdue, *The Course in Buddhist Reasoning and Debate* (Boston: Snow Lion Publications, 2014).

# 1

## Systems of Classification

WHEN ANCIENT INDIAN Buddhist texts present the topic of reality, they rely on numerous systems of classification. For example, in the *Attainment of Knowledge* (*Jñānaprasthāna*), which belongs to the Seven Treatises of Abhidharma,[28] as well as its commentary the *Great Treatise on Differentiation* (*Mahāvibhāṣāśāstra*), three groupings are used to classify all phenomena: (1) aggregates, (2) elements, and (3) bases. First, all conditioned phenomena are systematically presented in the five aggregates (*skandha*): the aggregates of form, feeling, discernment, formation, and consciousness. When external and internal phenomena are classified by means of object, sense faculty, and consciousness, phenomena are presented in terms of the eighteen elements and the twelve bases. The eighteen elements (*dhātu*) consist of six objective elements—form, sound, smell, taste, tactility, and mental objects; [140] six sense-faculty elements that support consciousness—the eye sense faculty, ear sense faculty, nose sense faculty, tongue sense faculty, body sense faculty, and mental sense faculty; and six subjective elements that are supported consciousnesses—eye consciousness, ear consciousness, nose consciousness, tongue consciousness, body consciousness, and mental consciousness. When the term *base* (*āyatana*) is applied as part of this classification system, all phenomena are categorized in twelve bases: the six objects comprising the sense base of form, sound, smell, taste, tactility, and the base of mental objects; and the six sense faculties comprising the eye sense base, ear sense base, nose sense base, tongue sense base, body sense base, and mental sense base. In these classification systems, unconditioned phenomena are included in the mental-object element from the list of eighteen elements and the mental-object base[29] in the enumeration of the twelve bases.

Alternatively, in Vaibhāṣika works such as Vasubandhu's *Treasury of*

*Knowledge* (*Abhidharmakośa*), the following framework of five basic categories of reality is presented: (1) evident material form, (2) primary minds, (3) concomitant mental factors, (4) nonassociated formative factors, and (5) unconditioned phenomena.

According to this classification all external and internal physical phenomena—the five sense objects such as material form and so forth and the five sense faculties such as the eyes—are included in the basic category of *evident material form*. The phenomena that constitute the inner world of consciousness are subsumed in the two categories of *primary minds* and *concomitant mental factors*. Those conditioned phenomena that belong to neither the class of material form nor consciousness—such as the three attributes of conditioned phenomena, namely, (1) arising, (2) enduring, and (3) disintegrating, as well as the three times of past, present, and future, and so forth—encompass the basic category of *nonassociated formative factors*. Finally, those phenomena that are not produced by causes and conditions constitute the basic category of *unconditioned phenomena*. In this way, all knowable objects are systematically classified in terms of the five basic categories of reality. [141]

In Indian Buddhist texts on epistemology, however, the presentation of reality is framed in terms of the following three topics: (1) the objects of knowledge, (2) the mind that knows, and (3) the way the mind engages objects. In contrast, in the texts of the Mind Only school, reality is presented in terms of the theory of the three natures: (1) dependent nature, (2) perfected nature, and (3) imputed nature.[30] In the texts of the Middle Way (Madhyamaka) school, phenomena are presented in terms of (1) the basis, which comprises the two truths,[31] (2) the path encompassing both method and wisdom,[32] and (3) the result, namely, the truth body and the form body of a buddha.[33] In the texts of tantra the presentation is arranged in terms of (1) the *ground*, that is, the nature of reality, (2) the stages for proceeding on the *path*, and (3) how the *result* is actualized. Thus, in brief, different formats were employed in classical Indian texts to systematically present the nature of reality.

In this two-volume compendium, we have organized our presentation on reality in terms of five major themes:

1. The presentation of knowable objects
2. The presentation of the cognizing mind
3. How the mind engages its objects

4. The means by which the mind ascertains objects
5. The presentation of the person, the agent who ascertains objects

One might ask, "Since there are a vast number of Indian Buddhist treatises, what are the principal sources for this compendium?" Here the presentation of physical phenomena in general, the primary elements and derivative form in particular, and the subtle and coarse particles that comprise material entities are based [142] by and large on the textual tradition of Abhidharma. We take as our sources: (1) the texts of the upper Abhidharma system, such as Asaṅga's *Compendium of Abhidharma* (*Abhidharmasamuccaya*) and its associated commentaries, (2) the Seven Treatises of Abhidharma as well as the *Great Treatise on Differentiation*, and (3) the treatises that expound these works, such as Vasubandhu's *Treasury of Knowledge* as well as its associated writings belonging to the lower Abhidharma system.[34] At times we have also consulted Bhāvaviveka's *Blaze of Reasoning* (*Tarkajvālā*) and its commentarial tradition. In the specific section where the question of whether indivisible particles exist or not is analyzed, we have principally selected the logical arguments negating the existence of partless entities presented by Nagārjūna, Āryadeva, Candrakīrti, and so on. We have also drawn from Vasubandhu's *Twenty Verses* (*Viṃśatikā*). As for the formation and destruction of external world systems and their inhabitants, and the progressive development of the fetus in the womb and so on, by and large we have relied on *Nanda's Sūtra on Entering the Womb*, the texts of the upper and lower Abhidharma systems, the corpus of the Kālacakra tantra, and traditional medical texts. Information concerning the channels, winds, and drops of the subtle body is derived from the texts of highest yoga tantra as well as works on medical science. The presentation on the brain too is based on traditional medical texts. [143]

Regarding the presentation on the subjective mind, we have drawn primarily on Asaṅga's Five Treatises on the Grounds as well as Yogācāra texts associated with these works, Vasubandhu's *Treasury of Knowledge* and its *Autocommentary*, and in particular the texts of the Buddhist epistemological tradition, such as the works of Dignāga and Dharmakīrti, as well as their subsequent commentators. The presentation on mental factors is compiled from the texts of the upper and lower Abhidharma systems. With regard to the levels of subtlety of mind, in terms of the notion of "coarse" and "subtle" minds contingent on (the subtlety of the mental states of) higher and lower realms of existence, we principally rely on the texts of the upper and lower Abhidharma systems as well as works pertaining to the Yogācāra school. As for the levels of subtlety of mind, the stages of dissolution of wind and mind, and the stages of death and so on as described in the highest yoga tantra system, we have drawn principally from the textual corpus of Guhyasamāja tantra as well as the cycle of texts belonging to the Kālacakra system.

On the way in which the mind engages its objects, we have taken the texts of the Buddhist epistemological tradition as our principal source while also drawing on the Yogācāra texts of Asaṅga and his followers. With respect to the means by which the mind engages its object, in terms of the application of formal reasoning, we have based our presentation on the Buddhist sūtras, Asaṅga's *Śrāvaka Grounds* (*Śrāvakabhūmi*), and so on, as well as Buddhist epistemological writings. On the essential points pertaining to the methods of training the mind, we have relied mainly on the following works: the *Unraveling the Intention Sūtra* (*Saṃdhinirmocana-sūtra*), the texts of the upper and lower Abhidharma systems, and works belonging to the Yogācāra school. In particular, we have utilized these specific texts: Śāntideva's [144] *Engaging in the Bodhisattva's Deeds* (*Bodhicāryāvatara*) as well as his *Compendium of Training* (*Śikṣāsamuccaya*), Dharmakīrti's *Exposition of Valid Cognition* (*Pramāṇavārttika*), and Kamalaśīla's *Stages of Meditation* (*Bhāvanākrama*). Finally, on the presentation of the person, the one who ascertains objects, we have drawn from treatises such as those of the Middle Way school.

# 2

# Methods of Inquiry

## Critical Analysis and the "Four Reliances"

QUESTION: When early Indian Buddhist thinkers in general and those of Nālandā University[35] in particular established their presentations on reality through various approaches, such as in terms of the five aggregates, the elements, and the sense bases, did they do so on the basis of a literal acceptance of what is found in the scriptures, or did they develop their presentations based primarily on logical reasoning?

Response: In general, according to Buddhism, the way one should engage in the analysis of reality was stated in a sūtra by the Buddha himself: [145]

> Monks and scholars, just as you test gold
> by burning, cutting, and polishing it,
> so too well examine my speech.
> Do not accept it merely out of respect.[36]

Thus the Buddha advised his followers that when they examine the meaning of the scriptures, they should rigorously analyze the relevant topic just as a skilled goldsmith tests gold by burning, cutting, and polishing it. And only when one has gained confidence in its words should one have a conviction about a scriptural statement. It is inappropriate to have a conviction about something without critical analysis, simply because it has been stated by one's teacher. So this sūtra statement demonstrates that when analyzing the nature of reality, it is evidence derived from reasoning that is primary, not scripture.

In Buddhist texts in general and those of the Nālandā thinkers in particular, when a specific issue is analyzed, greater importance is accorded to direct experience compared with logical reasoning and scripture. Moreover, if one were to take scripture to be the sole authority when inquiring into the nature of reality, [146] that scripture too would require validation by another scripture. This second scripture, in turn, would need the support of another scripture, and so on, resulting in an infinite regress. Thus these words of the Buddha demonstrate that validation of scripture must ultimately rest on stainless reasoning, and such flawless logic must be grounded, in turn, in valid experience. Such an approach is commended by the Buddha.

Furthermore, if we were to investigate the treatises composed by eminent Nālandā thinkers, we find that readers should approach such works on the basis of critical inquiry pertaining to the subject matter of the text, not through faith and conviction from the beginning. When positive doubt (that is, doubt tending to the fact)[37] has arisen, this is the stage when one engages the subject matter of the treatise. These thinkers also say that if the subject is presented in terms of the *five elements of exposition*, it helps to make the contents of the text easily perceivable to the reader's mind. The five elements of exposition are:

1. Any treatise should have a *purpose* for its composition.
2. There should be a *summary presentation* of the main body of the treatise so that the essential points of the subject matter of the text may easily appear to the mind of the reader.
3. Then there should be a *word meaning* that constitutes the exposition of the text.
4. There should be *objections and rebuttals* that critically analyze difficult points and areas of doubt.
5. There must also be *transition markers* that establish the connection between earlier and later sections of the text.

The custom thus emerged of emphasizing the importance of all five elements of exposition in any treatise. Vasubandhu's [147] *Rational System of Exposition*, for example, states:

> Those who would elaborate the meaning of sūtras
> should do so in terms of *purpose*, *summary*,
> *word meaning*, and *transitions*,
> as well as the *responses to objections*.[38]

When engaged in critical inquiry into the nature of reality, four reliances constitute the minimum condition required of the person. So when an interested person examines a treatise, he should not heed how famous the author is, nor the eloquence of his words. He should instead look at the composition of the treatise and examine if it is relevant or not. Even with respect to the contents, he should not heed the literal meaning but should focus instead on the definitive or the ultimate meaning of the text. And he should not be satisfied with understanding the meaning of the text by study and deliberation that employs only the immediate (ordinary) mind. He should instead combine critical reflection and meditation to generate a deep and profound realization of that meaning. This approach is referred to in the classical Buddhist texts as the *principles of the four reliances*. The *Questions of the Householder Ugra Sūtra*, for example, states:

> Rely on the meaning, not on the letters.
> Rely on transcendent wisdom, not on ordinary consciousness.
> Rely on the teaching, [148] not on the person.
> Rely on sūtras of definitive meaning, not on sūtras of provisional meaning.[39]

"Rely" should be understood to mean "have conviction in." The four things that should be relied on are the teaching, the meaning, the definitive meaning, and transcendent wisdom. The four things that should not be relied on are the person, the words, the provisional meaning, and ordinary consciousness.

How to rely on the first four and not rely on the latter four is as follows. "Rely on the teaching, not on the person" means that one should not accept or reject a teaching without critical analysis simply because of the fame of the person. Instead, on the basis of analyzing the teaching one should respond, "This I shall accept, for it's well taught" and "This I shall reject, for it's not well taught."

"Rely on the meaning, not on the words" means that one should not focus on the manner in which a statement is conveyed; instead one should focus on the meaning itself and apply oneself to understanding a clear presentation of its significance. In contrast, if the statement is flawed then even if the manner in which it is conveyed is finely crafted it must be rejected.

"Rely on the definitive meaning, not on the provisional meaning" means to recognize those statements of the Buddha whose literal meaning can be undermined by logical reasoning to be instruction given to help his followers. Such teachings, though not representing the Buddha's final intention, were taught for the sake of others for a specific purpose. [149] In contrast, those statements whose literal meaning cannot be undermined by logical reasoning—instruction that expresses the ultimate nature of reality representing the Buddha's final intention—one should trust as definitive in their meaning.

"Rely on transcendent wisdom, not on ordinary consciousness" means that when one seeks the definitive meaning, one should not place confidence in an ordinary person's sensory cognition, such as eye consciousness, nor should one trust what appears to the mental conception, a cognition that is mistaken with respect to the ultimate nature of reality. Instead one should generate conviction either in the nonconceptual transcendent wisdom of the sublime beings who see the way things really are, or in the rational cognition understanding suchness that gives rise to that transcendent wisdom. Alternatively, the principle of reliance states that when it comes to the definitive meaning, one should not take what is known by ordinary consciousness, which is laden with dualistic appearance, to be primary. Rather one should trust the truth as perceived by the mind of equipoise in which all forms of dualism have ceased. These points are explained in Asaṅga's *Bodhisattva Grounds*:

> Again, the bodhisattva[40] correctly and fully understands what is defective[41] or black instruction and what is great or pure instruction, and having well understood this he relies on logical reasoning. He does not rely on the person on the basis of the statement "This truth was explained by an elder, or a learned person, or the Tathāgata,[42] or the Saṅgha." Thus, relying on logical reasoning and not on the person, he will not deviate from the truth of suchness (*tattva*), and he will also not be swayed by any other factors.[43]

"Presenting defective instruction" refers to those treatises that promote only the causes of suffering, while "presenting great instruction" refer to those treatises that promote ways to cultivate the antidotes that help eliminate suffering.

With respect to [150] relying on the meaning, not on the words, *Bodhisattva Grounds* states:

> A bodhisattva listens to teachings because he desires the meaning, not because he desires well-crafted words. So the bodhisattva who relies on meaning will listen respectfully even when instruction is given in everyday common language.[44]

As for relying on the definitive meaning, not on the provisional meaning, the same text states:

> The bodhisattva who cherishes faith and joy in the Tathāgata and definitely delights in his words will rely on the Tathāgata's definitive sūtras, not on those of provisional meaning. For if he relies on the Tathāgata's sūtras of definitive meaning he will not be robbed of the teaching's capacity to tame. Because of the diversity of facts they relate, the sūtras of provisional meaning lack certainty and give rise to doubt. If the bodhisattva were to lack certainty about the sūtras of definitive meaning, then he would be robbed of the teaching's capacity to tame.[45] [151]

As for a reliance on transcendent wisdom, not on ordinary consciousness, the text states:

> A bodhisattva views realization that is transcendent wisdom to be the heart essence, not mere ordinary cognition of the teachings and its meaning derived from study and critical reflection. Therefore recognize that any phenomena understood through meditation cannot be understood by mere ordinary consciousness that engages its object through study and contemplation, and when you hear instruction uttered by the Tathāgata on the doctrine that is supremely profound, neither reject nor denegrate it.[46]

## Ascertaining Reality with Three Types of Object

All objects of critical inquiry are encompassed within the three types of ascertainable object (*prameya*). In turn, the three types of ascertainable object are differentiated by whether the mind cognizing them relies on reason-based inference or valid testimony and so on. Thus there are (1) those objects experienced through direct perception such as material form observed by eye consciousness. There are also (2) those objects not directly established by experience but established through inference based on objective fact, such as the presence of fire behind a distant mountain pass that is apprehended through taking smoke [152] as a logical sign. Finally there is (3) the category of objects that cannot be ascertained by direct sensory experience nor through inference based on objective fact but are established by inference based on conviction dependent on another's valid testimony. Thus there are three categories:

1. Evident perceptible phenomena
2. Slightly hidden phenomena
3. Extremely hidden phenomena

Regarding extremely hidden phenomena, there are even common, everyday facts that can be apprehended only in dependence on valid testimony, such as ascertaining one's birth date by the say so of one's parents, or different events that occurred in the past that may be known only through historical accounts. This being said, there are certain types of extremely hidden phenomena that can be established only through reference to the valid scripture of the Buddha.

In such contexts one may apprehend whether the testimony of the Buddha is valid. The point is not to cite the testimony itself as a logical proof, such as the statement "Wealth arises from generosity because it is stated so in scripture." Rather, the inference proceeds by establishing that (1) the content of the scriptural statement is not contradicted by direct perception or logical reasoning, (2) the words of scripture are not directly or indirectly contradicted by earlier or later scriptural statements, and (3) the author of the statement had no ulterior motive and so on. The critical point is that such conditions must be met when taking scriptural testimony as a source of inference. This matter will be discussed in detail in a later section.

In brief, when the presentation on the nature of reality was settled within the tradition of the Nālandā masters, it was achieved through reference to these three epistemological strategies. In their system, inference reliant on reasoning based on objective fact was accorded greater importance than inference [153] based on valid scripture or testimony, yet when compared with inference based on objective fact, the valid cognition of direct perceptual experience was granted higher priority.

## Analysis Grounded in the Four Principles

The Bodhisattva canon[47] (Bodhisattvapiṭaka) mentions four principles of reasoning as a mode of inquiry into the nature of reality:

1. The *principle of nature* takes the mode of being of things as its basis and refers to the specific character, or natural essence, or reality of things.
2. The *principle of function* takes the principle of nature as its basis and refers to any function performed that is in harmony with the basic state of its own essential nature.
3. The *principle of dependence* takes the principle of function as its basis and refers to how one thing depends on another as revealed in their natural modes of reliance, such as the relation of cause and effect, of part and the part possessor, and of the tripartite relation of action, agent, and the object of the action and so on.
4. The *principle of evidence* takes the first three principles (of nature, function, and dependence) as its basis, and refers to the capacity to prove a thesis through formulating formal reasons such as: "Since it is this it must be that," "Since this exists that must exist," "Since it is not this it must not be that," or "Since this does not exist that must not exist," and so on. This is called the *principle of evidence.*

Therefore the four principles of reasoning rely on the mode of being of things as they are, based on (1) the individual natural essence of a thing, (2) any function supported by that natural essence, [154] (3) how one thing depends on other things in reliance on that function, and (4) the reason why they are reliant in that way. These principles of reasoning will be elaborated in the context of explaining how the mind discerns objects.

## Causality and Dependent Origination

The law of causality and the system of dependent origination represent, out of the four principles of reasoning, the *principle of dependence* and the *principle of function*. These two matters are accorded crucial importance in the Buddhist tradition. It is on the basis of causality and dependent origination that the origin of everything is explained, from the larger evolution of the cosmos and its inhabitants to specific discrete events such as the fall of rain in a particular place at a particular time. Likewise, within the Buddhists' own system, everything from the manner in which sentient beings cycle within saṃsāric existence[48] because of the power of their karma[49] and affliction[50] to the process by which the causes and conditions that lead to birth in saṃsāra are gradually brought to an end, leading to attainment of highest excellence,[51] are explained as accomplished exclusively through the law of cause and effect. Thus, whether explained from the point of view of fact or from the standpoint of historical accounts, the logic of causality constitutes a critically important feature of Buddhist philosophy.

With respect to the Buddha himself, when he turned the wheel of Dharma[52] in Vārāṇasī pertaining to the four noble truths, he clearly set forth the principles of the four noble truths by describing two types of causal process: one that leads to suffering, the other to happiness. He spoke of the truth of suffering (the effect) [155] and the truth of its origin (karma and afflictions that give rise to suffering). He spoke of the truth of pacification or cessation of suffering (the effect) and the truth of the path (that which leads to the attainment of such cessation).

When elaborating in detail the causality of happiness and suffering, not only did the Buddha do so in terms of the twelve links of dependent origination,[53] but he also spoke of how each of these twelve links arise from their preceding link and how the origination of a subsequent link in the chain depends on the preceding link. Similarly, the Buddha taught that through the cessation of the preceding link a subsequent link ceases and that the cessation of the subsequent link depends on the cessation of the link that precedes it in the chain. Thus he taught with extreme clarity the law of causality in terms of how effects invariably follow their causes. The followers of the Buddha accorded great importance to the way the Buddha proclaimed causality and dependent origination. There is, for example, the following verse, known as the essence of dependent origination:

All phenomena originate from causes;
the Tathagata spoke of their causes;
also that which brings about cessation of these causes.
This too the Great Shramana has taught.[54] [156]

A tradition spread from ancient times that this sūtra verse should be inscribed above the doors and beams of temples and so on, a fact clearly discernible from closely examining ancient artifacts and archeological sites. But how is this system of dependent origination, in terms of causality, stated in sūtra? The *Rice Seedling Sūtra* states:

> What are the external phenomena of dependent origination that are related to causes? It is like this, from a seed comes a sprout, from a sprout a leaf, from a leaf a stem, from a stem a node, from a node the bud, from the bud a flower, from the flower the fruit. If there is no seed the sprout will not arise . . . up to . . . if there is no flower then the fruit will not arise. If there is a seed, the sprout will come to manifest. So too, if there is a flower, then everything up to the fruit will also come to manifest.

And:

> In what way should we view the external phenomena of dependent origination related to conditions? They arise from the assembly of the six elements. What are those six elements that assemble? It is like this. It's through the assembly of the elements of earth, water, fire, wind, space, and time, [157] and you should regard external dependently arisen phenomena that arise from their assembly as dependent on their conditions.[55]

Thus this sūtra explains the way external dependently arisen phenomena such as sprouts arise in dependence on causes and conditions. Furthermore, the same sūtra states that, just like external dependently arisen phenomena, internal dependently arisen phenomena, such as mental formations[56] and intention, also arise from causes and conditions: "So too internal dependently arisen phenomena have two types."[57] And when this system of causal dependent origination is investigated in detail, it may be summarized in terms of three conditions: (1) the condition of the absence of the prior design of a creator, (2) the condition of impermanence, and (3) the condition of potentiality. Given that these three conditions are more relevant to the discussion on the evolution and formation of external world systems and their inhabitants, they will be addressed in that later section.[58]

In general there are two ways of classifying causes: in terms of the division of direct cause and indirect cause, or in terms of substantial cause and cooperative condition. This system of classifying causes is a crucial point in many contexts when the Buddhist tradition explains its presentation on reality, such as when examining the essential nature, causation, and function of things. There are also other factors, such as how among causes there is a difference of importance, a difference of temporal proximity in relation to the effect, and how different aspects of the effect represent the imprint of the specific attributes of the cause, and so on. These points will be discussed in detail in a later section.[59] [158]

In brief, in the Buddhist tradition:

- One cannot accept that an effect arises without a cause.
- One cannot assert that the cosmos originates from the prior design of a creator.
- The arising of things from a permanent cause is illogical.
- Cause and effect must be commensurate.
- That specific results arise from specific causes is the nature of reality and cannot be otherwise.
- The relationship of cause and effect is a natural state of phenomena.
- Effects arise from various causes and conditions due to the law of dependent origination.

Now this system of causal dependent origination represents the understanding of dependent arising common to all Buddhist philosophical schools. There is, however, another meaning of *dependent origination* and *reliance* in terms of dependence on parts, which is subtler than the notion of *causal dependence*. There is also another meaning of dependent origination even subtler than dependence on parts that is found in texts of the Middle Way school. These will be discussed in the section on philosophy.[60] [159]

## Reasoning with the Laws of Contradiction and Relation

In the above sections on causal dependent origination and the four principles of reason, two modes of dependence emerged: the nonsimultaneous mode of reliance, such as an effect's reliance on its cause, and the simultaneous mode of reliance, such as between parts and the whole that possesses those parts, or the composite whole and the components that are its constituents and so on. In the treatises of classical Indian logicians these kinds of relationship were referred to as "contingent relations," and the two types of relation were called, respectively, causal relation and intrinsic relation. The first reveals the relationship of a cause to an effect, which are temporally sequential, whereas the latter indicates a relationship of reliance that is temporally simultaneous. Intrinsic relation, in turn, has different types. There is, for example, the relationship between the whole and its parts, such as between a jar and the components of a jar, the relationship between an instance and the class that it belongs to, such as a cypress tree and trees that share the same intrinsic nature, and the relationship of identical intrinsic nature that exists for phenomena that are mutually inclusive, such as being a product and being impermanent.

Phenomena that are intrinsically related are differentiated by means of language and concept, but in terms of their reality no such differentiation exists. For example, the thesis "sound is impermanent" can be inferred on the basis of the logical reason "because it's a product" [160] primarily because there exists a natural intrinsic relationship between being produced and being impermanent. Thus when engaging in refined rational analysis on the nature of reality, classical Indian logicians declare logical relationship to be a crucial point.

There are certain kinds of phenomena that possess this type of internal contingent logical relationship, in that if one does not exist the other would not exist. There are other kinds of phenomena that have relationships of internal mutual exclusion. For example, through the logical reason "because it is produced," one is able to establish that "sound is impermanent" because being permanent and being impermanent are directly contradictory in the sense of excluding each other. The logical reason "being produced" excludes "being permanent"—which is the property of the negandum—and there is a logical entailment that whatever is a product is necessarily impermanent. Similarly, when the logical reason "because intense fire exists at that site" negates the presence of such things as goosebumps—which result from the sensation of being cold—at that same site, the fact that one can negate the existence of the effects of cold temperature is because cold temperature that is the cause of goosebumps and the presence of intense fire are contradictory through incompatibility. In the texts of logic and epistemology, this type of proof is referred to as *a logical reason based on the observation of excluding contradictory phenomena.*[61]

In brief, in the process of establishing the basic view of reality, it is essential that matters that are obscure be clarified by formal reasoning. [161] In such contexts, if one lacks an understanding of how the principle of contradiction operates—the logical reason and the property of the negandum are mutually exclusive and the inferential cognition undermines the contrary perspective and so on—one will not be able to establish a given thesis. Understanding these points is therefore crucial. We shall, however, defer the more detailed discussion of contradiction and relation to a later section.[62]

Earlier we said that the system of the Nālandā masters emphasized the use of reason. We also spoke of how—starting from the laws of the natural world, such as the principles of nature, reliance, and function—using the principle of evidence, they established inference in relation to those facts that remained obscure. It is, however, also important to recognize the limit or scope of such inference based on objective facts, that is, what it can negate and what it cannot. Recognizing this point, Buddhist logicians state that one needs to differentiate between what a specific reason *does not prove as existent* and what it *proves as nonexistent.* Likewise, one must distinguish between the following: what a specific valid cognition *does not find* versus what that valid cognition *finds as being nonexistent*; what a specific reason *does not prove as being the case* versus what that reason *proves as not being the case*; and what specific valid cognition *does not establish as being the case* versus what that valid cognition *establishes as not being the case.* [162]

# 3

# Reasoning in the Collected Topics

Based on essential points concerning how negation and affirmation are established through logical reasoning, and in particular drawing on Dharmakīrti's seven treatises[63] and their commentaries—such as the essential points established when examining how language and concepts relate to their objects, or the presentations on logical reasons, which are the basis of inferential cognition, or repudiation or consequential reasoning that reveal internal contradictions within the assertions of others—a written tradition known as the Collected Topics, which organized these points in a curriculum on training in logic for beginners, became widespread among Tibetan logicians. We shall present a brief outline of this system of logic as an introduction.

Collected Topics (*düra*) emerged in Tibet through the introduction of a unique system of logic initiated by the twelfth-century Kadampa master Chapa Chökyi Sengé. Extremely learned in logic and epistemology, this master logician composed *Root and Commentary of the Condensed Epistemology Eliminating the Darkness of the Mind* and *Epistemology Eliminating the Darkness of the Mind*.[64]

In these texts Chapa summarized the subjects of logic and epistemology, such as valid cognition and the objects it ascertains, objective worlds and subjective minds, [163] identity and difference, universals and particulars, substantial phenomena and abstract phenomena, contradiction and relation, cause and effect, and the tripartite of definition—the definiens, definiendum, and instance, and so on. He systematized the presentation of these topics within a threefold framework: refutation of the positions of others, presentation of one's own position, and rebuttal of objections.

Furthermore, texts such as *Epistemology Eliminating the Darkness of the Mind* inspired other introductory texts, including those on the science of cognition (*lorik*) and the science of logical reasoning (*tarik*) that were aimed at beginner-level study. Even today this tradition of studying the science of cognition, logical reasoning, and the Collected Topics continues without decline in Tibetan centers of learning.

On the basis of Chapa's *Condensed Epistemology*, different abbreviated presentations of Collected Topics appeared bearing such names as Abbreviated Application (*bsdus sbyor*), Collected Topics (*bsdus grwa*), the Summaries (*bsdus pa*), and so on. The organization of the points explicitly presented in *Condensed Epistemology* into eighteen separate topics[65] became widely established: (1) color, white and red; and (2) substantial phenomena and abstract phenomena; (3) contradictory and noncontradictory; and (4) universals and particulars; (5) relation and absence of relation; and (6) distinction and nondistinction; (7) presence and absence; and (8) cause and effect; (9) first stage, intermediate stage, later stage; and (10) definiens and definiendum; (11) multiple reasons and multiple predicates; and (12) forward negation and reverse negation; (13) direct contradiction and indirect contradiction; and (14) two types of logical entailment; [164] (15) being and not being; and (16) the negation of being and not being; (17) understanding existence and understanding nonexistence; and (18) understanding permanent phenomena and understanding things.[66]

In later systems of calculating the topics of Chapa Chökyi Sengé's *Condensed Epistemology*, different methods of enumeration emerged, such as twenty-one, twenty-five, twenty-seven, and so on.

## Training in Logical Reasoning

The summary topics are traditionally presented within a threefold approach of refuting the position of others, positing one's own standpoint, and rebutting the objections to one's position. This approach is an important means of training the mind in logical reasoning. When one first analyzes a given topic, one cites the views of others that are flawed and refutes them, then one puts forth one's own views, and when objections are raised against one's standpoint one then dispels these critical objections. Regardless of the topic, when engaging in critical analysis by way of this threefold method of refuting, positing, and rebuttal, one will be able to develop a comprehensive understanding of the issues involved. Therefore the logical reasoning of the Collected Topics is an extremely useful system for engaging in critical inquiry. [165]

Now when one enters the path of reason of the Collected Topics, there are three main components of speech: (1) the proof, namely, the logical reason, (2) the predicate, and (3) the subject of the dispute.

Dividing one's statement in terms of these three components, the task of negation or affirmation is done through formulating an argument that reveals a consequence, thus framing a form of discussion between the proponent and opponent that highlights the essential points. To illustrate these three components, *reason*, *predicate*, and *the subject*, let us consider the following argument as an example: "Take sound as the subject under investigation. It follows that it is not produced because, according to you, it is permanent."[67] In this trimodal consequence, "sound" is the subject of dispute, "not being produced" is the predicate, and "being permanent" is the logical reason. The fact that sound is not produced—which is the combination of "sound," the subject of the dispute, and "being not produced," which is the predicate—constitutes the *thesis* of that trimodal consequence. The entailment that "if something is permanent it is necessarily the case that it is not produced" is the *actual pervasion*. "Being produced" is the opposite of the predicate "being not produced," and "being impermanent," which is the opposite of the current logical reason "being permanent," is the reverse reason cited in that trimodal consequence. This understanding of the components of the arguments should be extended to other similar formulations of reasoning.

In the chart at the top of the next page, the reason, predicate, and subject are presented for the trimodal consequence "Take sound; it follows that it is not produced because it is permanent."

When giving responses in the context of a debate, there are different forms corresponding to the components of the statement. In general there are two responses corresponding to the predicate component of the statement: "I agree" (yes) and "Why?" (no). Corresponding to the reason component of the statement there are two possible responses: "The reason is not established" (unestablished) and "There is no logical pervasion" (inconclusive). Thus there are four in total. This said, there are also numerous other

| *Subject of dispute* | *Predicate* | *Reason* | *Thesis* | *Reverse predicate* | *Reverse reason* |
|---|---|---|---|---|---|
| sound | being not produced | being permanent | sound is not produced | being produced | being impermanent [166] |

responses, such as "There is a contrary pervasion" or "I have doubt about this," and so on.

(1) Now among these types of responses, "I agree" means "Yes, I agree that the subject concurs with the predicate," indicating that one accepts the stated thesis. For example, in the formulation "Take sound; it follows that it is impermanent because it is produced," if one responds "I agree," then one responds by agreeing that "sound is impermanent." Also, if a predicate alone is formulated without any specific subject and one responds with "I agree," then one accepts the predicate. For example, to the statement "It follows that it exists" one responds, "Yes I agree that it exists."

(2) "Why?" is uttered when the thesis of a consequential reasoning as formulated is not established by valid cognition. Here the response "Why?" is uttered to indicate that one does not accept the thesis and is inquiring further about the reason stated. For example, to the formulation "Take sound; it follows that it is permanent" if one responds "Why?" it means "Why is sound permanent?"

(3) "The reason is not established" means "That subject is not established to be in agreement with that stated reason." For example, to the formulation "Take sound; it follows that it is not produced because it is permanent" one responds, "The reason is not established that sound is permanent." Again, if just a reason alone is stated without a subject and one responds "The reason is not established," one is stating, "The reason does [167] not establish that it is so." For example, to the formulation "Because it is a vase" one responds, "The reason does not establish it is a vase."

(4) "There is no logical pervasion" means that "There is no logical entailment that if something is the reason it is necessarily the predicate." For example, to the formulation "Take sound; it follows that it is per-

manent because it is a knowable object" one responds, "If something is a knowable object it is not necessarily the case that it is permanent."

"There is a contrary pervasion" means that it is the opposite, that if something is the reason it must not be the predicate. For example, to the formulation "Take sound; it follows that it is permanent because it is an audible object" one responds, "If something is an audible object it is necessarily not permanent."

"I have doubt" means "I have doubt as to whether the predicate exists in relation to the subject." For example, to the formulation "Take the child in front of me; it follows that he will live to the age of eighty" one responds, "I have doubt about the predicate." [168]

Here is a chart of responses to questions when engaged in debate:

<table>
<tr><td>1. "I agree"</td><td>This means accepting the stated subject concurs with the stated predicate.</td><td rowspan="2">Addresses the relationship between that subject (or basis of dispute) and that predicate</td></tr>
<tr><td>2. "Why?"</td><td>This indicates that one does not accept that the stated subject concurs with the stated predicate and inquires for further proof.</td></tr>
<tr><td>3. "The reason is not established"</td><td>This means one doesn't accept that the stated subject concurs with the stated reason.</td><td>Addresses the relationship between that subject and that reason</td></tr>
<tr><td>4. "There is no logical pervasion"</td><td>This means that if something is the stated reason it is not necessarily the stated predicate.</td><td rowspan="2">Addresses the relationship between that reason and that predicate</td></tr>
<tr><td>5. "There is a contrary pervasion"</td><td>This means that if something is the stated reason it must not be the stated predicate.</td></tr>
<tr><td>6. "I have doubt"</td><td>This means that I have doubt or hesitation about the stated predicate, reason, and so on.</td><td>Addresses the relationship between any two: the subject, predicate, or reason</td></tr>
</table>

In the early Indian Buddhist texts, however, four types of response were set out as the means of replying to a question: (1) categorical statement, (2) qualified statement, (3) response in the form of a question, and (4) response in the form of silence. [169]

For instance, Vasubandhu's *Treasury of Knowledge* states:

> Stated categorically, through qualifying,
> through questioning, through remaining silent.[68]

To give examples, then, with regard to the question "Are all material things impermanent?" the categorical response "Yes" is an example of the first type. To the question "Are all impermanent entities material?" the qualified response "Those with obstructive spatial properties are material, but those with the nature of inner experience are not" is an example of the second type. To the question "Is this tree tall or short?" one could respond by asking a question, "In relation to what do you ask that?" Then if the questioner points to a shorter tree and says, "In relation to that tree," and one responds "It's tall," and if he points to a taller one, and one responds, "It's short," this is an example of the third kind. There are also certain contexts where for a specific purpose and reason one may need to respond by adopting silence. This is an example of the fourth type.

One extremely vital method of engaging in a comprehensive analysis of a given topic employing the logical reasoning of the Collected Topics is the system of discerning conceptual relationships in terms of the trilemma and the tetralemma.[69] On this model four kinds of permutation are identified: (1) trilemma, (2) tetralemma, (3) contradiction, and (4) equivalence. The system of enumerating the trilemma or tetralemma and so on is a simple method for examining in a most comprehensive manner what might be the conceptual difference between two things. [170]

The *trilemma*: There are three possibilities, say, regarding the difference between a person and a Tibetan: (1) a Tibetan child is both, the first possibility, (2) an Indian is a person but not a Tibetan, the second possibility, and (3) a jar is neither of the two, the third possibility.

The *tetralemma*: There are four possibilities, say, between a carpenter and a Tibetan: (1) a Tibetan carpenter is both, the first possibility, (2) an Indian carpenter is the former but not the latter, the second possibility, and (3) a Tibetan child is the latter but not the former, the third possibility, and (4) a pillar is neither, the fourth possibility.

*Contradiction*: Mutually exclusive phenomena refer to distinct entities that share no common locus, such as the pair a horse and an ox, or white and red, or being permanent and impermanent. For these pairs of combinations, no example can be found that is both.

*Equivalence*: Mutually inclusive phenomena are those that not only have a common locus but also possess eight modes of pervasion. Examples include pairs like being produced and being impermanent, and fire and something that is hot and burning. The eight modes of pervasion are the two modes of entailment pertaining to being, two modes of entailment pertaining to not being, two modes of entailment pertaining to existing, and, finally, two modes of entailment pertaining to not existing. To illustrate these in their respective order:

(1–2) For example, if something is a product it is necessarily the case that it is impermanent; and if something is impermanent it is necessarily the case that it is a product.

(3–4) For example, if something is not a product it is necessarily the case that it is not impermanent; and if something is not impermanent [171] it is necessarily the case that it is not a product.

(5–6) For example, if something that is a product exists it is necessarily the case that something that is impermanent exists; and if something that is impermanent exists it is necessarily the case that a product exists.

(7–8) For example, if something that is a product does not exist it is necessarily the case that something that is impermanent does not exist; and if something that is impermanent does not exist it is necessarily the case that something that is a product does not exist.

In brief, this tradition of logic of the Collected Topics invented by the Tibetan master Chapa Chökyi Sengé is an excellent method for opening wide the doors of critical reasoning when engaged in the various disciplines of knowledge. Not only is it relevant to the study of traditional fields of knowledge, but it constitutes a remarkably effective system of reasoning that is applicable to contemporary subjects as well. Here we have offered only a brief introduction. Those who wish to study this system of reasoning in detail should consult the texts of the *Collected Topics* themselves.[70] [172]

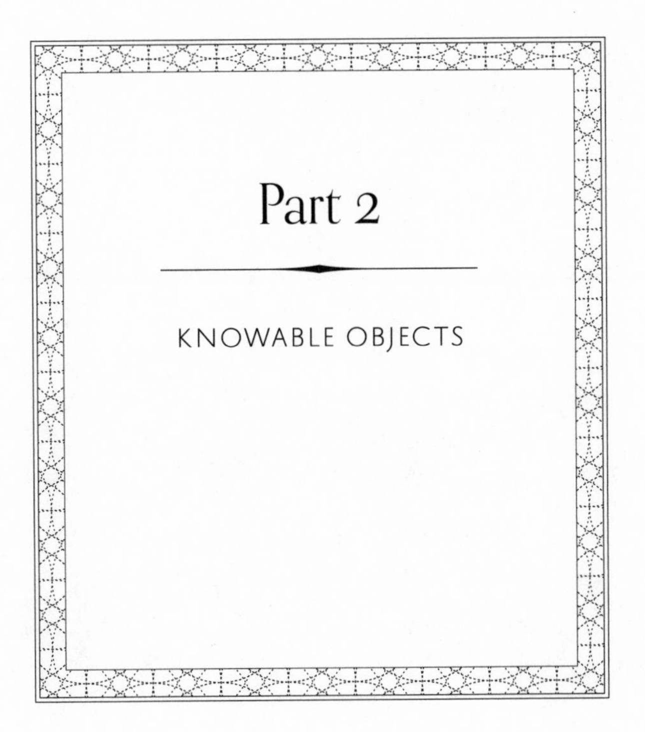

# Part 2

## KNOWABLE OBJECTS

## Defining the Physical

Setting aside differences among Buddhist schools, the picture of the natural world presented in Buddhist texts is something like the following. The world, or at least the human experience of it, has two primary components—*matter* and *consciousness*. A third component, *nonassociated conditioning factors*, qualifies material and mental phenomena. This Buddhist differentiation of the conditioned world into three classes is strikingly similar to what Karl Popper refers to as "the three worlds": (1) the world of physical objects, (2) the world of subjective experience, and (3) "the world of statements in themselves" (Popper 1982, 211). Buddhists have in addition the further category of the *unconditioned*, which is essentially the absence or negation of conditioned things.

The first task then for Buddhist theorists is to clearly define what is meant when something is said to be physical or material. (Consciousness is defined in volume 2.) In the very early sources what is material (*rūpa*) is described simply as comprising the "four great natural elements" (*mahābhūta*) and things derived from such elements. The earth element performs the function of supporting, the water element cohesion, the fire element maturation, and the wind element performs the function of extension. Apart from this, one sees little in the earliest texts in the way of rigorous definition of what matter is. They are "great" (*mahā*) in that they are found in all matter, and they are characterized as "elements" (*bhūta*) for they are the ultimate constituents of everything within the physical world. Since this concept of the great elements comes to be closely connected with Buddhist atomic theory, I shall discuss the evolution of this concept below in my introduction to part 3, which investigates Buddhist theories of subtle particles.

Over time, a more systematic approach emerged in defining matter in terms of four distinctive characteristics: (1) that which accumulates, (2) that which occupies space, (3) that which is visible, and (4) that which can be modified through contact. These features of matter came to be narrowed further to two main characteristics—resistance and visibility—

with the former becoming the primary. Vasubandhu, for example, states that matter is characterized by resistance because it hinders the arising of another thing in its own location. Here in part 2, the defining features of matter, drawing from authoritative Abhidharma sources, are presented at length and examined.

A key problem in the early Buddhist attempt to define matter lies in the tension between trying to capture simultaneously the *phenomenological* character of material entities—their resistance and visibility—and their *ontological* existence in terms of their atomic constituents. Early Ābhidharmikas wished to define matter to include the objects conceived as forms during meditation, for they occupy space in the mind and hinder other things from arising in that space. Other Buddhist thinkers, such as the Sautrāntikas, reject this, defining matter as consisting of atomically composed things. For these later thinkers, the material world is composed of the objects of the five senses, that is (1) visible forms, (2) sounds, (3) smells, (4) tastes, and (5) tactile objects; the five sense organs, defined not as the gross organs but as some kind of refined inner form; and the four great elements. They reject the materiality of the so-called *mental-object forms*, which include the objects of meditative states as well as Sarvāstivāda Abhidharma's "nonrevelatory" or "nonindicative" form, a special category of form that is not characterized by resistance. For the Sautrāntikas, these so-called forms are nothing but constructs of the mind and have no physical reality. In any case, the status of objects known as *kṛtsnāyatana* (*kaṣina* in Pali) perceived in advanced meditative states, which are thought to have real material effects, posed a daunting challenge to the early Buddhist theorists, with some broadening their definition of the physical to accommodate these and others rejecting their materiality altogether.

## Conditioning Factors

As observed above, Buddhist theorists postulate, in addition to matter and consciousness, an important third category of conditioned phenomena. Although it was primarily the Sarvāstivāda school that advocated a distinct category of *nonassociated conditioning factors* (or *nonassociated formative factors*, as it is rendered in this volume) with their own intrinsic reality, the need to recognize some conditioning factors that are neither material nor mental states seemed to be broadly shared by most Buddhist schools. They are "nonassociated" in that they belong neither to the category of matter nor to the category of consciousness; they are "conditioning factors" because, like matter and consciousness, they have the capacity to act as conditions for the arising of experience. As we can see (pages 142–43), the Sarvāstivāda school postulates fourteen such factors, each defined in terms of their unique functions. *Possession*, for instance, is a factor that makes it possible for one to acquire and maintain an attainment, *homogeneity* provides a basis for shared characteristics within a group of sentient beings, and so on. Asaṅga's *Abhidharmasamuccaya* adds nine more to the list, making twenty-three in all. The attribution of a specific activity to a particular factor as well as how many such factors should be on the list became a matter of debate and controversy among Buddhist theorists.

While the Sarvāstivādins conceived these dissociated forces to possess intrinsic and substantial realities of their own, Asaṅga explicitly speaks of them as "constructs" that have reality only in name and thought (*prajñapti*). Later Buddhist theorists are also clear in maintaining that the list—whether numbering fourteen or twenty-three—should not be treated as exhaustive. These differences aside, having a category of conditioned things distinct from matter and mind helps even the critics of Sarvāstivāda Abhidharma. It gives them a way to speak about the ontological status of such abstract entities as time and persons as well as characteristics of conditioned phenomena like arising, abiding, and cessation. For these theorists, then, the category of nonassociated conditioning factors becomes equivalent to the class of phenomena referred to as *prajñaptisat*, those that are real only in name and thought.

On the rationale for positing this distinct category of existence, a contemporary researcher on Sarvāstivāda Abhidharma writes, "Given the diversity of activities explained and doctrinal constraints satisfied, the category of dissociated forces appears to be a derivative category with no single integrating principle. Instead, it is a miscellany containing functionally unrelated factors that are unified only by their successful operation demanding their separation both from form and thought" (Cox 1995, 73). In other words, the category is best understood as defined primarily in terms of things that do not fit in the other two categories. From the perspective of the history of ideas, what one appreciates from the Sarvāstivāda's sustained attempt to systematize the category of nonassociated factors, including its subsequent expansion, is the need early Buddhist thinkers had for a comprehensive account of the world, one that is not just composed of static objects—whether material or mental—but is dynamic, with complex causal activity.

## CAUSALITY

As in contemporary science, one of the most important explanatory principles in Buddhist exploration of reality is the law of causality. Whether it is the framework of the four noble truths or dependent origination, the principle of causality lies at the heart of the Buddhist worldview, from the natural world to the process of mental purification. A well-known statement attributed to the Buddha reads, "When this exists, that exists; from the arising of this, that arises," which encapsulates the principle of dependent origination (*pratītyasamutpāda*) and explicitly presents causation in terms of conditionedness. "This conditionedness" (*idaṃpratyaya*) is a ubiquitous phrase in Buddhist texts, tantamount to a slogan, especially to draw a contrast with those who explain the evolution of the cosmos in terms of a transcendent force such as God (Īśvara in Sanskrit).

In explaining the causal process inherent in the arising of things, we see a distinction drawn between two crucial terms, *causes* (*hetu*) on the one hand and *conditions* (*pratyaya*) on the other. When distinguished, *cause* is primary while *conditions* are complementary in that they help the cause produce its effect. To give an example popular in Buddhist sources, the arising of a rice sprout, the rice seed is the cause while factors such as soil, moisture, warmth, and the farmer's efforts are conditions. In some contexts *hetu* and *pratyaya* are used interchangeably, thus making cause and condition in fact equivalent. When used in this way, a distinction is drawn between what is a *primary* or *substantial* cause and what is a *cooperative* cause, with the former being unique—that is, having a one-to-one relation to a given effect—while those in the latter category are shared and common with other effects. In other words, a cooperative cause can have multiple effects.

In the early Buddhist scriptures, four conditions are proposed for the arising of a perceptual event, such as a visual perception. There is (1) the *causal* condition, all the factors that are coexistent and concomitant with the perception itself, (2) the *objective* condition, the external object, (3) the *dominant* condition, the sensory faculty, and (4) the *immediately preceding* condition, the preceding moment of consciousness. Later thinkers, such as the Sautrāntikas, modified this view and understood the first condition to be in fact the general category with the remaining three as specific subclasses of causal conditions. They explain the effects of the three conditions on the basis of three specific features of the perceptual experience, such as the perception of a flower, for example. That the perception possesses the form of a flower as its content is the imprint of the *objective condition*, that it is visual as opposed to some other sensory experience is the imprint of the *dominant condition*, and that the perception is subjective experience is the imprint of the *immediately preceding condition*.

The Sarvāstivāda school of Abhidharma developed the doctrine of the four conditions into its theory of six causes: (1) efficient cause, (2) coexistent cause, (3) homogeneous cause, (4) concomitant cause, (5) omnipresent cause, and (6) fruitional cause. In this theory, the notion of cause is understood in very general terms such that to be a cause is to exist without hindering the arising of the given thing. Furthermore, the second—the coexistent cause—implies the acceptance of two simultaneous things conditioning each other in a mutual causal relationship. Asaṅga's Yogācāra school too seems to accept a notion of simultaneous causation, but most subsequent Buddhist schools, including especially the Sautrāntikas, reject the notion of simultaneous causation. For the Sautrāntika, to be a cause of something is to have an impact on producing that something, and given that nothing remains static even for a single moment, the idea of two temporally simultaneous things having an effect on each other is logically untenable. Causation, by this logic, must be sequential.

An important related issue in Buddhist thought is the distinction between coincidence and causation. An example of a mere coincidence often cited is the case where a coconut falls down from a tree the very moment a crow lands on that tree. What is required to establish causality is more than mere coincidence; there needs to an invariable relation such that if *x* then *y*. In Dharmakīrti's language, this invariability is due to a "necessary relationship" between cause and effect. In brief, (1) a cause must be temporally antecedent, (2) it must be an agent of the change effected, and (3) the arising of a given effect must be related to it invariably. Furthermore, in this Buddhist view, all conditioned things are causally efficacious—that is, they are capable of producing effects.

In Western thought, especially in contemporary philosophy, much of the inquiry into the nature of causality tends to focus on the question of necessary versus sufficient conditions. If *x* is a *necessary* condition for *y*, then the presence of *y* implies the existence of *x*. If, on the other hand, *x* is a *sufficient* condition of *y*, then *y* does not imply the existence of *x*; alternate conditions could also result in *y*. In classical Buddhist sources, as mentioned earlier, a relationship of invariability (necessity in Dharmakīrti's language) is assumed for, say, *y* to be the effect of *x*. There seems to be hardly any discussion of the matter of sufficient condition in the manner in which the concept is understood in the Western thought. Where the notion of *sufficiency* is introduced in Buddhist sources in relation to causation, it tends to be understood in terms of the aggregation of the multiple conditions that invariably lead to the effect. For Buddhist thinkers, given that all effects arise from multiple conditions, individual things can never be by themselves sufficient causes for any effects.

## Reality or Mental Construct?

The final section of part 2 relates to a host of constructs central to Buddhism's systematic approach to defining and describing what is real. These include the concepts associated with definition itself, notions of identity and difference, universals and particulars, substance and attributes, and relationships, as well as the distinctions drawn earlier between the three types of facts—evident, concealed, and extremely concealed. For non-Buddhist Indian philosophers like the Naiyāyikas, many of these categories—especially universals, substance, and relationships—constitute objective features of reality. Buddhist thinkers reject the objective reality of these categories and recognize them as constructs of the mind, albeit important ones. They are not totally nonexistent, like the horn of a rabbit or a flower growing in the sky; nor do they have the robust reality of matter and mental states. In what sense, then, can they be said to exist at all? Do these constructs based on binary conceptual distinctions belong to the category of nonassociated conditioning factors? In Sarvāstivāda thought, one finds no serious discussion of these questions. In Sautrāntika, however, arguably the most epistemologically oriented Buddhist school, especially as understood through Dharmakīrti, the above questions are addressed explicitly. For Sautrāntikas, these differentiations are constructs of the human mind created for expedient purposes through shared language and convention. Not only are they not material or mental states, they cannot be accepted even as nonassociated conditioning factors, for they have no causal efficacy. In brief, they belong to category of the unconditioned.

One final point needs clarification in connection with part 2. This has to do with the way in which the Sanskrit terms for negation (*pratiṣedha*) and affirmation (*sādhana/vidhi*) are often rendered in Tibetan as nouns rather than verbs, thus referring to classes of phenomena—in other words, negative phenomena and positive phenomena. Some Tibetan thinkers thus speak of a tree as a positive phenomenon and "non-nontree" as a negative phenomenon. This is because the cognition of the former does not entail explicit negation of something, while the latter evidently does. Within negative phenomena, too, one can distinguish between a non-brahman as an implicative negative phenomenon—that is, it implies that he belongs to some other caste—whereas the absence of an elephant in the room would be a nonimplicative negative phenomenon in that its understanding entails no implicit affirmation.

In many of the original Indian Buddhist sources, however, the usage of these two terms seems better understood as involving what contemporary Western philosophers call *speech acts*. On this view then the statement "This is a tree" would be a speech act of affirmation, while the statement "Buddhist monks do not drink alcohol" would be a speech act of negation. Understood thus, negation and affirmation do not necessarily refer to some negatively characterized or positively characterized facts of the world. The distinction between nonimplicative (*prasajya*) and implicative (*paryudāsa*) negations would simply be a function of whether a given negation implies anything else other than the simple act of negating. This more linguistic and epistemic interpretation of the theory of negation also seems to fit better when analyzing what the negation is being applied to. Thus the statement "Buddhist monks do not drink" can be represented as "–*x*," where the negative particle is attached to the verb "drink," while in the statement "he is not a brahman," the negative particle is attached to the noun "brahman." Given this distinction, some contemporary scholars render these two types of negation as *verbally bound negation* and *nominally bound negation*.

Within the verbally bound negation, too, a distinction can be drawn between those that directly negate a proposition versus those that negate only the statement of the proposition, as illustrated in the following two sentences: "Buddhist monks do not drink alcohol" versus "I do not say that Buddhist monks drink alcohol." The American philosopher John Searle characterized the first type as "propositional negation" and the second kind as "illocutionary negation."

In brief, keeping in mind these two ways of interpreting the role of negation and affirmation—one more ontological and the other linguistic and epistemological—would be helpful as the reader engages with the texts presented here in part 2.

## Further Reading in English

For a critical analysis of Abhidharma ontology including its conception of matter, see Paul Williams, "On the Abhidharma Ontology," *Journal of Indian Philosophy* 9 (1981): 227–57.

For a detailed analysis of the concept of nonassociated conditioning factors according to Abhidharma, see chapters 4–9 in Collet Cox, *Disputed Dharmas: Early Buddhist Theories of Existence* (Tokyo: The International Institute of Buddhist Studies, 1995).

For a detailed analysis of the theory of causality in Buddhism, see David J. Kalupahana, *Causality: The Central Philosophy of Buddhism* (Honolulu: University of Hawaii, 1975); and Kenneth K. Inada, "A Review of David J. Kalupahana's *Causality: The Central Philosophy of Buddhism*," *Philosophy East and West* 23 (1976): 339–45.

On the concept of negation in classical Indian philosophy in general, see Bimal K. Matilal, *Epistemology, Logic, and Grammar in Indian Philosophical Analysis* (The Hague: Mouton, 1971; reprinted by Oxford University Press, 2015); on the analysis of negation in Indian Buddhist thought, see Yuichi Kajiyama, "Three Kinds of Affirmation and Two Kinds of Negation in Buddhist Philosophy," *Wiener Zeitschrift für de Kunde Südasien* 17 (1973): 161–75; and for Tibetan interpretation of the concept of negation, see chapter 13 of Georges B. Dreyfus, *Recognizing Reality* (Albany: State University of New York Press, 1997).

# 4

# Phenomena in General

EARLIER WE SPOKE of how prior Buddhist thinkers employed different systems of classification when establishing their presentations on reality. Each of these systems of classification possesses unique structures within which the subject matter of the treatises may be made more accessible to the minds of their readers. Here in this compendium we take the best features of these classical systems of codification, but we develop our presentation principally with the aim of making the subject matter of the treatises accessible to the contemporary reader. We thus organize the subject matter of the two volumes by five major themes:

1. The presentation of knowable objects[71]
2. The presentation of the cognizing mind
3. How the mind engages its objects
4. The means by which the mind ascertains objects
5. The presentation of the person, the agent who ascertains objects

The essential natures of specific phenomena are discussed in the presentation of knowable objects. The nature and classification of mental phenomena that are characterized by subjective experience are discussed in the presentation of the cognizing mind. In the section on how the mind engages its objects, topics such as whether the mind engages its objects by way of elimination or affirmation and what mind possesses what aspects of those objects and so on are addressed. The section on the means by which the mind ascertains objects presents the following topics: the four principles of reasoning, as well as repudiation, proof statements, formal reasoning, [174] and the ancillary topic of the stages of training the mind. In the final section on the person who is the perceiver of objects, the views of the Buddhist and non-Buddhist schools of classical India on the identity of the person are presented.

On the presentation of knowable objects, in general existence and nonexistence are defined on the basis of whether something is established by valid cognition. If something is established by valid cognition it must be posited as existent, and if it is not established by valid cognition it must be posited as nonexistent. A vase, for example, exists since it is established by valid cognition, whereas the horn of a rabbit does not exist since it is not established by valid cognition. Thus the following is stated in Dharmakīrti's *Exposition of Valid Cognition*:

> In this regard, those fit to be observed exist,
> others that are not observed do not exist.[72]

So the meaning of existence is "that which is observed by valid cognition." "Existent," "established basis," "object," "knowable object," "ascertainable object,"[73] and "phenomenon" are equivalent with respect to their reference. "That which is established by valid cognition" is the definition of an *established basis*, "that which is cognized by awareness" is the definition of an *object*, "that which is fit to be an object of awareness" is the definition of a *knowable object*, "that which is comprehended by valid cognition" is the definition of an *ascertainable object*, "that which retains its nature" is posited as the definition of a *phenomenon* (*dharma*). [175]

With respect to knowable objects, as mentioned earlier, there are numerous systems of classification, such as the eighteen elements and twelve bases and so on, whereby their specific natures, characteristics, functions, and types are explained. However, when all of these phenomena are classified in terms of their essential natures they fall within two categories: those that are produced by causes and conditions that are subject to change, and those that are not produced by causes and conditions that are not subject to change. The first kind is characterized as impermanent or conditioned phenomena, whereas the second type is characterized as permanent or unconditioned phenomena. Those belonging to the first category are referred to as *functional things* since they perform the function of generating their result. They are called *products* since they arise from their causes and conditions. They are called *conditioned phenomena* since they are produced from the aggregation and convergence of causes and conditions. And they are called *impermanent* since they are states that are subject to disintegration moment by moment. Thus "functional thing," "product," "conditioned phenomenon," and "impermanent" are equivalent with respect to their reference. "That which is capable of function" is the definition of a *functional thing.* "That which has arisen" is the definition of a *product.* "That which is fit to arise, cease, and endure" is the definition of a conditioned phenomena. And "that which is momentary" is the definition of impermanent.

If the category of "conditioned, impermanent phenomena" is further differentiated, it can be classified in terms of three types: (1) material phenomena that are characterized as obstructive, such as what can be seen by the eyes or touched by the hand; (2) those characterized as consciousness, which are nonphysical and constitute internal subjective experience alone; and (3) factors that are neither matter nor consciousness but whose ground of imputation is either material or mental phenomena, such as time. [176] The categorization of conditioned phenomena into three classes remained a standard approach among classical Buddhist thinkers.

Phenomena that are not produced by causes, and conditions that are not subject to change, are designated as *nonthings* since they do not perform the function of generating a result. They are *unproduced* since they are not generated by causes and conditions. They are *unconditioned* since for them the three features of arising, ceasing, and enduring are incompatible. And they are *permanent* since these are phenomena that do not transform from moment to moment. Terms such as these are used to describe unconditioned phenomena. Thus "nonthing," "unproduced phenomena," "unconditioned phenomena," and "permanent" are equivalent with respect to their reference. "Being devoid of the capacity to perform a function" is the definition of nonthing, "not arisen" is the definition of being unproduced, "not fit to arise, cease, or abide" is the definition of unconditioned phenomena, and "being both a phenomenon as well as being not momentary" is the definition of a permanent entity. Permanent phenomena are of two types: eternal permanent phenomena and occasional permanent phenomena. Unconditioned space, for instance, is posited as an example of an eternal permanent phenomenon since it is a permanent phenomenon that exists forever throughout all time. In contrast, the space inside a vase, for example, is posited as an occasional permanent phenomenon since it exists only when the vase exists and ceases to exist when the vase no longer exists. [177]

# 5

# The Essential Nature of Physical Entities

COARSE MATERIAL PHENOMENA must be defined in terms of their character of obstructive resistance. In general, however, when the essential nature or unique attributes of physical entities are explained in various Buddhist texts, the term "that which is capable of materiality"[74] is mentioned. In Candrakīrti's *Entering the Middle Way*, for example, one reads:

> Form possesses the characteristic of that capable of materiality.[75]

The meaning of "capable of materiality" is explained by Vasubandhu in his *Treasury of Knowledge Autocommentary*:

> What is called *capable of materiality* ultimately means "that which is capable of being damaged."[76]

For example, when one touches a newly arisen flower petal, the mere touch of the hand affects the new flower petal, and the site touched withers and darkens. *Damage*, in this context, means "to induce change," and this is explained in the *Autocommentary*:

> What harms a physical entity? That which produces total change.[77]

In essence this notion of "damage" means that since various changes occur because of the mutual interaction between material things, they are characterized as "capable of being damaged." [178] In the *Autocommentary* three questions are posed in responding to objections raised against explaining the phrase "capable of materiality" in terms of "capable of being damaged by other physical phenomena."

1. The first objection: If "capable of materiality" is explained in terms of "capable of being damaged," then subtle particles cannot be material entities because they cannot be harmed by other material phenomena.

The text responds by stating that whereas a subtle particle that resides in isolation cannot be damaged, it is capable of being harmed when combined with other subtle particles.

2. The second objection: In that case, according to you material things of the past and of the future cannot be material entities because they are not capable of being damaged by other material phenomena.

In response the text states that as for both past and future material forms, since one has already existed (in the past) as a material entity and the other will exist (in future) as a material entity, and furthermore since both are similar in type with present material form, they can be posited as material phenomena as well. For example, because some trees in a forest are firewood the entire forest may be labeled as *firewood*.

3. The third objection: Given your definition, nonindicative form[78] would not qualify as physical form because it cannot be harmed by other material phenomena.

The text responds that nonindicative form is characterized as material form since it is supported by indicative form that itself exists as material form. Analogously, when a tree moves its shadow also moves.

Thus the *Autocommentary* [179] states:

> Objection: But subtle particles are not material form because they don't exist as material form.
>
> Response: Even though the discrete form of a subtle particle cannot exist in an isolated state, those residing as an aggregation are capable of materiality.
>
> Objection: But that which exists in the past and future is not material form.
>
> Response: That which has already existed as material form and that which will become material form are material form because they are similar in type to present material form such as firewood.
>
> Objection: But that which is nonindicative is not material form.

> Response: That which is indicative exists as material form, that which is nonindicative exists as material form, just as the shadow moves when the tree moves.[79]

In Asaṅga's *Compendium of Abhidharma*, however, when the nature of material form is posited it is defined in terms of "that which exists as form" and it is classified in two different types: that which exists as material form since it has the property of obstructive resistance, and that which exists as material form since it is apprehended conceptually[80] by discernment that takes physical objects and their attributes as *some such* object of mental cognition. The *Compendium of Abhidharma* states:

> What are the characteristics of the form aggregate? The characteristic of that existing as form is "that existing as form through either aspect," which refers to (1) that existing as form owing to making contact, [180] and (2) that existing as form owing to imagining the object. What is existing as form owing to making contact? It is that which is capable of materiality when contacted by a hand, a stone, a stick, a weapon, cold, heat, hunger, thirst, bees, gadflies, wind, sun, scorpions, and snakes.[81]

The same text states:

> What is that which exists as form owing to imagining the object? It refers to any object apprehended by meditative equipoise[82] as such-and-such material form or some such material form. Or it is that which is apprehended conceptually by nonequipoise mental conception and its concomitant mental factors.[83]

"Meditative equipoise" in the *Compendium of Abhidharma* refers to the mind of meditative equipoise that is single-pointed concentration cultivating repulsiveness,[84] and the mind of nonequipoise refers to concept-laden mental consciousness unmistaken with respect to conceived objects that it apprehends, which are external objects in relation to which conception proliferates.

When commenting on passages cited earlier from the *Compendium of*

*Abhidharma*, Jinaputra/Yaśomitra states in his *Elucidation of the Compendium of Abhidharma*:

> "Any object" refers to "manifest states." "Such-and-such" [181] refers to physical images corresponding to actual things such as skeletons and so on. "Some such" refers to particular subsets of color and shape. "Apprehended conceptually" refers to the discernment of material form by conceptual means.[85]

Although there is a slight difference between the two brothers who are masters of Abhidharma in how they define form, namely, *what is capable of materiality* or *what exists as form*, their views converge in the fact that in general there are two main categories of form: one type of form that is an object of sense consciousness and a second that is an object of mental cognition alone. Of these two, the first class of form, namely, those that have the property of obstructive resistance, include the five sense objects such as visible form, sound, and so on. The second category of form, those that are the objects of mental cognition alone, in turn has two kinds: those that are obstructive and those that are nonobstructive. The examples of the first kind are the five sense faculties, such as the eyes and so on, which are internal form, whereas the second type, which are nonobstructive, are known as "mental-object form."[86]

Thus in general when the presentation of physical form is made in Buddhist texts, it is crucial to understand the difference between physical form and visible material form. "Physical form" is a term applied in general to all material phenomena, whereas "visible material form" (form base) is a subcategory of physical form. The latter must be understood in terms of visible form such as material entities possessing shape and color that are perceived by eye consciousness. Hence one must make the distinction that sound, smell, taste, and tactility are material forms, but they are not visible material form (form base). [182]

Moreover, according to Asaṅga's *Compendium of Abhidharma*, one should differentiate between physical form defined in terms of *that capable of materiality* and form defined in terms of the property of obstruction. What that text refers to as "existing as form owing to making physical contact" must be understood as referring to obstructive material phenomena. Conversely, "that existing as form through imagining the object," such as mental-object form, is not compelled to possess the property of obstructive resistance. Therefore, although these forms (such as mental-object forms) are characterized as physical they need not be posited as obstructive.

Again, these Abhidharma masters speak of three specific characteristics and three general characteristics of form. For example, Asaṅga's *Compendium of Ascertainment* states:

> There are three specific characteristics of form: (1) the translucent sense organs,[87] (2) the perceived objects of the sense organs, and (3) the perceived objects of mental cognition.
>
> The five types of translucent form that support the five sense consciousnesses belonging to sense bases such as the eyes, and causally derived from the four great elements, are called "sense organs."
>
> The five objects such as visible form that are objects of their five congruent physical sense organs are called "form perceived by the sense organs." The sense organs (*rūpaprasāda*) associated with consciousness are called "homogeneous" because they are congruent with their respective consciousness and its objects. Those [sense organs] that are devoid of consciousness [183] are called "partially homogeneous" because they eventually become congruent in the same way as those that are homogeneous. The form of a physical image that is a referent field of single-pointed concentration is called "form that is an experiential field of mental cognition."[88]
>
> One should recognize the three general characteristics of form:
>
> 1. All form impedes objects and always exists separate from such objects. This is the first general characteristic.
> 2. All form perceived by the sense organs and the perceived form of the sense organs exhibit increase and decrease. This is the second general characteristic.
> 3. All form is capable of tactility, such as that contacted by the hand, a stone, a stick, a weapon, cold, heat, hunger, thirst, gadflies, bees, mosquitoes, wind, sun, scorpions, and snakes. This should be recognized as the third characteristic.[89]

The three specific characteristics referred to in this text pertain to the specific nature of different types of form, such as:

1. Translucent form of the five sense faculties, such as the eyes
2. Form perceived by the sense organs is the five objects, such as visible form
3. Form perceived by mental cognition is the referent field of mental consciousness alone, such as the image that is an object of single-pointed concentration

By "general characteristics" the text means those properties that are common to and shared by many types of physical phenomena. Thus "impeding objects" [184] refers to the property of many physical phenomena whereby they display the nature of impeding the presence of other physical objects. "Exhibit increase and decrease" refers to the character of material phenomena that is subject to increment and decrement. "Capable of tactility" means that many physical phenomena are characterized as having the property of tangible contact, either external or internal.

In general when classifications of material form are made, the texts tend to speak in terms of the twofold division of (1) outer physical form that is an object of the five senses consciousnesses and (2) inner physical form such as the five sense faculties. Here in our compendium we shall follow the approach of Asaṅga's *Compendium of Abhidharma* and present the classifications of form in four sections:

1. Outer physical form, such as the five sense objects
2. Inner physical form, such as the five sense faculties
3. Mental-object form
4. The primary elements that act as the causes of the above three types of form

# 6

## The Five Sense Objects

THE FIVE SENSE OBJECTS, such as visible form, are posited as objects of consciousness of the five sense doors. Thus the objects that are perceived by eye consciousness, ear consciousness, nose consciousness, tongue consciousness, and body consciousness are defined as visible form, sound, smell, taste, and [185] tactility bases, respectively. The *Compendium of Abhidharma*, for example, states:

> What is form? It is an object derived from the four great elements and constitutes the field of experience of the eye sense faculty. . . .
>
> What is sound? It is an object derived from the four great elements and constitutes the perceived object of the ear sense faculty. . . .
>
> What is smell? It is an object derived from the four great elements and constitutes the perceived object of the nose [sense] faculty. . . .
>
> What is taste? It is an object derived from the four great elements and constitutes the perceived object of the tongue sense faculty. . . .
>
> What is tactility? It is an object derived from the four great elements and constitutes the perceived object of the body sense faculty.[90]

The phrase "the field of experience of the eye sense faculty" indicates that, just as the image of an object appears in a mirror, it is on the basis of physical form appearing to the eye sense faculty that eye consciousness apprehends physical form.

## Visible Form

To explain the topic of visible form (form base), the perceived object of eye consciousness is called *visible form*, such as blue or yellow color. If classified, visible form is twofold: color and shape. The first refers to hues like blue, yellow, and so on, which are referred to as *colors*, while configurations, such as long, short, square, and so on, are called *shapes*. If one were to take a house as an example, [186] its white or yellow hue is posited as its color while its proportions or dimensions are posited as its shape.

### *Color*

Colors fall into two classes: primary colors and secondary colors. "That which is fit to be a primary hue" is the definition of primary color. When differentiated there are four: blue, yellow, white, and red. These four are called *primary colors* since the secondary colors are derived from combinations of these four. For example, yellow combined with a predominance of blue makes green. Therefore the term *primary* should be understood in terms of being the basis or the cause of the secondary colors. Alternatively, these four are called *primary colors* because they conform with the colors of the four primary elements that constitute the basis or the cause of resultant form.

"That which is fit to be a secondary hue" is the definition of secondary color. When differentiated, eight secondary colors are identified as light, darkness, cloud, smoke, dust, shadow, mist, and sunlight. These colors are called *secondary* since it is in dependence on the combination of the four primary colors in varying measures that the secondary colors emerge. In his *Commentary on the Treasury of Knowledge* Pūrṇavardhana explains that "secondary" here means "individuated," for he states:

> There are four primary colors and there are eight secondary colors such as "cloud-color" that are individually related to combinations of those four, in greater or lesser measure.[91] [187]

Although there is no need for a secondary color to be any of these eight, the reason why secondary colors are explained in terms of these eight in Abhidharma texts is this. For instance, the four—cloud, smoke, dust, and mist—can, like the horse or ox of a magical illusion, appear like a solid wall when viewed from a great distance, yet no such solidity can be seen when observed close-up. So to counter the possible misconception that they do not exist at all these four are listed. Similarly, with respect to the remaining four—light, darkness, shadow, and sunlight—these are mentioned to counter the misconception that light and so on do not exist separately from the sites where they are perceived. In Jinaputra/Yaśomitra's *Explanation of the Treasury of Knowledge*, for example, it is said:

> Here this should be analyzed. If some say that secondary colors such as cloud-color are specific subsets of the four primary colors such as blue and there are no others, then cloud-color and so on should not be listed as secondary colors, just as green and so on are not mentioned.
>
> Response: Those who state this have not analyzed the issue properly. Such attributes as "green" can be listed, and though *cloud-color* and so on are specific subsets of blue and so forth, they are explained separately because they are like illusory form. Thus when one approaches colors that appear to be like a solid wall in the distance, they stop appearing, and some, upon reflection, think they do not exist in the least. But it is inappropriate to be of two minds about this, thinking "Whatever it is exists" or, conversely, "It is merely an illusion." Therefore secondary colors such as cloud-color [188] are explained for the purpose of stopping such distorted conceptions and doubts, for it is said the essential nature of these form bases does exist.[92]

### *Shapes*

"That which is fit to be a configuration" is the definition of shape. When differentiated there are eight types: long, short, square, round, high, low, even or smooth, and uneven or rough. Of these, "square" refers to that which is four-sided, "round" refers to that which is circular or spherical in shape, "even" refers to a shape that is smooth or broad, while "uneven" refers to that which is neither smooth nor broad. The *Autocommentary*, for example, states:

> Even is a smooth shape. Uneven is a shape that is not smooth.[93]

Thus when the divisions of visible form (form base) are added up, there are twenty in total. For example, the *Autocommentary* states:

> Also the form base has twenty types: blue, yellow, red, and white; long, short, square, round, high, low, even or smooth, and uneven or rough; cloud, smoke, dust, mist, shadow, sun, light, and darkness.[94] [189]

This said, given that colors such as green, black, and so on also belong to the category of "visible form," it is not necessarily the case that if it is a visible form it must be any of these twenty.

In Asaṅga's *Compendium of Abhidharma*, however, visible form is classified as twenty-five types. Color is differentiated, in terms of its essential nature, into the four primary colors: (1) blue, (2) yellow, (3) white, and (4) red. From the perspective of how an array of color is located within a specific shape, there are ten: (5) long, (6) short, (7) square, (8) round, (9) fine particles, (10) coarse particles, (11) even or smooth, (12) uneven or rough, (13) high, (14) low. From the point of view of whether they are augmented or diminished, there are eight: (15) shadow, (16) sun, (17) light, (18) darkness, (19) cloud, (20) smoke, (21) dust, and (22) mist. In addition, there is (23) the form of open space that serves as a necessary basis for eye consciousness to see distant forms, (24) indicative form, which refers to the specific shape of the body that has the characteristic of indicating a person's motivation, and (25) the uniform color of the sky as a factor of ornamentation. Jinaputra/Yaśomitra, for example, writes in *Elucidation of the Compendium of Abhidharma*:

> The twenty-five types of form, such as blue and so on, should be understood in terms of the presentation of six features: (1) characteristics, (2) location, (3) augmented or diminished, (4) basis of activity, (5) characteristic of activity, and (6) ornamentation. In their respective order, they consist of four, ten, eight, one, one, and one types of form.[95] [190]

"The form of open space" refers to the whitish vacuity of vacant space that is the absence of obstruction whose existence is essential for eye consciousness to see distant physical objects. "The uniform color of the sky" refers to the form appearing uniformly to eye consciousness as the clear blue color of the vacant sky in the distance. "Indicative form" refers to the configurations of the body that reveal another's inner motivation, such as, for example, the expression of joy in the form of a smiling face. *Elucidation of the Compendium of Abhidharma*, for example, says:

> "Form of open space" refers to any space free of obstructive contact that would otherwise prevent the existence of a physical entity there. "Sky" is the appearance of blue anywhere above.[96]

Furthermore, the fact that visible form can be classified in terms of three primary categories and, when elaborated, in terms of thirty-two types is explained by Asaṅga in his *Yogācāra Ground*:

> What is the object of focus of eye consciousness? It is physical form that is demonstrable and possesses obstructive resistance. This, in turn, consists of numerous kinds. In brief, there is color, shape, and indicative form. What is color? It is these: blue, yellow, red, white, shade, sun, light, darkness, cloud, smoke, dust, [191] mist, and sky. What is shape? It is these: long, short, square, round, minute particles, coarse, even, uneven, high, low. What is indicative [form]? It is these: taking up, setting down, contracting, extending, standing, sitting, lying, going, stopping, and so on.[97]

## SOUND

To explain the sense object of sound, then, the field of experience of ear consciousness, what is audible to the ear, is sound. Examples include the sound of water flowing or the melody of a flute. Sound is classified in terms of eight types, on which the *Autocommentary* states:

> Sound consists of eight types: Four types are causally derived from the great elements either conjoined or not conjoined with mind, which convey or do not convey [meaning] to sentient beings. Within these there exist sounds that are either appealing or unappealing, thus making it eightfold.[98] [192]

The four sounds derived from elements conjoined with the mindstream are:

1. Pleasant sound that conveys meaning to sentient beings
2. Unpleasant sound that conveys meaning to sentient beings
3. Pleasant sound that does not convey meaning to sentient beings
4. Unpleasant sound that does not convey meaning to sentient beings

Examples of these four are, respectively, the pleasant sound of a person singing, the sound of a man speaking harsh words, the song of a cuckoo,[99] and the sound of a fist striking something.

In contrast, the four sounds that are derived from the great elements not conjoined with the mindstream are:

5. Pleasant sound that conveys meaning to sentient beings
6. Unpleasant sound that conveys meaning to sentient beings
7. Pleasant sound that does not convey meaning to sentient beings
8. Unpleasant sound that does not convey meaning to sentient beings

In their respective order, examples of these four are: pleasant words issuing from a television, harsh words issuing from a television, the sound of a lute,[100] and the sound of rock cracking. The *Autocommentary* states:

> Sound derived from the four great elements conjoined with the sense faculties as their cause is, for example, the sound of hands clapping and the voice. Sound derived from the four great elements not so conjoined as their cause is, for example, the sound of wind, of trees, and of water. That which conveys meaning to sentient beings is, for example, indicative sound in the form of language. The remainder is sound that does not convey meaning to sentient beings.[101]

"That which conveys or does not convey meaning to sentient beings" indicates whether the sound communicates some level of meaning, while "being conjoined or not conjoined" should be understood as whether that sound is conjoined with the physical sense faculties. "Being conjoined with a sense faculty," in turn, means [the mind] is harmed or benefited when that physical sense faculty is harmed or benefited, [193] and "not conjoined with a sense faculty" means [the mind] is not harmed or benefited when that physical sense faculty is harmed or benefited. That hair and nails are not conjoined with the sense faculties, except at their roots, is explained in the *Autocommentary*:

> The others are not conjoined, such as the hair of the head, body hair, nails, teeth, excrement, urine, saliva, mucus, blood, and so on, as well as earth, water, and so forth. What is the meaning of conjoined? They are affected equally by benefit or harm.[102]

In Asaṅga's *Compendium of Abhidharma*, however, sound is classified into eleven types:

1–3. Three types of sound that are pleasant and so on based on their benefit or harm
4–6. Three types of sound that are derived from the great elements based on being conjoined with the sense faculties and so on
7–9. Three types of sound, such as worldly conventions and so on, based on their revelation and presentation
10–11. Two sounds uttered by āryas and nonāryas based on their designation[103]

In that very text one reads, for example:

> What is sound? It is any perceived object of the ear sense faculty derived from the four great elements that is pleasant, unpleasant, or neither. It is derived from the great elements conjoined with the sense faculties as its cause, or derived from the great elements not conjoined with the sense faculties as its cause, or both. It is uttered within everyday language, revealed by tenet systems, or fictitious. It is uttered by āryas or [194] uttered by nonāryas.[104]

Of these, the sound derived both from the primary elements conjoined and not conjoined with the sense faculties includes, for example, the sound produced from the contact of the hand and a drum. The "conventions of the world" are terms such as "pillar" and "vase" uttered within everyday language. "Conventions revealed by tenet systems" are expressions used by tenet systems, such as, for example, "All conditioned phenomena are impermanent." "Fictitious conventions" are expressions that superimpose or denigrate, such as the statements "sound is permanent" and "there are no earlier or later rebirths."

"Conventions used by āryas" have eight types and refer to the eight kinds of true statements uttered by āryas. They are: to state correctly "I have seen" when it is seen, "I have heard" when it is heard, "I have comprehended" when it is comprehended, "I have understood completely" when it is understood completely, "I have not seen" when it is not seen, "I have not heard" when it is not heard, "I have not comprehended" when it is not comprehended, "I have not understood completely" when it is not understood completely.

"Conventions used by nonāryas" have eight types and refer to the following false statements: "I have not seen" when it is seen, "I have not heard" when it is heard, "I have not comprehended" when it is comprehended, "I have not understood completely" when it is understood completely, "I have seen" when it is not seen, "I have heard" when it is not heard, "I have comprehended" when it is not comprehended, "I have understood completely" when it is not understood completely. The *Autocommentary* states:

> Regarding the statements comprising the sixteen declarations in sūtra, these eight statements—"I have seen" [195] when it is not seen, "I have heard" when it is not heard, "I have comprehended" when it is not comprehended, "I have understood" when it is not understood, and "I have not seen" when it is seen, up to "I have not understood" when it is understood—are the declarations of nonāryas. The eight from "I have seen" when it is seen, up to "I have understood" when it is understood, and from "I have not seen" when it is not seen, up to "I have not understood" when it is not understood are the declarations of āryas.
>
> What are the characteristics of seeing, hearing, comprehending, and understanding? What is experienced by the eye, the ear, mental consciousness, and the three sense bases are, respectively, that which is seen, that which is heard, that which is understood, and that which is comprehended. What is experienced by eye consciousness is "seen," what is experienced by

ear consciousness is "heard," what is experienced by mental consciousness is "understood," and what is experienced by nose, tongue, and body consciousnesses are "comprehended."[105] [196]

### *Linguistic Sounds*

Linguistic sound is a specific type of sound that makes known its content. Here linguistic sound can be defined as "an audible sound making known its content through the power of symbols," such as, for example, the sound "vase." When differentiating linguistic sound there are four ways in which it can be classified: in terms of its essential nature, its content, its mode of expression, and what it excludes.

Linguistic sound in terms of its essential nature is threefold: nouns, predicated phrases,[106] and letters. A noun refers to "that which expresses or denotes the essential nature of the thing itself," a predicated phrase refers to "that which expresses the attributes of an object," and letters refer to "linguistic sounds that do not express either of these two." The word "house," for example, is a noun expressing merely the entity "house," the utterance "white house" is a phrase expressing an attribute of a house, and "h" is a letter that expresses neither the thing nor its attributes. These points are explained in Guṇaprabha's *Exposition of the Five Aggregates*:

> What is the collection of nouns? It consists of appellations pertaining to the essential nature of phenomena, such as the utterance "eyes" and so on. What is the collection of phrases? It consists of appellations pertaining to the attributes of phenomena, such as "Alas, all conditioned phenomena are impermanent" and so on. They are complete with respect to what the person wishes to express, for they lead to the understanding of the associated agent, attribute, and time of a given act. [197] For example, the statement "a vase has been produced" makes known attributes related to *production*; the statements "Black Devadatta," "White Devadatta," and "Light Blue Devadatta" express attributes related to *qualities*; and the statements "Devadatta cooks," "Devadatta will cook," "Devadatta has cooked" make known attributes related to *time*. They are referred to as "expressing specific characteristics" and thus they are called

> "phrases expressing attributes related to production, qualities, and time." What is the collection of letters? It is said, "Letters are for the purpose of illuminating both nouns and phrases." "Illuminating" means "making manifest."[107]

In view of the above, a noun can be defined as "a sound expressing the mere entity itself." When classified it consists of two types: proper nouns and figurative nouns. Alternatively, nouns can be differentiated in terms of arbitrary nouns initially formulated and subsequent derived nouns.

"Proper noun," "primary name," and "arbitrary name" are similar in their meaning, as are "figurative noun," "secondary name," and "subsequent name." Regardless of whether there exists a specific reason or purpose, the name that was initially applied to signify an object is a *proper noun*. For example, [198] the word "fire" is applied to that which is hot and burning. The name that is subsequently applied to something because of its resemblance or its relationship to another object that is signified by the proper name is called a *figurative name*. When differentiated there are two types: figurative names based on relationship and figurative names based on resemblance.

This first type, in turn, is of two kinds: figurative names based on causal relationship and figurative names based on intrinsic relationship. The example of causal relationship is when one refers to the rays of the sun as "the sun." The example of intrinsic relationship is when a collective name is given to a part, such as referring to a cushion that has been scorched in one corner as "a burned cushion."

An example of this second type, a figurative name based on resemblance, is calling a man with large ears and a broad nose a lion. Dharmakīrti's *Exposition of Valid Cognition*, therefore, states:

> For the mind that engages its object undistortedly
> it is always the proper name. Through exaggeration
> the word "lion" is applied thus to a brahman boy,
> for such expressions exist as worldly conventions.
> If it has no other object, and beings
> formulate this term for it by way of worldly convention,
> then it is the proper name for it. Because it bears resemblance,
> the other is the secondary name through erroneous
> conceptualization.[108]

The term "lion" is, for the king of beasts, always a proper name, whereas it is applied as a figurative name to a brahman boy whose features resemble a lion. And such a custom does exist as part of worldly convention. So although no objective independent basis exists for the application of a term, such as the eternal universal asserted by the Vaiśeṣikas, a term can be described as a proper name if it is the name initially formulated for a specific object [199] by way of worldly convention. Similarly, owing to some resemblance to the object to which it is a proper name, the same term—when subsequently applied to some other thing—becomes a secondary figurative name associated with erroneous conceptualization.

A predicated phrase is defined as a "sound expressing a thing qualified by an attribute," such as, for example, the expression, "Alas all conditioned phenomena are impermanent."

A letter is defined as "the audial tones that form the basic constituents of both nouns and phrases." Since the nature of letters is asserted to be "audial tones," the forms of written letters are not the actual letters. Letters, when classified, are of two types: vowels and consonants.

The definitions of nouns, predicated phrases, and letters should be understood in the specific context of differentiating these three. They need not be taken as definitions of nouns, predicated phrases, and letters in general.

One critical issue that is related to the discussion of sound is the statement found in the Buddhist texts about the absence of type continuity in relation to sound. In general, as shall be explained later, no beginning can be asserted for any conditioned thing with regard to its substantial continuum. So just like any other material phenomena, even for sound, one must accept its substantial continuum. Similarly, given that sound is a conditioned phenomenon that possesses a string of moments as its components, it has to be something that possesses a continuum. For if sound does [200] not possess a continuum, then since it would merely endure for the briefest moment, such consequences as not being directly perceivable by ear consciousness and so on would ensue. Yet the fact that sound does not possess a type continuum is stated in many Buddhist texts. In *Elucidation of the Compendium of Abhidharma*, for example:

> Given that its continuum ceases, sound does not proceed from one place to another through the medium of the primary

> elements. As such, it exists in its own place and arises for an instant.[109]

The *Treasury of Knowledge* states:

> The four external elements
> are those that dissect and are dissected.
> So too they are burned and weighed.[110]

In conformity, the *Autocommentary* also says:

> Just as the four external elements are those that dissect and are dissected, so too they are burned and weighed. Those that are burned are those that are weighed. The sense faculties are not like that because they are translucent like light. Also sound is not like that because its continuum is severed.[111]

Now if one were to investigate, the statement that "sound does not possess a type continuum" appears to be a plausible explanation. With respect to other material things, it is through the accumulation of similar-type atoms that coarse-level composite entities [201] come into being. One could posit a type-continuum for these entities on the grounds that preceding moments transforming into subsequent moments occur because these physical phenomena emerge through the accumulation of subtle particles and increase in size as they develop into coarse phenomena. In contrast, even when there is an accumulation of similar-type particles, sound does not increase in material size. Regardless of however loud or soft its volume may be, sound exists in dependence on the size of the airwaves that constitute its supportive medium. Thus although there is no basis for speaking about the size of the sound itself, still such questions require further examination. In any case, sound possesses the essential nature of a wave. For example, sound is indeed like ocean waves that appear to ordinary perception to move as a progression of earlier and later waves. It appears that the force of earlier waves newly generates later waves and that earlier waves themselves transform into later waves, but in reality such waves never reach the shore. So too, if one takes a loud sound such as the one produced from striking a gong, the force of the sound of the first moment determines how long the sound will endure. That the sound of the later moments arises from the force of the sound of earlier moments in the oscillation of air—its medium—may be inferred from the example of the wave, but this should be examined. [202]

## SMELL

Smell is "that which is the field of experience or that which is sensed by olfactory consciousness." Examples include the aroma of sandalwood and saffron and so on. If differentiated, smell consists of four types: pleasant aroma and unpleasant aroma are, in turn, differentiated in terms of uniform aroma and nonuniform aroma. Vasubandhu's *Autocommentary*, for example, states:

> Smell has four types: both pleasant aroma and unpleasant aroma, and uniform and nonuniform aroma.[112]

"Uniformity" is explained as a small measure of a pleasant or unpleasant aroma, whereas nonuniform aroma is described as a greater measure of either type. In Asaṅga's *Compendium of Abhidharma*, however, there is a system of classifying smell in terms of six types: pleasant aroma, unpleasant aroma, uniform aroma, natural aroma, fabricated aroma, and aroma derived from transformation.[113] As for examples of the last three, *Elucidation of the Compendium of Abhidharma* states:

> Natural aroma refers to aroma such as that of sandalwood and so on. Fabricated aroma refers to the aroma produced by incense and so on. Aroma derived from transformation refers to the aroma of ripened fruit.[114] [203]

Furthermore, in Pṛthivībandhu's *Explication of the Five Aggregates* one reads:

> If differentiated, it is posited in terms of three types:
>
> Pleasant aroma is that which benefits the faculties and the four great elements, such as the aroma of musk, sandalwood, saffron, and so on.

> Unpleasant aroma is that which harms the faculties and four great elements, such as the aroma of vomit and so on.
>
> Other aromas that do not benefit or harm the five sense faculties and the four great elements are neither pleasant aroma nor unpleasant aroma. Further, there is the aroma of a conch shell and oyster shell and so on.
>
> Pleasant aroma also has three types: naturally arisen, fabricated, and ripe. The aroma of sandalwood and saffron and so on are innately arisen. Those such as a ball of incense and so on are applied aroma. The aroma of a ripe mango fruit and so on are ripe aroma. Also, unpleasant aroma is any of the three types: naturally arisen, fabricated, and ripe.[115]

Thus the text speaks of various types of smell, such as pleasant, unpleasant, and neither pleasant nor unpleasant aroma that respectively benefits, harms, or neither harms nor benefits the sense faculties and four great elements. [204]

## TASTE

Regarding taste, "that which is the field of experience of tongue consciousness" is called the *sensory base of taste*. The taste of molasses is an example. Vasubandhu's *Autocommentary* states:

> Taste is of six types: sweet, sour, salty, pungent, bitter, and astringent.[116]

As for their examples, Pṛthivībandhu's *Explication of the Five Aggregates* says:

> Sweet includes the taste of sugar, molasses, honey, and so on. Sour refers to the taste of citrus and that of fermenting agents. Salty refers to a flavor that includes the taste of the five types of salt. Bitter is bile, the *tikta* herb,[117] and so on. Pungent is that of long peppers, peppers, ginger, and so on. Astringent is that of myrobalan, the fruit of the myrobalan, and so on.[118]

In *Explanation of the Treasury of Knowledge*, however, it says:

> "Taste is of six types" is stated from the perspective of six types of basic tastes, for many types of taste emerge from different combinations of these six tastes.[119]

Thus one may differentiate 6 tastes within each of the 6 basic tastes—sweet, sour, and so on—such that there are 36 different tastes. Also, these 36 tastes can be further differentiated on the basis of (1) tastes that are beneficial to the tongue when experienced, (2) those that do not harm body consciousness when food is digested, and (3) those that bring benefit to the body and mind when digested. [205] Thus 36 times 3 generates 108 types of taste.

Furthermore, in Asaṅga's *Compendium of Abhidharma* twelve types of taste are mentioned:

> Bitter, sour, pungent, astringent, salty, sweet, pleasant, unpleasant, neither, naturally arisen, fabricated, and derived from transformation.

This twelvefold classification is made on the following grounds: the first six in terms of their essential nature, the next three—pleasant, unpleasant, and neutral—from the standpoint of their benefit or harm, and the last three from the perspective of their source. The examples of the last three include the taste of arura[121] that is a naturally arisen taste, the taste of a soup that is derived from preparation, and the taste of [ripe] mango that has emerged from maturation.

Classical Indian medical texts also contain explanations of how these six tastes emerge and their functions and so on. For example, Candranandana's work on medicine, *Moonlight Commentary on the Essence of the Eight Branches*, states:

> "Earth and water . . ." is mentioned. Here the phrase "earth and water" indicates that it is through a predominance of the primary elements of earth and water that sweet taste arises. "Fire and earth" indicates that it is from a predominance of

> the primary elements of fire and earth that sour taste arises. "Water and fire" indicates that it is from a predominance of the primary elements of water and fire that pungent taste arises. "Space and wind" indicates that it is from a predominance of the primary elements of space and wind that bitter taste arises. [206] "Fire and wind" indicates that it is from a predominance of the primary elements of fire and wind that hot taste arises. And "earth and wind" indicates that it is from a predominance of the primary elements of earth and wind that astringent taste arises. Thus these tastes emerge from two elements each. The term "predominance" signifies that all tastes are derivative in nature.[122]

With regard to the functions of these six tastes, the same text states:

> Sweet eliminates wind and bile and generates phlegm.
> Sour eliminates wind and generates phlegm and bile.
> Salt eliminates wind and generates phlegm and bile.
> Pungent eliminates phlegm and generates wind and bile.
> Bitter eliminates phlegm and bile and generates wind.
> Astringent eliminates phlegm and bile and generates wind.[123]

## Tactile Entities

As for tactile entities, "that which is the field of experience or tactile object of body consciousness" is called the *tactile sense base*, such as, for example, the four primary elements. Two main classifications exist for tactile objects, as Sthiramati's *Specific Explanation of the Five Aggregates* says:

> Objects of contact are of two types: the primary elements and material forms [207] derived from these primary elements.[124]

Thus tactile entities are classified in terms of a twofold division: primary elements that are causal tactile entities and resultant tactile entities that are their derivatives. As for the primary elements, these will be discussed in a later section.

As for derivative form, in general one can refer to anything that is produced on the basis of being derived from the primary elements as "derivative." In some commentaries on Asaṅga's *Compendium*, three types of derivative form are mentioned: (1) derivative form caused by the primary elements, (2) derivative form whose designations ultimately depend on the primary elements, and (3) derivative form that is perceived in dependence on taking the primary elements as objects.

Examples of the first include the resultant forms, such as a vase or pillar, that come into being on the basis of having the four primary elements as their cause. Examples of the second type, those that ultimately depend on the primary elements for their designation, include the subtle particles. For when the primary elements exist, the possibility exists of designating something as a material form, and when that is possible, designations such as subtle particles become possible. Therefore even these subtle particles are called *derivative form* since ultimately it is on the basis of the primary elements that the designation of subtle particles exists. Nevertheless, they are not derivative forms in the manner of composite entities, such as a vase, which exist as a collection of primary elements. The manner in which mental-object forms, such as open space and the form derived through a rite, are posited as derivative is similar to the previous example of subtle particles. [208]

Examples of the third type, which are derivative in the sense of being perceived on the basis of taking the primary elements as objects of focus, include the form derived through meditative expertise. On this, Asaṅga's *Compendium of Ascertainment*, for example, states:

> Any physical entity belonging to a mental-object form that is a result of meditative attainment should be recognized as being dependent on meditative attainment alone. As such, it is not dependent on the great elements because it is established from [single-pointed] concentration that focuses on images similar in type. It is said to be "derived from the great elements" but mental-object form does not arise in dependence on them.[125]

Thus the form arising from meditative expertise is described as "derivative form" since it appears owing to meditative concentration that takes the primary elements as its object of focus, and it is not derivative in the sense of being defined in dependence on the great elements.

That there are seven types of derivative tangible forms is mentioned in Vasubandhu's *Autocommentary*:

> Smoothness, roughness, heaviness, lightness, cold, hunger, and thirst.[126]

As for the essential nature of these seven tangible forms, Guṇaprabha's *Exposition of the Five Aggregates* states:

> Smoothness is suppleness and pliability and this is perceived through rubbing a tactile object with the skin. Roughness is coarseness and this is perceived from merely touching or holding something. Heaviness and lightness [209] are perceived by weighing something with scales. Cold is experienced through coming into contact with wind, and it has the function to produce desire for warmth. Hunger is the desire to eat. Thirst is the desire to drink.[127]

These seven derivative tactile forms are merely imputed as attributes of the tactile qualities of the four great elements; apart from the tactile aspects of the four great elements they have no separate nature of their own. In fact, it's stated that all tactile bases can be subsumed in the threefold division of smoothness, roughness, and an intermediate state.

As to how these seven derivative tactile objects arise from the primary elements, their causes, is stated in *Specific Explanation of the Five Aggregates:*

> Smoothness and so on are, owing to their distinct locations, distinct from one another. If there is a predominance of water and fire elements, smoothness emerges; a predominance of earth and wind elements, roughness; a predominance of earth and water elements, heaviness; and a predominance of fire and wind elements, lightness. This is why there is a predominance of heaviness with respect to a corpse. A predominance of the water element indicates cold, a predominance of the wind element indicates hunger, and a predominance of the fire element indicates thirst.[128]

Also, Asaṅga's *Compendium of Abhidharma* states:

> Smoothness, roughness, lightness, heaviness, suppleness, [210] softness, rigidity, cold, hunger, thirst, satisfaction, robustness, weakness, fainting, shuddering, slipping, sickness, aging, death, fatigue, rejuvenation, resilience.[129]

These twenty-two derivative tangible forms are, in one way or another, encompassed within the seven types of tactility. Suppleness refers, for example, to the texture of silk, which appears smooth when touched. Softness refers, for example, to the texture of wool that is not hard when pressed. So these tangible forms are included in either smoothness or lightness.

Rigidity refers, for example, to the tactile sensation of iron or stone that does not depress when squeezed. Satisfaction refers, for example, to the tactile sensation felt when satisfied by food or drink, or the tactile sensation experienced when the elements are balanced. Robustness refers, for example, to the tactile sensation felt when the body is stout and strong. Weakness refers, for example, to the tactile sensation experienced when the body is thin. So these four sensations are included in either roughness or heaviness.

Fainting refers, for example, to the tactile sensation derived from a vital point in the body being penetrated through the cessation of a coarse level of movement of mind and mental factors. It is hence included in either roughness or smoothness. Shuddering refers, for example, to the tactile sensation of physical trembling from the sound of iron scraping against iron that generates mental intolerance owing to irritation. So it is included in the texture of roughness.

Slipping arises as a specific tactile experience from the coalescence [211] of water and earth, and it is included in heaviness. Sickness refers, for example, to the pressure of a headache, and is therefore included in roughness. The tactility of aging is heaviness of the body, therefore it is included in heaviness. The tactility of death refers, for example, to the tactile sense of severing the vital essence. Fatigue refers, for example, to the tactile sense when the body lacks flexibility, as in physical exhaustion. These two are therefore included in either roughness or heaviness.

Rejuvenation refers, for example, to the tactile sense when the body is flexible and free of fatigue, and is therefore included in lightness. Resilience refers, for example, to the tactile sense when the body has good strength and good complexion, and is therefore included in either lightness or heaviness. In this regard, Sthiramati's *Specific Explanation of the Five Aggregates*, for example, says:

> Fainting, robustness, weakness, and so on that are mentioned elsewhere as subsets of tactility, these too are included in these seven. Thus they are not mentioned separately. Fainting does not exist separately from smoothness, while robustness does not exist separately from roughness and heaviness. Therefore even those specific types of tactility mentioned in other sources should be subsumed into these seven in an appropriate manner.[130]

Thus in the texts of the lower Abhidharma system visible form is enumerated in terms of twenty types, sound in eight types, smell in four, taste in six, and tactility in eleven—the four primary elements and seven derivative forms. In the texts of the upper Abhidharma system, however, visible form is [212] enumerated in terms of twenty-five kinds, sound in terms of eleven, smell in terms of six, taste in terms of twelve, and tactility in terms of twenty-six—the four great elements and twenty-two derivative forms.

# 7

# The Five Sense Faculties

THE FIVE FACULTIES, such as the eyes, exist as clear internal sense organs that cannot be dissected into parts nor weighed by scales since they are, like light, extremely translucent. They are dependent on the skeletal cavity supporting the individual sense faculty, and, as their effect, they generate sensory consciousness. Just as, for example, a reflection of an object appears in a mirror, images of color and shape appear to the eye sense faculty and visual consciousness arises. Thus "the clear internal sense organ that acts as the specific dominant condition of eye consciousness, which is its result" is the definition of the eye sense faculty. On the basis of this one can understand how to define the four remaining sense faculties, such as the ear sense faculty and so on. On this point Asaṅga's *Compendium of Abhidharma*, for example, states:

> What is the eye sense faculty? It is translucent form (*rūpa-prasāda*) derived from the four great elements that is the basis of eye consciousness. What is the ear sense faculty? It is the translucent form derived from the four great elements that is the basis of ear consciousness. What is the nose sense faculty? It is the translucent form derived from the four great elements [213] that is the basis of nose consciousness. What is the tongue sense faculty? It is the translucent form derived from the four great elements that is the basis of tongue consciousness. What is the body sense faculty? It is the translucent form derived from the four great elements that is the basis of body consciousness.[131]

With each of these five sense faculties, such as the eye, there is an active homogeneous or partially homogeneous function. The eye sense faculty at the time of seeing a form is homogeneous, whereas the eye faculty at the time of falling asleep and closing the eyes is partially homogeneous. Pūrṇavardhana's *Investigating Characteristics* elaborates:

> The statement "the subtle particles of the eye sense faculty" occurs widely in the texts. At times all sense faculties are homogeneous, such as when viewing things that one desires to apprehend owing to one's desire to engage them. On these occasions all such sense faculties remain homogeneous. On other occasions, however, they are partially homogeneous, such as when one is asleep and one's eyes are closed. At times some of the sense faculties remain homogeneous, such as when a person spontaneously views objects without the desire to engage them; during such moments some of the sense faculties remain partially homogeneous.[132] [214]

This way of differentiating between homogeneous and partially homogeneous in relation to the eye sense faculty should be extended to the other sense faculties.

With respect to the essential nature and functions of the five sense faculties, including their specific shapes and so on, these are clearly explained in Pṛthivībandhu's *Explication of the Five Aggregates*:

> The objects of the eye sense faculty are blue, yellow, white, red, and so on. And just as, for example, images appear in a mirror, a crystal, or water and so on due to their being translucent and clear, similarly the eye sense faculty is translucent and clear. So when the sense faculty comes into contact with a form, the image of the form appears and generates eye consciousness that is consonant with just that physical object. The sense faculty is therefore referred to as a translucent form.
>
> Some say: The discrete physical units[133] of the eye sense faculty reside within the retina like the opening buds of a sesame flower.
>
> Some say: In that case describing it as translucent or clear would indeed be sufficient, therefore what need is there to call it form?

Response: If "translucent or clear" alone is stated and "form" is not mentioned, then there is the danger of confusing the explanation in the context of mental factors: "What is faith? It is clear mind" may be seen as referring to the eye sense faculty. Hence the term "form" is mentioned. Although faith is clear, intention does not exist as a material phenomenon, whereas the eye sense faculty is by nature physical [215] and it has the nature of being translucent or clear as well.

Objection: In that case it would be sufficient to mention "form," so what need is there to mention "translucent or clear" at all?

Response: If "form" alone is stated and "clear" is not stated, there is the danger of mistakenly thinking that material entities, such as stones, plants, and so on, also constitute the eye sense faculty. There is also the potential flawed consequence that those affected by cataracts and blindness also possess the eye sense faculty. Hence the word "clear" is mentioned. Therefore, although stones, plants, and so on as well as those sense organs affected by cataracts and blindness and so forth exist as material forms, they are not the eye sense faculty, for they lack clarity.

Question: Again some might say, "In that case, it is sufficient to mention 'clear form,' but what need is there to state 'whose object is color' in your definition of the eye sense faculty?"

Response: It is necessary to include the phrase "whose object is color" for two reasons. For if "translucent form" alone is mentioned and the phrase "whose object is color" is not present, then, given that ears and so on too are translucent form, there is the possibility of mistakenly thinking that they might be the eye faculties as well. Similarly, since a mirror, crystal, and water are also translucent form, there is the possibility of mistakenly thinking that they might be the eye sense faculty. Even though these things are translucent form they do not have colors as their objects. Therefore the definition of eye faculty is "that which is a material form and clear as well and whose object is color." This then is the meaning. [216] In the same manner, the text [Asaṅga's *Compendium*] extends this explanation in appropriate ways to the remaining sense faculties, such as the ears and so on.

The *Compendium* asks: "What is ear sense faculty?" which means, "What is the definition of the ear sense faculty?" To present its definition, the *Compendium* states: "Its object is sound and it is translucent form." So the object of the ear sense faculty is sound, such as pleasant or unpleasant sound and so on. And its form is translucent and clear in a manner similar to crystal or a mirror, and due to the force of such clarity it acts as the cause for generating ear consciousness. Its discrete physical units are like twisted knots of birch located inside the ear.

The *Compendium* inquires: "What is the nose sense faculty?" which means, "What is the definition of the nose sense faculty?" In response, the text states: "Its object is smell and it is translucent form." The object of the nose sense faculty is smell, pleasant and unpleasant aroma and so on. Like a crystal or a mirror and the like, it is translucent and clear form, thus it acts as the cause for generating nose consciousness. Its discrete physical units are like metal spoons for smearing eye medicine and exist throughout the back of the nose.

The *Compendium* asks: "What is the tongue sense faculty?" which means, "What is the definition of the tongue sense faculty?" In response, the text states: "Its object is taste and its form translucent." The object of [217] the tongue sense faculty is taste—sweet, sour, and so on. Its form is translucent and clear like crystal or a mirror, and because of this it acts as the cause for generating tongue consciousness. The discrete physical units or molecular structures of the tongue sense faculty resemble half moons, yet the tongue sense faculty does not permeate the core of the tongue even to the width of a hair tip. For if it did pervade the tongue one would feel revulsion when swallowing since saliva would continuously flow as it descended from its root source, the cerebrum.

The *Compendium* asks: "What is the body sense faculty?" which means, "What is the definition of the body sense faculty?" In response, the text states: "Its object is tactility and its form translucent." The object of the body sense faculty is tactile

> sensation, such as smoothness, roughness, and so on. Its form is translucent and clear like crystal or a mirror and, because of this, it acts as the cause for generating body consciousness. Its discrete physical units are like skin or hide and permeate the entire body.[134]

In *Explanation of the Five Aggregates*, the shape of the five sense faculties are described, in their respective order, as resembling flax flowers,[135] twisted knots of birch, rows of metal spoons for eye medicine, half moons, and skin. [218]

# 8

## Mental-Object Forms

In general, as discussed earlier, there are two types of form—those that are objects of sense consciousness and those that are objects of mental cognition. Within this twofold classification, mental-object forms belong to the second category.[136] For example, when a coarse material form such as a composite mass of particles is reduced to its parts, it becomes progressively subtler until it transcends the domain of sensory cognition. But even then it does not become a nonmaterial entity. Since such form cannot be any of the five sensory objects or the five sense faculties, it has to be posited as a mental-object form. Thus the definition of mental-object form is: "That which is capable of materiality that appears to mental consciousness alone but that is not a supportive sense element." The phrase "mental consciousness alone" precludes it being the object of cognition of any of the five sense doors. And the phrase "not a supportive sense element" is used in order to preclude the five sense faculties, such as the eyes, from being mental-object forms even though they are forms perceptible to mental consciousness alone. That it is necessary for mental-object forms to be perceptible to mental consciousness alone and that they cannot be a supportive sense element is stated also in Vasubandhu's *Rational System of Exposition* (*Vyākhyāykti*):

> It is within the objects of mind that what is called *mental-object form* is found. Such a thing is an object [219] of mental cognition alone and it is only an object and not a supportive basis.[137]

Asaṅga speaks of five types of mental-object form in his *Compendium of Abhidharma*:

> What is a mental-object form? It should be viewed in terms of five types: (1) form emerging from a process of deconstruction, (2) form of open space, (3) form of vows derived through a rite, (4) form that is imagined, and (5) form arising from meditative expertise.[138]

Of these, "form emerging from a process of deconstruction" refers to subtle particles that emerge from the process of mentally deconstructing coarse material form, which can be perceived by mental cognition alone. Thus when a coarse everyday object is examined and reduced into its constituent parts it becomes progressively smaller, ultimately resulting in the form of a subtle particle that can only be an object of mental cognition. It is this that is posited as the first type of mental-object form. It is called "form emerging from a process of deconstruction" since it is the smallest discrete unit of matter identified from deconstructing a composite mass into its component parts by mental cognition.

The "form of open space" refers, for example, to the vacant space that appears as a whitish vacuity to mental consciousness alone. In general, although the whitish vacuity of vacant space that is essential for eye consciousness to perceive a distant mountain is posited as "the form of open space," here what is being referred to as a mental-object form is specifically the whitish vacuity of vacant space [220] that appears to mental cognition alone. This is because the whitish vacuity of vacant space that appears to eye consciousness is a visible form.

"Form of vows derived through a rite" refers, for example, to nonindicative form. Such nonindicative form includes the form derived from taking *pratimokṣa* vows,[139] negative commitments,[140] and interim precepts.[141] They are called "nonindicative form" in that they do not reveal the person's intention, that is, what motivated them, to the minds of others. The pratimokṣa vows, for instance, are received through a rite in the presence of a preceptor, and from that angle they are described as a form derived through a rite. Alternately, this term should be applied to such form that is derived through a rite and so on."[142]

"Imagined form" refers, for example, to the skeletons that are clearly perceived as pervading the face of the earth by meditative states focused on repulsiveness,[143] or the vivid perception of five sensory objects, such as horses, oxen, houses, and so on to the mind in a dream. Such things are called "imagined forms" because, apart from the perception of skeletons pervading the earth by the meditative attention focused on repulsiveness, and, similarly, apart from the appearance of horses, oxen, and so on to the dreaming mind, in reality skeletons do not pervade the face of the earth, and [dream] horses, oxen, houses, and so on do not exist at all. [221]

"Form arising from meditative expertise" refers, for example, to fire, water, and so on that are clearly perceived in meditative concentration that mentally projects fire and water and so on in every direction. They are called "forms arising from meditative expertise" since it is through the power of obtaining mastery over concentration that everything is pervaded by fire and water and everything is turned into fire and water. *Elucidation of the Compendium of Abhidharma* states:

> "Form emerging from a process of deconstruction" refers to material forms like subtle particles. "Open space" refers to space that represents the absence of obstructive contact. "Those derived through a rite" refers to nonindicative forms. "Those that are imagined" refer to mental image forms. "That which emerges from meditative expertise" refers to any form that is the domain of experience of meditative absorption aimed at liberation.[144]

Thus mental cognition being referred to in the context of the phrase "it is to mental cognition alone that the five types of mental-object form appear" should be understood as being nonconceptual mental consciousness. Given that the appearing object of the conceptual mind is permanent (i.e., the object-universal), it would be difficult to posit it as a mental-object form.[145] [222]

Generally speaking, there is a consensus among classical Buddhist scholars on how mental-object form must be understood in terms of a form that is the object of mental cognition alone. This said, there is a divergence of opinion on whether it should be further differentiated into five types and also whether all five are material forms or not. For example, in the Vaibhāṣika system, apart from nonindicative form no other mental-object form is mentioned. The *Treasury of Knowledge*, for example, states:

> Form is the five sense faculties, the five referent fields,
> and nonindicative form alone.[146]

Thus, explaining the form aggregate in terms of eleven categories—the five sense faculties, the five sense objects, and nonindicative form—the eleventh, nonindicative form, is characterized as mental-object form, and no other mental-object form is mentioned. The same is found in *Investigating Characteristics*:

> Since nonindicative form belongs to mental objects it is not included among the five sensory objects, so it is presented separately. The term "alone" is a term of emphasis, indicating that it is nonindicative form alone that is mental-object form. The significance of this is to preclude other types of mental-object form postulated by the Sautrāntikas and others.[147] [223]

Thus the text explains that mental-object form consists only of nonindicative form. Even so, according to the Sautrāntika position presented by Vasubandhu himself in his *Autocommentary*, it appears that even this nonindicative form is not accepted as constituting material form. The *Autocommentary* states:

> Sautrāntikas assert that this nonindicative form does not have substantial reality because it lacks any function after the vows are received. It is imputed as well in dependence on the great elements that belong to the past: they lack any essential nature and lack the characteristics of material form.
>
> Vaibhāṣikas say nonindicative form has substantial reality. How is this known? *Treasury* states: "Scripture declares three types: stainless form, increase, and the path not committed and so on."[148]

This assertion of the Sautrāntikas is mentioned also in the Korean master Wonch'uk's *Commentary to the Unraveling the Intention Sūtra*, where he writes:

> With regard to these eight types of material form, in the Sarvāstivāda texts it is asserted that, of these eight aspects of form, the four great elements are subsumed in tactile objects, while the four, such as blue and so on, are subsumed in the class of visible form. [224] And their texts assert mental-object form consists

> only of nonindicative form. In contrast, the Sautrāntika texts state, just as in the treatise *Establishment of Reality*, that nonindicative form is neither material form nor mind at all.[149]

For the Mind Only school, however, the vows received through a rite are accepted either as the mental factor of intention or its seed in the form of an imprint. They do not assert it to be an actual physical form. This is explained in Vasubandhu's *Treatise on the Establishment of Karma*:

> If intention alone were physical karma, then when intention is not present for those who have distracted mind or dwell in mindless states, how would either vows or negative commitments come to exist? Both vows and negative commitments exist because the propensity or latent potency of specific instances of karma remain undestroyed.[150]

Candrakīrti states, however, in his *Autocommentary*:

> Form that belongs to the class of mental objects that is an object perceived by mental consciousness, such form does exist in a dream.[151]

Thus he states that imagined forms such as a dream horse or oxen, which appear vividly [225] to mental consciousness, are mental-object forms. Moreover, it is said in the *Treatise on the Five Aggregates*, a work widely attributed to the same master:

> What is a nonindicative form? It is any form that is a mental object, which is undemonstrable, nonobstructive, and known by mental consciousness alone. Thus vows, negative commitments, and interim precepts, which are by their very nature part of the continuum of virtuous or nonvirtuous karma, are nonindicative forms.[152]

Thus Candrakīrti states that nonindicative forms, such as vows and negative commitments and so, on are mental-object forms. It also appears that he accepts that the other three mental-object forms constitute fully qualified material forms.

# 9

## The Causal Primary Elements

In many Buddhist texts, when presentation on material form is made, it is done so in terms of the twofold division of causal primary elements and derivative forms that originate from these primary elements as their cause. For example, Asaṅga's *Compendium of Abhidharma* states:

> What is it that is presented as the aggregate of form? It refers to anything that is capable of materiality, everything including the four great elements (*mahābhūta*) and those that are derived from the four great elements.[153] [226]

Thus these Buddhist masters unanimously maintain that all physical phenomena are included in these two classes of causal material form, namely, the four primary elements and resultant or derivative forms that arise from their accumulation, where the latter consists of the five sensory objects, such as visible form, and the five sense faculties, such as the eyes and so on.

In brief, Nālandā scholars take as their shared standpoint that material entities emerge through the aggregation of subtle particles forming coarser levels of reality and that it's through the combination of numerous atoms that the diverse forms of coarse everyday objects emerge. However, to understand this essential point about how physical phenomena arise, it is crucial to gain a proper understanding of the primary causal elements. So we shall present below, in a more extensive way, how the topic of the primary elements is addressed in Indian Buddhist texts.

In general such texts relate numerous ways in which the primary elements act as the cause of their derivative forms. For example, they speak of the following five types, namely: (1) the generating cause, (2) the supporting cause, (3) the stabilizing cause, (4) the dependent cause, and (5) the enhancing cause. The primary elements are the *generating cause* in that if the primary elements do not exist no derivative forms will emerge, just as, for example, if there is no seed no sprout will arise. The primary elements are the *supporting cause* in that without the support of the primary elements derivative resultant forms cannot separately arise, just as, for example, if there is no bowl the food supported by it cannot exist. The primary elements are the *stabilizing cause* in that if the primary elements degenerate then the form caused by them will also decline, just as, for example, when a tree is cut down [227] so too its shadow ceases. The primary elements are the *dependent cause* in that it is due to the primary elements that the temporal stages of resultant derivative form arise in succession, just as, for example, an old person moves with the support of a walking stick. The primary elements are the *enhancing cause* in that if the primary elements are enhanced their resultant form also displays enhancement. Pṛthivībandhu's *Explication of the Five Aggregates*, for example, states:

> The four great elements constitute five types of cause that act as producers of derivative form. These five causal types include the generating cause, the supporting cause, the stabilizing cause, the dependent cause, and the enhancing cause. The four great elements act as the *generating cause* since if the four great elements do not exist then their derivative form will not arise, just as, if there is no seed a sprout will not emerge. The great elements act as the *supporting cause* since their derivative form cannot exist unsupported by, and separate from, the great elements. They act as the *stabilizing cause* in that if the four great elements degenerate or decline then their derivative form will degenerate, just as if a tree is cut down its shadow ceases. They act as the *dependent cause* since the four great elements—that are, for example, like an old person who moves in dependence on a walking stick—arise in a stream of moments, and they also produce an uninterrupted stream of moments in their resultant derivative form. They act as the *enhancing cause* [228] since if the four great elements develop and mature then also their derivative form will develop.[154]

With regard to the etymology of the term "primary elements" (*bhūta*), *Investigating Characteristics* says:

> "Primary elements" are so called because when various types of form arise, they emerge from them. . . . Alternatively, they are called "primary elements" since such primary elements proliferate as entities. This means that it is the primary elements that bring such entities into existence and propagate the birth of sentient beings.[155]

Thus two different etymologies are provided on the meaning of the term "primary elements." One is that when the diverse resultant material forms come to exist, they do so with characteristics such as solidity and so on. Alternatively, they are so called because they increase the subsequent stream of the primary elements themselves and propagate births of sentient beings.

As for the meaning of the phrase "great elements" (*mahābhūta*), these elements are called *great* because each of them—earth, water, fire, and wind—possesses a greater force of the function of supporting, binding, maturing, and moving. *Exposition of the Five Aggregates*, for example, says:

> Further, they are *great* since these aggregates of earth, water, fire, and wind predominantly function to support, [229] bind, mature, and move.[156]

Another meaning of the phrase "great element" is found in Sthiramati's *Specific Explanation of the Five Aggregates*:

> "Greatness" of the primary elements (*bhūta*) lies in their being great in magnitude, since they are the support of all material form caused by them. Alternatively, they are great because solidity and so on exist in all composite form, or because the scope of their presence extends far and wide.[157]

In brief, what are referred to as "the four great elements" that act as the constructive cause of coarse physical phenomena, such as the five sense objects of visible form and so on, must be understood in terms of the function of supporting, binding, maturing, and moving. The reason these primary elements (*bhūta*) are called *elements* (*dhātu*) is because these constituent elements such as earth maintain their own natural essence such as solidity and so on, and, serving as constituent elements, they generate the eyes and so forth. On this Jinaputra/Yaśomitra's *Explanation of the Treasury of Knowledge* states:

> They are called *elements* (*dhātu*) because they maintain their specific characteristics [230] and they maintain their derivative form. This means they are elements (*dhātu*) because they maintain their specific characteristics such as solidity and so on and they maintain the form of the eyes and so on derived from them.
>
> Because the eighteen elements maintain specific and general characteristics or because in reliance on those characteristics they maintain type continuity, they are *elements*. When teaching the six elements (*dhātu*), the meaning of "element" is the "seed of existence" because they expand or proliferate existent entities.[158]

Thus the meaning of the word "element" (*dhātu*) pertains both to essential nature and to cause. There are two subdivisions of the primary elements (*bhūta*): external primary elements not associated with the mindstreams of beings and internal primary elements associated with the mindstreams of beings. Each of these, in turn, can be differentiated in terms of subtle and coarse primary elements. There are also contexts where differentiation is made, for example, between earth, water, fire, and wind understood in conventional terms of everyday common usage, and the four primary elements of earth, water, fire, and wind that are said to be the cause of physical entities. In this sense, the first kind are known simply as earth, water, fire, and wind, whereas the latter are referred to by the additional qualifying term "element," such as "earth element" or "water element" and so on. For example, the *Autocommentary* states:

> What subsets are there of earth and so on, and the earth element (*dhātu*) and so forth? The *Treasury* states: "In common usage, what is called *earth* is color and shape." Thus when earth is taught, color and shape are taught. And in just the way earth

> is taught, [231] "also water and fire." In worldly usage [water and fire] are called *color and shape*. "Or else it is similar" states that in just the way color and shape are called *earth*, so too wind is called *blue wind* or *circular wind*.[159]

Thus by calling them "earth," "water," "fire," and "wind" they are differentiated from one another as well as from their resultant forms, such as the eyes and so on. Referring to them as "elements" (*dhātu*) refutes the assertion that material forms such as color and shape commonly perceived in the everyday world are primary elements, and demonstrates that these primary elements are, in the parlance of the treatises, tactile objects. That in everyday worldly conventions one refers to color and shape as "earth" is revealed by the example of how commonly one speaks of "blue earth," "square earth," and so on.

In terms of their essential nature, the primary elements are classified as the four primary elements of earth, water, fire, and wind. Given that there exist four modes of causation with respect to their resultant material forms, the primary elements are fixed in number as four. Just these four primary elements alone are capable of completing the function of the primary elements, such as supporting, binding, maturing, and expanding. And if any of these primary elements are lacking, the functions of the four primary elements will not be complete. On this point, the *Great Treatise on Differentiation* states:

> Some say that if there are less than four their function will be incomplete, but if there are more than four [232] they will have no function; just as, for example, a square throne has only four legs.
>
> Question: What is the meaning of the term "primary element"?
>
> Response: That which diminishes, enhances, harms, benefits, and by nature accommodates both emergence and disintegration—this is the meaning of "primary element." In terms of entity, type, and mass they pervade every direction and accomplish great efficacy—this is the meaning of the term "great."[160]

In *Investigating Characteristics* it says:

> It is said the primary elements are just four, neither more nor less, because derivative form cannot be generated from the primary elements if any are absent.[161]

Similarly, in *Specific Explanation of the Five Aggregates* one reads:

> They are just four in number. This is neither too many nor too few, otherwise they will lack potency or be redundant. The activity of the primary elements consists just of supporting, binding, maturing, and expanding. Given that it's within these that their functions are complete and they have no other function, the primary elements remain just four in number.[162]

As for the specific natures of each of the primary elements, in texts such as Vasubandhu's *Treatise on the Five Aggregates*, solidity, moistness, heat, and lightness and mobility are explained, respectively, [233] as the essential nature of the four primary elements of earth, water, fire, and wind. That only a single function is mentioned for the first three, but two for wind, namely, lightness and mobility, is explained thus: to preclude the lightness of tactility, the word "mobility" is mentioned, and to preclude the mental factor of intention, "lightness" is stated. *Explication of the Five Aggregates*, for example, states:

> In order to reveal the characteristics of the four great elements *Compendium of Abhidharma* states: "What is the earth element? It is solidity" and so on. Here the question "What is the earth element?" ultimately means "What are the characteristics of the earth element?" and the response given is that solidity, hardness, and heaviness are the characteristics of the earth element. When it states: "What is the water element?" which means "What are the characteristics of the water element?" the response given is "moistness," for moistness, oiliness, and wetness are the characteristics of the water element. When it states: "What is the fire element?" which means "What are the characteristics of the fire element?" it responds with "hotness,"

for heat, warmth, and burning are the characteristics of the fire element. When it states: "What is the wind element?" which means "What are the characteristics of the wind element?" it responds that lightness and mobility are the characteristics of wind. Lightness causes things to stir and mobility causes a thing to move and relocate over time.

The three—earth, water, and fire—are said to have one characteristic each, since these characteristics are beyond dispute, but the wind element has two characteristics: lightness and [234] mobility. Why does the wind have two characteristics? If "lightness" alone were stated without mentioning "mobility," then causal form that is an object of the body sense faculty would also exist as form caused by smoothness, roughness, heaviness, and lightness, and the doubt would arise as to whether it is the wind element. If "mobility" alone were stated without mentioning "lightness," then when the phenomena of mental factors is explained, and when it is said "What is intention? It is that which moves the mind," doubt may arise as to whether intention is the wind element. Therefore that which has both the property of lightness and mobility possesses the characteristics of the wind element.[163]

The essential nature and function of each of the four primary causal elements may be illustrated in a chart:

| *Name* | *Essential nature* | *Function* |
|---|---|---|
| 1. Earth primary element | Solidity | Supporting |
| 2. Water primary element | Moistness | Binding |
| 3. Fire primary element | Warmth | Maturing |
| 4. Wind primary element | Lightness and mobility | Increasing [235] |

On the difference of subtlety among the four primary elements, Asaṅga explains how the earlier primary elements are coarser compared with the later. For example, in his *Compendium of Ascertainment*, he writes:

> The four great elements, such as earth and so on, are recognized as progressively coarser states. In this regard, the earth element (*dhātu*) and its resultant form are the main factor in terms of supportive function. The functions of water, fire, and wind are to give rise to moistening, blazing, and moving, and these functions manifest in dependence on the earth element.[164]

On the function of the four primary elements, Pṛthivībandhu's *Explication of the Five Aggregates* states:

> When the characteristics of the four great elements are explained their function is revealed. As such, the function of the four great elements is stabilizing, binding, maturing, and increasing. Furthermore, the function of the earth element (*dhātu*) is the activity of stabilizing, the function of the water element is the activity of binding, the function of the fire element is the activity of maturing, and the function of the wind element is the activity of increasing. But what is the meaning of "element"? The meaning of *element* is to support, since the elements support their eleven resultant derivative forms. Thus they are called *earth element*, *water element*, *fire element*, and *wind element*.[165] [236]

In brief, if an explanation is given of the four elements by combining their natures and functions, then that which is solid and hardens is earth, that which is wet and moistens is water, that which is hot and burns is fire, and that which is light and moves is wind.

In the texts of the lower Abhidharma system, such as Vasubandhu's *Treasury of Knowledge*, for example, it is stated that when each of the four primary elements combine, though the other primary elements are present, the most forceful and dominant primary element manifests while the others exist in an unmanifest state. Take, for example, something like a rock. Since it has the property of cohesion, dryness, and movement, the three elements of water, fire, and wind are present. Similarly, water supports wood and matures flowers, and has movement; fire supports flames, it has cohesion and movement; and wind gives support and has cold and warm tactile sensation. Therefore these three primary elements are each asserted to coexist with three other primary elements.

Thus since the activities of all four elements exist in all composite material form, the four elements are universally present. They are, however, referred to by a specific element based on whatever function happens to be dominant. A blazing red flame, for example, is called "fire" since fire is dominant even though the other three primary elements exist. Thus the four elements are posited on the basis of their function.

The *Great Treatise on Differentiation* explains this point:

> Question: How do the four great elements always manifest together without separation from one another?
>
> Response: Because their characteristics and functions are observed in every compound. [237] Thus the characteristics of earth exist in a solid composite mass because they appear to perception. If that compound lacked the water element, then any gold, silver, and lead and so on existing within it would not melt; moreover if it was devoid of water it would fall apart. If it lacked the fire element then sparks would not fly when it was struck by stones and the like; moreover if such a compound was devoid of fire it would not mature and decay, but such compounds do decay. If it lacked the wind element it would not possess movement; furthermore, in lacking wind it would not possess expansion or growth.
>
> The characteristics of the water element exist in a liquid compound because they appear to direct perception. If such a compound lacked the earth element, then ice would not form in the presence of extreme cold; moreover if that mass of water was devoid of the earth element even ships and so on would sink. If it lacked the fire element it would never warm up; and devoid of fire, it would decay. If it lacked the wind element it would not accommodate movement; furthermore, being devoid of wind it would not expand.
>
> The characteristics of the fire element exist in a warm compound because they appear to direct perception. If such a compound lacked the earth element, tongues of flame such as those in butter lamps and so on could not be extinguished; moreover if it lacked earth it would not support entities. If it were devoid of the water element it would not arise as a continuum; moreover in lacking water tongues of flame would not collect

> together. If it were devoid of the wind element [238] it would be without movement; furthermore, without wind there would be no expansion or growth.
>
> The characteristics of the wind element exist in a moving compound because they appear to direct perception. If that compound lacked the earth element then when such a compound came into contact with something obstructive, such as a wall, it would not be stopped; moreover without earth it could not support anything. If it was devoid of the water element cold wind would not arise; furthermore, without water it would disperse or disintegrate. If it were devoid of the fire element there would be no warm wind; moreover, lacking fire it would decay.[166]

Thus, as explained in the texts of the lower Abhidharma system, Vaibhāṣikas assert that even one aggregated atom must possess all eight substantial particles—the particles of the four primary elements plus the four substantial particles of form, smell, taste, and tactility. They do not merely assert that these substantial particles constitute fully qualified earth, water, and so on. Sautrāntikas, on the other hand, do assert that material phenomena are composed of the accumulation of the eight substantial particles, and they consider these eight substantial particles to be more like potentials or seeds. They maintain that when a particular primary element is perceived it is manifest, and when not perceptible it remains unmanifest, and it is not necessarily the case that the primary elements remain manifest at all times. Therefore although earth, water, and so on are fully present in all physical entities as potentials or seeds, there is no need for actual earth, water, and so on to be present. *Explanation of the Treasury of Knowledge* states, for example:

> "Others say that they exist as seeds," this is a reference to the Sautrāntikas. [239] "In the manner of seeds" means in the form of potentials, which ultimately means "as potencies." "They do not exist by means of their essential nature" ultimately means "they do not exist substantially."[167]

In the texts of Asaṅga and his followers, the names of the individual primary elements (*bhūta*) are used when the elements strongly manifest and are dominant, such as "this or that primary element," and when they are weak in strength and potency they are referred to simply as "elements" (*dhātu*). Through such differentiation it is maintained that although all four primary elements are present as elements (*dhātu*) in all composite matter, it does not follow that all four primary elements of earth, water, fire, and wind are present. For example, Asaṅga's *Facts of the Grounds* states:

> All composite form possesses the elements (*dhātu*) of all four great elements at all times. Thus fire arises from rubbing dry wood, and so too the fire primary element (*bhūta*) arises from striking stones together. So too, if iron, silver, gold, and so on are heated with a gentle fire, they stretch. So too the water primary element arises from the moon gem.[168]

So too *Explication of the Five Aggregates* mentions:

> For example, wherever earth exists, water, fire, and wind also exist there. Wherever water exists, [240] earth, fire, and wind also exist there. If that which is stronger and more dominant in a specific place is explicitly described as *earth*, then water, fire, and wind reside weakly and recessively and these are referred to as *elements* (*dhātu*). Further, for example, though a mixture of a handful of sand and a handful of salt may explicitly be referred to as *salt* since salt is stronger and more dominant, sand also exists there, for it is not the case that the characteristics of the elements do not exist there.[169]

Further:

> When this is critically examined, it is evident that wherever one [element] exists all exist, for when rock and wood and so on cohere without separation then it is evident that the earth element possesses the water element. When rock and wood and so on come to move and stir then they also possess the wind element. When sparks arise from two stones being struck together they also possess the fire element. When the water element supports ships and petals and so on then it is evident that the water element possesses the earth element. When the leaves of trees

> and grass fall into water and rot, and the water is observed to be warm, it is evident that the water element possesses the fire element; when they fall and move it is evident that the wind element exists; when the wind element supports leaves and grass it is evident that it possesses the earth element. When we observe warm wind and a moist substance dries [241] when exposed to it, it is evident it possesses the fire element. When a vortex and wind converge as one and transform, it is evident it possesses the water element. When tongues of fire and the fire element support each other, and tongues of fire support and hold up leaves and grass and so on, it is evident that they possess the earth element; when tongues of fire blaze together without separating, it is evident that they possess the water element; and when they shake and move it is also evident that they possess the wind element.[170]

Furthermore, it is explained that composite form possesses from one up to four fully qualified primary elements such as earth and so on, and it possesses from the particle of form alone to up to the particles of all five—form, sound, smell, taste, and tactility. Asaṅga's *Compendium of Abhidharma* says, for example:

> We say that any primary element that is observed in any composite thing, there it exists. There are composite things made of a single primary element, two primary elements, up to that of all primary elements.[171]

*Explanation of the Treasury of Knowledge* also states:

> Those who think of themselves as Yogācāras speak of compound states as definitely consisting of the primary elements and their derivative forms. [242] Why? They say that there exists some thing that is derived from the accretion of one primary element, dry clods of earth, for example. Also there are some things that arise from two primary elements, moist clods of earth, for example. Also there are those things that arise from three primary elements, such as warm, moist clods of earth, for

> example. Furthermore, there are those things that arise from all four primary elements, such as moist, warm clods of earth when they exhibit movement.
>
> One should understand that in terms of derived form, derived form is observed in any composite and thus is said to exist there. A composite that possesses a single derived form exists, such as light, for example. A composite that possesses two derived forms exists, such as wind that possesses sound and aroma. A composite that possesses three derived forms exists, such as wind that also possesses smoke, for it is categorized by specific form, aroma, and tactility, and its attribute of tactility should be understood as lightness. Also a composite that possesses four derived forms exists, such as an agura tree. A composite that possesses five derived forms also exists, such as those plus sound and so on.[172]

In the texts of Middle Way school too, there are statements about how all four primary elements (*bhūta*) are present as elements (*dhātu*) of composite material form. For example, in Nāgārjuna's *Precious Garland*, one reads: [243]

> The primary elements are false like the self.
> Also earth, water, fire, and wind
> individually lack any essential nature.
> When any three do not exist, one does not exist.
> When one does not exist, also three do not exist.
> If when three do not exist one does not exist,
> then they cannot exist separately.
> How then could a composite be produced?[173]

Thus the primary elements are demonstrated to be fictitious and devoid of inherent existence. Nāgārjuna explains that within each of the four primary elements, such as earth, the elements (*dhātu*) of the other three primary elements (*bhūta*) must be present. For if only one of the four primary elements is present or if all four are not present, there would be no emergence of composite material form. Candrakīrti too says in *Commentary on the Four Hundred Stanzas*:

> Therefore if the kalala embryo was about to descend, the earth element would support it. If it lacked the water element, then it would not become a consolidated mass at that time, but since it coalesces owing to the water element, it becomes a consolidated mass and thus doesn't disintegrate into its subtle component particles like grains of sand. If it lacked the fire element, then at that time the kalala embryo would decay, but since the fire element dries [the embryo], it does not decay. And if it lacked the wind element the kalala embryo would not develop, [244] increase, or expand, for the wind element is that which enhances.[174]

To summarize, in the texts of the lower Abhidharma system it is said there is no difference between the four primary elements (*bhūta*) and the elements (*dhātu*) of the four primary elements, and that all composite matter possesses the four primary elements since they possess all the functions of the four primary elements. In contrast, the texts of the higher Abhidharma system as well as those of the Middle Way school differentiate between the four primary elements and the elements of the four primary elements. Thus they explain that even though all composite form possesses the elements of the four primary elements, they do not necessarily possess the four primary elements themselves. The substance of water in a composite of subtle particles, for example, is the water element or potentiality, but not water itself. The fact that the great elements reside through mutual dependence on one another is mentioned in the *Condensed Perfection of Wisdom Sūtra*:

> Wind depends on space, the water aggregate depends on that;
> on this depends the great earth, and moving beings depend on
> that.[175]

Also in the following sūtra cited in *Explanation of the Treasury of Knowledge* one reads:

> "O Gautama, on what does earth depend?"
> "Brahman, earth depends on water."
> "O Gautama, on what does water depend?"

"Brahman, water depends on wind."
"O Gautama, on what does wind depend?"
"Brahman, wind depends on [245] space."
"O Gautama, on what does space depend?"
"Brahman, you are now going too far. Because space has no support, it has no appearance and cannot be objectified."[176]

Similarly, when refuting the view of some philosophers[177] who assert that the spherical shape of a bean, the sharpness of a thorn, and so on come about naturally without any cause, Āryadeva states in *Establishing the Proofs Refuting Distorted Views*:

In empty space devoid of support,
if beans and thorns were to exist,
one might see their very nature as uncaused
but they grow in dependence on the earth and so on.
Also if water were to flow upward
and the sun were to shine equally in all four directions,
one may verify their very nature to be causeless,
but water flows downhill owing to the earth's gravity.[178]

Here Āryadeva explains how beans, with their spherical shape and so on, grow in dependence on the earth, and how water flows downward because of the earth's gravity. These things do not come into being causelessly.

Vajrayāna texts such as those of the Kālacakra system speak of five primary elements—earth, water, fire, wind, and space. Similarly, some medical texts explain six external elements—wood, along with earth, water, fire, wind, and space—and how they give rise to the illnesses of sentient beings. Such texts explain how the illness of paralysis and atrophy of the limbs are caused by the poison of the wood element entering the nerves, the illnesses of heaviness and torpor [246] are caused by the poison of the primary element of earth entering the flesh and bones, the illness of cold is caused by the poison of the primary element of water entering the blood and lymph, the illness of heat is caused by the poison of the primary element of fire entering the warmth of one's body, the illness of wind is caused by the poison of the primary element of wind entering the breath, and the illness of disturbed mental states such as hallucinations are caused by the poison of the great element of space entering the mind. Thus one can understand that different presentations of the essential nature and enumeration of the primary elements exist.

Furthermore, in highest yoga tantra texts, such as those of the Guhyasamāja system, explanations exist about subtle and gross levels of outer primary elements, and also there is discussion of five inner primary elements of earth, water, fire, wind, and space. Particularly in the texts of the Kālacakra system, there are highly explicit explanations presenting the relationship between the outer and inner primary elements. Not only that, these texts also describe how the movements of sun and moon affect changes within the bodies of sentient beings, especially the breath patterns of inhalation and exhalation. Similarly, these texts say that by gaining mastery over the inner primary elements through the power of yogic practices, mastery over outer primary elements can occur and so on.

In brief, with respect to material form—the first of the five basic categories of reality—the important points to understand are: (1) the nature or definition of matter, (2) how the ten obstructive physical entities and mental-object form are posited, (3) the distinction between primary elements and their derivative forms, and (4) how the substances of the four primary elements need not be understood in terms of actual material entities but can also exist as potentials. [247]

# 10

## Nonassociated Formative Factors

EARLIER WE SPOKE of how conditioned phenomena consist of three main types: (1) material phenomena, (2) consciousness, which is a nonphysical phenomenon, and (3) nonassociated formative factors, which belong neither to matter nor consciousness.

To present the third category, namely, nonassociated formative factors, its definition is posited as "conditioned phenomena that are neither material form nor consciousness." They are called "nonassociated formative factors" since they are the aggregate of formation that is not concomitant with the mind. Instances include the three characteristics of the arising, disintegrating, and enduring of a vase; time in terms of years, months, and days; persons, propensities, things, impermanence; functions such as a vase holding water; and directions such as east, south, west, and north, and so on.

The arising, enduring, and disintegration of a vase refers to, for example, the new arising of a vase that did exist before, the vase enduring in its own time, and the nonendurance of the vase in the second moment in the aftermath of its own temporal existence. Since these characteristics are effects produced by the same single collection of causes that gave rise to the vase, they too are conditioned phenomena. Now if these three characteristics were material form there is no way other than for them to be the material form of the vase itself. However, they do not constitute the material form of the vase nor are they forms of consciousness. Hence they must be posited as attributes of the vase and nonassociated formative factors. [248]

The topic of time, years, months, and so on will be addressed below. As for "persons," it is a phenomenon imputed on four or five aggregates. Given that all three—material form, consciousness, and nonassociated formative factors—are the basis for the designation of the term "person," it would be difficult to posit the person as either material form or consciousness.

Hence nonassociated formative factors is the only viable candidate. The topic of the person too shall be discussed in a later section.[179]

Propensities too are posited as nonassociated formative factors. For example, after the mind cognizes an object such as a person, a latent potential is left behind that has the capacity to give rise to recollection of that person even though the mental event itself has ceased. This kind of latent potency is referred to as a propensity. Such an imprint is not a material form, nor is it a form of awareness; it's merely a potential left by an imprinting cognition. Hence it must be characterized as a nonassociated formative factor. Thus "products," "impermanent phenomena," and "functional things" (all of which are mutually inclusive) contain the three categories of material form, consciousness, and nonassociated formative factors, yet they themselves must be characterized as nonassociated formative factors.

Furthermore, there are countless conditioned things that are neither matter nor awareness, such as the act of a woodcutter chopping wood, the emergence of cause and effect in a sequence, and the possession of attributes by a person and so on. When classified, nonassociated formative factors consist of fourteen types: (1) obtainment, (2) nonobtainment, (3) homogeneity, (4) state of nondiscernment, (5) meditative attainment of nondiscernment, (6) meditative attainment of cessation, (7) faculty of life-force, (8–11) the four characteristics of conditioned things: arising, enduring, aging, and disintegrating, (12) the collection of [249] names, (13) the collection of phrases, and (14) the collection of letters.

The *Treasury of Knowledge*, for example, states:

> Formative factors dissociated from mind
> are obtainment, nonobtainment, homogeneity,
> state of nondiscernment, meditative attainment,
> life-force, characteristics,
> and the collection of nouns and so on.[180]

Of these, *obtainment* is explained as a substance enabling the possession of that which is obtained within one's mindstream. This too is considered to be something separate from the thing obtained, just as, for example, the rope securing a load exists separately from the load. In this way, unlike higher philosophical schools (Sautrāntika, Mind Only, and the Middle

Way), the Vaibhāṣika school holds all remaining nonassociated formative factors to be not only imputed as properties of other phenomena but also existing with a separate identifiable nature of their own. *Nonobtainment* is asserted to be the substance that constitutes the nonpossession of something that is not obtained. *Homogeneity* is the substance that causes the similarity of conduct, thought, and characteristics within sentient beings. *State of nondiscernment* is the substance that temporarily brings to an end the minds and mental factors in the mindstreams of gods born in the realm of nondiscernment. *Meditative attainment of nondiscernment* is the substance that acts as the cause of nondiscernment as its result, and brings about the cessation of mind and mental factors until one rises from that meditative attainment. *Meditative attainment* [250] *of cessation* is the substance that temporarily stops minds and mental factors in the mindstreams of ārya beings in dependence on the mind of the summit of existence, which represents the subtlest mind within the three realms of existence. *Life-force* is taken to be the life of beings or persons and is said to be the basis of either heat or consciousness. For example, the *Treasury of Knowledge* says:

> Life-force is life, which is
> the support of heat and consciousness.[181]

And Vasubandhu's *Autocommentary* states:

> The phenomenon called "life" is not well understood. Therefore to rectify this *Treasury* says: "The support of heat and consciousness." And the Bhagavan spoke to this point when he stated in *Bhikṣuṇī Dharmadinnā Sūtra*: "However long this life, heat, and consciousness may endure, the body will be destroyed. And once discarded, what remains lacks mind, like a tree." Thus the phenomena that acts as the basis of heat and consciousness and as the cause of continued abidance is life.[182]

Among classical Buddhist scholars, there are some who maintain a distinction between *life-force* and *life*. We find the following in the *Great Treatise on Differentiation*, for example:

> Some say because living is called the *factor of life-force*, hence death is called the *factor of life*. Some say that which is set in place is the factor of life-force and that which is relinquished [251] is the factor of life. Some say that phenomena suitable to arise are the factor of life-force and phenomena not suitable to arise are the factor of life. Some say that which resides intermittently is the factor of life-force and that which maintains type similarity is the factor of life. Some say that which is homogeneous is the factor of life-force and that which is partially homogeneous is the factor of life. Some say that which is the result of habituation is the factor of life-force and that which is the result of karma is the factor of life. Some say that which is the result of uncontaminated karma is the factor of life-force and that which is the result of contaminated karma is the factor of life. Some say that which is the result of cognition is the factor of life-force and that which is the result of ignorance is the factor of life. Some say that which is the result of new karma is the factor of life-force and that which is the result of old karma is the factor of life. Some say that which is the result of karma producing a result is the factor of life-force and that which is the result of karma not producing a result is the factor of life. Some say that which is the result of close karma is the factor of life-force and that which is the result of distant karma is the factor of life. Bhadanta Ghoṣaka says that the karmic result experienced in this life is [252] the factor of life-force, and the karmic result experienced after rebirth, the karmic result experienced in a third life, and the karmic result uncertain to be experienced are the factor of life. These indicate the difference between the factor of life-force and the factor of life.[183]

Sautrāntikas refute these assertions by raising the following question: "If you assert that the body and consciousness depend on life-force, then on what does life-force depend, since it cannot exist without a basis? If you say that life-force, in turn, depends on the body and consciousness, in that case without life ceasing the others would not cease and without the others ceasing life would not cease, which means that life-force will never come to cease. Such mutual dependence is untenable. In this regard *Autocommentary* states:

> Now what acts as the basis of this life? It is heat and consciousness, and that is because they are established through mutual reliance. But stopping one of them by any means would stop the others, and it would [absurdly] follow that they would never stop.[184]

Thus, upholding the Sautrāntika standpoint, Vasubandhu raises objections to citing the Vaibhāṣika view. As for the Sautrāntika position, the *Autocommentary* says:

> But what is it? That which is called "life" (*āyus*) is what is temporarily projected by karma of the three realms that maintain continuity of type. It exists [253] for a specific duration and maintains type continuity for as long as it is projected by karma.[185]

So Vasubandhu states that what is referred to as the faculty of life-force is that which, due to the force of karma, maintains type continuity of mental consciousness along with its seeds.

There are also philosophical systems that assert that life-force consists of wind. For example, Vasubandhu's *Autocommentary* states:

> What is referred to as *life-force* is the wind that functions on the basis of the body and mind.[186]

In Śākyamati's *Commentary on the Exposition of Valid Cognition* it says:

> Life-force is exhalation, the phrase "and so on" indicates inhalation and so forth.[187]

In Asaṅga's *Compendium of Ascertainment*, however, he says:

> What is the faculty of life-force? It refers to that projected by prior karma, which establishes a body in a specific place for a set duration.[188]

Similarly, in *Compendium of Abhidharma* Asaṅga states that the term *life-force* is given to the maintenance of type continuity of foundation consciousness (*ālayavijñāna*) owing to the power of one's karma.

Therefore in the texts cited earlier a divergence of opinion emerges on how the term *life* or *life-force* is defined. For example, life or life-force is defined as the support of heat and consciousness, as a wind that functions on the basis of body and mind, as continued survival, or as the continued existence of the type continuum of consciousness. [254]

Now some medical texts speak of the trio of vital essence, life-force, and life. The vitality of the body is *vital essence*, the period during which the union of the body and mind is maintained is *life*, and the fact of the inseparability of wind and mind is *life-force*. To illustrate these three through the metaphor of an oil lamp, the oil is the vital essence, the wick is the life-force, and the light of the flame is life.

The *four characteristics of conditioned phenomena* are so called because they illustrate a given phenomenon as a conditioned thing. *Arising* indicates generation, *enduring* indicates continued existence, *aging* indicates decay, and *impermanence* indicates destruction. The Vaibhāṣikas assert these four to arise simultaneously to the conditioned entity and that they are capable of being grasped as distinct in substance from the conditioned thing itself.

Of the three—*nouns*, *phrases*, and *letters*—nouns refer to that which expresses merely the thing itself, as discussed earlier in the section pertaining to linguistic sounds. The word "material form" is an example of this. *Phrases* refers to expressions that indicate their reference in terms of the predication of attributes, such as, for example, "impermanent form." *Letters* are the basis of nouns and words, for example, the Tibetan letter *ka* and so on. "Collection" refers to the bodies of these three things.

That the Vaibhāṣikas assert these three—nouns, phrases, and letters—to be nonassociated formative factors is mentioned in Vasubandhu's *Autocommentary*:

> They are not the natural essence of speech, for even if speech is sound, sound alone does not give rise to the understanding of meaning. Why? Speech engages nouns, [255] and nouns express meaning.[189]

When the word "vase" is uttered in speech, it is through the appearance of the referent, namely, the vase, in one's mind that the meaning of the word is understood. Speech alone is not capable of conveying the meaning, just as an oxen's bellow cannot convey any linguistic meaning. Therefore Vaibhāṣikas maintain that the three—nouns, phrases, and letters—as well as their three collections, are nonassociated formative factors since they constitute nothing but strings of word-universals that are perceived by a concept-laden mind. Sautrāntikas, on the other hand, assert these to be uniquely characterized indicative sounds of speech, hence they belong to material form.

In the texts of the upper Abhidharma system, such as the *Compendium of Abhidharma*, the following nine nonassociated formative factors are added to the earlier fourteen, thus making twenty-three in total: (1) continuity, (2) distinction, (3) union, (4) rapidity, (5) sequence, (6) time, (7) location, (8) enumeration, and (9) assembly.

In those texts, nonobtainment is listed by the name "ordinary being" and homogeneity is referred to by the term "type similarity," and these, in turn, are stated as being merely imputed on the aspects of other phenomena. For instance, the *Compendium of Abhidharma* states:

> What are formative factors not associated with mind? They are: (1) obtainment, (2) meditative attainment of nondiscernment, (3) meditative attainment of cessation, (4) state of nondiscernment, (5) the faculty of life-force, (6) type similarity, (7) arising, [256] (8) enduring, (9) aging, (10) disintegrating, (11) the collection of nouns, (12) the collection of phrases, (13) the collection of letters, (14) ordinary beings, (15) continuity, (16) distinction, (17) union, (18) rapidity, (19) sequence, (20) time, (21) location, (22) enumeration, and (23) assembly.[190]

As to how these twenty-three nonassociated formative factors are defined: (1) Obtainment is a label applied to the increment and decrement of any phenomenon. The three, (2) meditative attainment of nondiscernment, (3) meditative attainment of cessation, and (4) the state of nondiscernment, are labels applied to the cessation of awareness and cognition. (5) The faculty of life-force is a label applied to the sustenance of life. (6) Type similarity is a label applied to the aspect of likeness. The four, (7) arising,

(8) enduring, (9) aging, and (10) disintegrating, are labels applied to characteristics of phenomena. The three, (11) the collection of nouns, (12) the collection of phrases, and (13) the collection of letters, are labels applied to linguistic elements. (14) Ordinary beings is a label applied to those who have not obtained the ārya path. The nine, (15) continuity, (16) distinction, (17) union, (18) rapidity, (19) sequence, (20) time, (21) location, (22) enumeration, and (23) assembly, are labels applied in relation to causation. These are explained in some detail in the *Compendium of Abhidharma*:

> What is obtainment? It is designated as the obtainment, acquisition, and possession of phenomena that increase or decrease or are virtuous, nonvirtuous, or neutral. [257]
>
> What is the meditative attainment of nondiscernment? It is the cessation of unstable minds and mental factors owing to attention preceded by discernment that is free from attachment to the heaven of Śubhakṛtsna (Extensive Bliss) but not free from attachment to the highest level.
>
> What is the meditative attainment of cessation? It is the cessation of unstable—and even stable—minds and mental factors, owing to attention preceded by the discernment of the state of peace that is free from attachment to the base of nothingness and progresses beyond the summit of existence.
>
> What is the state of nondiscernment? It is the cessation of unstable minds and mental factors that arises in sentient beings who are gods devoid of the mental factor of discernment.
>
> What is the faculty of life-force? It is the fixed period of maintaining type similarity projected by prior karma.
>
> What is type similarity? It is the similarity of the bodies of various sentient beings within the various classes of sentient beings.
>
> What is arising? It is the formative factors that arise having not previously arisen, while maintaining type similarity.
>
> What is aging? It is the continuity of formative factors that transforms while maintaining type similarity.
>
> What is [258] enduring? It is the continuity of formative factors that abide without destruction while maintaining type similarity.

What is impermanence? It is the destruction of the continuum of type similarity of formative factors.

What is the collection of nouns? It is the appellations indicating the essential nature of phenomena.

What is the collection of phrases? It is the appellations indicating the attributes of phenomena.

What is the collection of letters? It is the letters that are the abode or basis of both names and phrases because they make both clear. Also they communicate meaning because they correctly express meaning. Letters cover diverse contexts and do not transform.

What are ordinary beings? It is those who do not obtain ārya qualities.

What is continuity? It is the continuum of cause and effect not being severed.

What is distinction? It is cause and effect being distinguished.

What is union? It is the effect being in harmony with its cause.

What is rapidity? It is cause and effect arising swiftly.

What is sequence? It is cause and effect each arising discretely.

What is time? It is cause and effect occurring in a continuous stream.

What is location? It is cause and effect existing in all ten directions: east, south, west, north, below, [259] above, and so on.

What is enumeration? It is distinct separate formative factors.

What is a assembly? It is the assembly of the conditions of cause and effect.[191]

Based on the explanation of these twenty-three, other nonassociated formative factors should be understood as well. That this number is not exhaustive of nonassociated formative factors is demonstrated by some of the commentarial treatises. The Mind Only school and others identify nonassociated formative factors as nominal existents imputed on states of form as well as on minds and mental factors and do not accept, as the Vaibhāṣikas do, that they possess a substantially distinct reality separate from their basis of designation. Nevertheless they are neither material forms nor mind. That such nonassociated formative factors are nominal existents is affirmed by the above passage from the *Compendium of Abhidharma*, and is stated also in *Treatise on the Five Aggregates*:

> What are formative factors not associated with mind? Those that are imputed on the basis of the states of form, mind, and mental factors. They are not imputed as identical or different from their bases.[192] [260]

# 11

# Causes and Effects

## Causes

ANYTHING THAT IS a conditioned phenomenon, such as material phenomena, consciousness, as well as nonassociated formative factors, necessarily comes into being in dependence on its causes and conditions. So here we shall briefly present this principle of causation.

The meaning of cause is "that which produces its effect." Furthermore, when a given cause exists there is the possibility that effects can come into being, and this very basis of dependence is what is called "causation." Dharmakīrti's *Exposition of Valid Cognition* states:

> Where *it* exists *that* arises,
> and when *it* changes *that* changes as well,
> this is referred to as the cause.[193]

In general the text speaks of various usages of the term "cause":

1. The *cause of origination*, such as the production of consciousness from the assembly of three conditions
2. The *cause of abiding*, such as the four types of nourishment that sustain the body
3. The *cause of supporting*, such as how the earth supports everything
4. The *cause of illumination*, such as in the illumination of objects by a lamp
5. The *cause of transformation*, such as how fire transforms firewood and how a goldsmith [261] transforms gold by beating it
6. The *cause of prevention*, such as the prevention of illness by medicine

7. The *cause of obtainment*, such as how medicine helps obtain the happy state of freedom from illness
8. The *proximate cause*, such as the direct cause
9. The *distant cause*, such as the indirect cause
10. The *evidential cause* (or explanatory reason), such as the inferential cognition of fire by taking the smoke as the reason, or discerning another's intended meaning on the basis of a proof statement

In general "evidential cause" means reason, and in some contexts a single cause may be posited as both an explanatory reason and a productive cause. For example, in the texts of Prāsaṅgika Madhyamaka the exhaustion of oil is taken as both the productive cause and the explanatory reason for the demise of an oil lamp flame.

Further, the Buddhist texts speak in terms of the twofold notions of "causes" and "conditions." So there are contexts in which when one speaks of the *cause* of a specific phenomenon it must be understood as referring to the primary agent that produced it, such as its substantial cause. In contrast, *condition* needs to be understood as referring to those factors that assist in the production of the effect, such as the cooperative conditions. Kamalaśīla, for example, explains in his *Extensive Commentary on the Rice Seedling Sūtra*:

> In this regard, "causes" refers to the substantial causes, for they are unique to the phenomenon. "Conditions" refers to the cooperative causes, for they are shared with others.[194]

Therefore when the causes and conditions of a specific phenomenon are differentiated, then cause and condition are not equivalent in meaning. In general, however, "causes" and "conditions" are equivalent with respect to their reference. [262]

Regarding the classification of causes and conditions, various presentations are found in the texts. For example, according to the Vaibhāṣika system there is a presentation of six causes and of four types of condition, such as the causal condition, objective condition, dominant condition, and immediately preceding condition. Given that this sixfold classification of causes is a specific viewpoint of the Vaibhāṣika school, we shall discuss this later in the section on philosophy. Similarly, since the presentation of the four conditions is more closely related to how cognitions arise, we shall discuss this later in the presentation of the subjective mind.

In brief, generally, when classified from the perspective of time, causes are twofold: the direct cause and the indirect cause. And from the perspective of the way in which they produce their effects, causes consist of both the substantial cause and the cooperative condition. A direct cause is defined as "the direct producer of its effect." For example, fire is the direct cause of smoke, which is its effect, and it is called the "direct producer of smoke that is its effect" because no other cause intervenes between that fire and its immediate effect, the smoke. This explanation should be extended to other similar phenomena.

An indirect cause is defined as "the indirect producer of its effect." For example, firewood is the indirect cause of smoke, which is its effect. Firewood cannot directly produce smoke, which is its effect, but produces it through fire intervening between the two.

A substantial cause is defined as "the principal producer of its substantial effect within its own substantial continuum." Alternatively, it can be defined as "that which primarily produces the essential nature of its effect as opposed to the attributes of the effect." For example, a rice seed is the substantial cause of a rice sprout, which is its effect, and clay is the substantial cause of a clay vase, which is the effect. Dharmakīrti's *Exposition of Valid Cognition* states: [263]

> Without transformation of the substantial cause
> changes in that which possesses a substantial cause
> cannot take place; for example, without transformation
> in the clay, no vases and so on arise.[195]

A cooperative condition is defined as "that which primarily produces its cooperative effects outside its own substantial continuum." Alternatively, it can be defined as "that which primarily produces the attributes as opposed to the essential nature of the effect." For example, water and fertilizer are the cooperative conditions of a sprout. They are called the "cooperative conditions" of a sprout since they act as conditions of the sprout together with the seed. Dharmakīrti's *Exposition of Valid Cognition* states:

> The effects arising from the cooperative causes
> reside together with them,
> just like fire and molten copper.[196]

Thus Dharmakīrti speaks of how the present body and consciousness reside together, each with its own distinct substantial causes while acting as each other's cooperative condition. Analogously, although the melted copper and the fire that have arisen from the preceding moment of fire and copper are not substantial causes of each other, they still reside together.

In brief, cooperative conditions act to complement substantial causes in their production of effects. Furthermore, not only is there is a difference between the two regarding whether they produce their effects within their own substantial continua, but substantial causes are also unique, whereas cooperative conditions are shared with other effects. [264] For example, a barley seed acts as the cause of a barley sprout but does not act as the cause of a rice sprout. Similarly, a rice seed acts as the cause of a rice sprout but does not act as the cause of a barley sprout. Therefore these seeds are posited as unique causes, whereas water and fertilizer are recognized as common causes since they serve as the causes of both effects. Alternatively, one could say that the substantial cause is the principal producer of the essential nature of the effect, whereas cooperative conditions are the principal producers of its attributes. For example, whether a barley sprout grows or a rice sprout grows is the function of the substantial cause, whereas the height and quality of the sprout are principally the function of the cooperative conditions.

One might ask, "What then is the measure of something becoming a cause of something else?" When something becomes endowed with the potency to produce a given thing, it can then be posited as having become a cause of that entity. For example, Dharmakīrti's *Exposition of Valid Cognition* states:

> When the nature of the earth and so on
> transform to produce a sprout
> it is a cause, for when it's nurtured well
> one can observe the effects.[197]

For example, when a field transforms from its winter state (when it does not produce any sprouts) the field then becomes a productive cause capable of generating sprouts. This is so because when the field is well nurtured, for example through ploughing, an excellent result, such as a good harvest [of grain], is observed. [265]

Thus whether something is an external or internal phenomena, it is through the combination of both a substantial cause and cooperative conditions that they must arise. For example, in order for a sprout to arise it must do so in dependence on the accumulation of many causes and conditions, such as its substantial cause, the seed, and its cooperative conditions, water, fertilizer, and so on. A solitary seed cannot produce a sprout, and without the seed, water and fertilizer alone cannot produce the sprout. This applies to other conditioned phenomena. *Eight Thousand Verse Perfection of Wisdom Sūtra*, for example, states:

> Children of good families, it is like this. Consider, for example, the sound of a lute. In arising it does not come from anywhere, in ceasing it does not go anywhere, nor does it relocate anywhere else, for it arises in dependence on the assembly of causes and conditions, and it relies on causes and it relies on conditions. Thus in dependence on the body of the lute, in dependence on the skin, in dependence on the strings, in dependence on the neck, in dependence on the bow, in dependence on the fretboard, in dependence on the exertion generated by the musician, sound arises from the lute. . . . So too, children of good families, establishing the body of a buddha relies on causes, and on conditions, for it is fully established by the manifold practice of the root of virtue. The body of a buddha is not distinguished by having just one cause, nor by one condition, nor by one root of virtue, and neither is it causeless. It arises from the accumulation of many causes and conditions.[198] [266]

So too, in Devendrabuddhi's *Commentary on the Exposition of Valid Cognition*, one reads:

> In that case, it would not be possible for multiple effects to emerge from a single thing. What is it then? "Multiple means

> from multiple causes," which is the position of Buddhist schools. If that is so why is it untenable for one to arise from one? Because ultimately "all arise from collections; from one collection another collection arises."[199]

## EFFECTS

An effect is defined as "something that is produced by its cause." For example, a sprout is posited as the effect of a seed and smoke the effect of fire. To define what an effect is Dharmakīrti states in his *Exposition of Valid Cognition*:

> Smoke is a result of fire
> because it follows as its consequence.
> If smoke were to arise without fire,
> it would then possess no cause at all.[200]

So Dharmakīrti explains how, because smoke is contingent on fire and follows as its consequence in that it would not exist if there were no fire, smoke constitutes an effect of fire. And in contrast, if smoke were to exist even if there is no fire, then smoke would become something that does not possess any cause at all. [267]

If differentiated, effects are classified from the perspective of time into two: direct effects and indirect effects. When differentiated from the perspective of their continuum, effects are classified into substantial effects and cooperative effects.

A *direct effect* is "that which is produced directly," for example, smoke is a direct effect of fire. It is so called because when smoke arises from fire it does so without any other effect intervening between the two.

An *indirect effect* is "that which is produced indirectly," for example, smoke that is an indirect effect of firewood. It is so called because when smoke arises from firewood it does so with another effect, namely, fire intervening between the two.

A *substantial effect* is "that which is primarily produced within its substantial continuum," for example, a clay vase is the substantial effect of the clay that is its cause. It is so called because it is through a change in the substantial cause, which is the clay, that its subsequent substantial continuum assumes the form of a clay vase.

A *cooperative effect* is "that which is primarily produced outside its substantial continuum," for example, the clay vase is a cooperative effect of the potter who is one of its causes. It is so called because when the effect, the clay vase, arises, it does so not in a manner whereby the potter himself turns into a clay vase, yet the effect, the clay vase, is produced.

In brief, substantial effects arising from their substantial causes include things such as cloth from wool and thread, clay vases from clay, gold vases from gold, and sprouts from seeds. Cooperative effects, on the other hand, arise from their cooperative conditions, such as the arising of the above-mentioned things (cloth, clay vase, gold vase, and sprouts) from the weaver, the potter, [268] the goldsmith, the person who planted the seed or the water, the fertilizer, and so on.

"What then is the difference between causes and effects?" one may ask. In general the two are considered to be equivalent. For example, a vase is an effect of its own cause and it is also the cause of its own effect. So too morning is the result of the night that preceded it and, at the same time, it is the cause of the afternoon that follows it, and so on. This is true of all conditioned phenomena. This said, in the context of a specific phenomenon, the two are mutually exclusive. For example, fire is the cause of smoke that is its effect, and not the effect of that smoke. Similarly, smoke is the effect of fire that is its cause, and not the cause of that fire. The same applies to all instances of cause and effect.

# 12

# Unconditioned Phenomena

IN GENERAL there are two main categories within knowable objects: (1) those that are conditioned by causes and conditions and are subject to change, and (2) those that are not conditioned by causes and conditions and are thus not subject to change. The first of these two are conditioned phenomena, whereas the second are unconditioned phenomena.

Unconditioned phenomena are defined as: "That which is devoid of the three characteristics of arising, ceasing, and enduring." Unconditioned phenomena and permanent phenomena are equivalent. [269] In the lower Abhidharma system, such as that of Vasubandhu's *Treasury of Knowledge*, for example, three types of unconditioned phenomena are mentioned: (1) space, (2) analytical cessation, and (3) nonanalytical cessation.

Vaibhāṣikas define *space* as "that which is free of obstruction and accommodates physical phenomena." They assert space to be a permanent substance that does not obstruct physical phenomena and in turn is not obstructed by physical phenomena. However, they assert the space element, which is part of the six elements, to be a visible form having the nature of either light or darkness. Other Buddhist schools, beginning with the Sautrāntika school, define what is referred to as *space* in terms of a nonimplicative negation, which is the mere absence of obstructive contact. *Analytical cessation* refers to the cessation of the afflictions relevant to one's stage on the path owing to analytical wisdom. An example would be the total elimination of hatred that is attributable to the power of cultivating the path. *Nonanalytical cessation* refers to the factor that prevents the arising of future phenomena because of the power of incomplete conditions. For example, there is the cessation of hatred permanently, not because of the power of cultivating the path but because the completion of the conditions giving rise to hatred have become impossible. Vasubandhu's *Treasury of Knowledge* states:

> The three unconditioned phenomena
> are space and the two cessations.
> Space does not obstruct.
> Analytical cessation is freedom.
> Each cessation is different.
> Nonanalytical cessation absolutely obstructs arising [270]
> and is different from analytical cessation.[201]

According to the Vaibhāṣikas, space accommodates material objects and acts to support wind, analytical cessation stops contaminated phenomena that are objects to be negated on the path, and nonanalytical cessation inhibits the arising of objects of negation, such as the afflictions. Thus they assert all three to be unconditioned phenomena with distinct substantial existence.

In Vasubandhu's *Treatise on the Five Aggregates*, however, four types of unconditioned phenomena are enumerated: (1) space, (2) analytical cessation, (3) nonanalytical cessation, and (4) suchness.

> What are unconditioned phenomena? They are space, nonanalytical cessation, analytical cessation, and suchness. What is space? It is that which provides a spatial locus for form. What is nonanalytical cessation? It is the absolute nonarising of the aggregates that is not free from what is to be negated since it lacks the antidote to the afflictions. What is analytical cessation? It is the cessation that is freedom and the absolute nonarising of the aggregates because of applying the antidotes to the afflictions. What is suchness (*tathātā*)? It is the reality of any phenomena and it [pertains to] phenomena and the person.[202] [271]

In Asaṅga's *Compendium of Abhidharma*, however, unconditioned phenomena are classified in terms of the following eight types: (1) suchness of auspicious things, (2) suchness of inauspicious things, (3) suchness of neutral things, (4) space, (5) analytical cessation, (6) nonanalytical cessation, (7) the immovable, and (8) cessation of discernment and feeling. The text continues:

What is the suchness of auspicious phenomena? It is the two selflessnesses, emptiness, signlessness, the reality limit, the ultimate state, and the expanse of reality. Why is suchness called suchness? Because it does not transform into something else. Why is it called emptiness? Because it does not act as the cause of that which is thoroughly afflicted. Why is it called signless? Because it fully pacifies signs. Why is it called the reality limit? Because it is a focal object that lacks distortion. Why is it called the ultimate state? Because it is the pure referent field of ārya transcendent wisdom. Why is it called the expanse of reality? Because it is the cause of all the Dharmas of śrāvakas, [272] solitary buddhas, and buddhas. So too you should view the suchness of inauspicious phenomena and the suchness of neutral phenomena like the suchness of auspicious phenomena.

What is space? That which is formless and accommodates all action. What is nonanalytical cessation? That which is not free from what is to be negated. What is analytical cessation? That which is free from what is to be negated. What is the immovable? It is the cessation of happiness and suffering that is free of attachment to the heaven called Extensive Virtue but not free from attachment to higher heavens. What is the cessation of discernment and feeling? The cessation of unstable—and even stable—minds and mental factors owing to attention and preceded by the discernment of the state of peace that is free from attachment to the base of nothingness and progresses beyond the summit of existence.[203]

Of these eight types of unconditioned phenomena, the first three, although not different with respect to all being suchness, are differentiated by their bases. There are two things that are the objects of cessation: the afflictions and the feelings that are the source of the afflictions. It is on the basis of the cessation of the first that analytical cessation is defined. Within feelings there are two kinds: (1) pleasurable and painful, which effect change in the body and mind, and (2) neutral feeling, which does not cause such change in body and mind. It is on the basis of the cessation of pleasure and pain that the immovable is defined, and on the basis of the exhaustion of neutral feeling that cessation of discernment and feeling is posited. [273]

The immovable is posited from the point of the cessation of feelings of pleasure and pain, which are objects to be negated, wherein one is free from attachment to the third absorption and below due to the mundane path but not free from attachment to the fourth absorption. It's so called because it remains unmoved by the eight errors, and in particular it is so called because it is not moved by pleasure and pain.

The cessation of discernment and feeling refers to the cessation of discernment and feeling on the basis of the mind of the *summit of existence* where one is free of attachment to the realm of nothingness and below.

The distinction between the meditative attainment of the cessation of discernment and feeling referred to earlier in *Compendium of Abhidharma* as a nonassociated formative factor, versus the cessation of discernment and feeling just referred to, is said to be this: the former is an implicative negation, whereas the latter is a type of nonimplicative negation.

Moreover, according to the views of the great masters of Buddhist epistemology, many of the factors that are defined from the perspective of conceptual thought, such as subject and predicate, common loci and universals, definiens and definiendum, identity and difference, thesis and proof—all these universal categories must be posited as imputed existents and unconditioned phenomena. Dharmakīrti's *Exposition of Valid Cognition*, for example, states:

> The presentations of subject and predicate,
> difference, nondifference, and the like,
> these exist without analysis into reality,
> and in accordance with conventions of the world.
> So in accordance with this [convention] alone,
> everything formulated, thesis, proof, and so on, [274]
> these were constructed by the learned ones
> to help lead to the ultimate truth.[204]

In view of the above, it must be understood that unconditioned phenomena need not be exhausted within the classification of the eight types enumerated earlier in Asaṅga's *Compendium of Abhidharma*.

# 13

## Other Presentations of Ascertainable Objects

### Definitions, the Defined, and Their Instances[205]

When logicians engage in their quest for discovering the nature of reality, as a means to understanding the ultimate truth they utilize categories such as subject and predicate, thesis and proof, definiens and definiendum, and so on—as listed earlier in the citation from Dharmakīrti's *Exposition of Valid Cognition*. Such categories are identified on the basis of examining even a single object of knowledge from the perspective of its conventional level of reality. Since these categories are more closely related to the way in which the mind engages an object, in the classical texts they are presented in the section on how the mind engages its object. Here, however, to make it easier for beginners to engage such topics, we shall present them in the section on the objects of knowledge.

To explain the definiens and definiendum, as Maitreya's [275] *Ornament of Mahāyāna Sūtras* says, there are three elements—definiens, definiendum, and the basis of the definition:

> The buddhas, to help sentient beings,
> gave perfect explanations by means of
> the distinction between the basis,
> the definiens, and the definiendum.[206]

The three elements of the definition are, respectively: (1) *definiendum*, that which is defined, (2) *definiens*, that which defines, and (3) the *basis*, that is, the basis on which the definition is established.[207] To illustrate this with an example: "vase" is the basis, "object of knowledge" is what is being defined, and "that which is fit to be an object of cognition" is the definiens. To elaborate, it is on the basis of knowing that a vase is fit to be an object of cognition that one can understand a vase as being a knowable object. That is to say, it is by means of that specific definition that a vase can illustrate an instance of a knowable object to someone's mind.

In general there are many different usages of the term "definition" (definiens), such as the following examples:

1. As a *specific characteristic*, such as heat in relation to fire, which is what distinguishes fire uniquely from other entities.
2. As a *general characteristic*, such as impermanence in relation to fire, which is a characteristic shared in common by both fire and other conditioned things that are not fire.
3. As a *differentiating characteristic*, such as the differentiations made on the basis of a single entity, for example, a vase, whereby it is characterized as impermanent on the basis of its exclusion of permanence and characterized as a product on the basis of its exclusion of nonproduct. [276]
4. As a *defining characteristic that eliminates dissimilar types*, such as positing "that which is capable of functioning as a bulbous, flat-based, water pourer" as the definition of a vase. Such a definition eliminates the twin fault of nonpervasion (inconclusiveness) and overpervasion (overextension).
5. As a *defining characteristic that eliminates wrong conception*, such as those that have been formulated as definitions for the purpose of eliminating doubt in one's mind.

From among those, the meaning of "definition" that is relevant in the context of our topic—differentiating the three elements of definiens, definiendum, and basis—is principally the fourth, namely, definition as a *defining characteristic that eliminates dissimilar types*. One should know that such definitions could also include definitions as a defining characteristic that eliminates misconception as well.[208]

For a definition to be a defining characteristic eliminating dissimilar types it must be free from the following three faults: (1) the fault of nonpervasion, (2) the fault of overpervasion, and (3) the fault of not being present in its instances. For example, if we were to posit "that which possesses branches and leaves" as the definition of a tree, then it would suffer the fault of nonpervasion due to underpervasion, since it would not pervade all classes of tree. Similarly, if one were to state "that which arises from the earth" as the definition of a tree, it would then suffer the fault of overpervasion (overextension), since it would pervade not only trees but all other things that arise from the earth as well. And if one were to posit [277] "that which is capable of functioning as a bulbous, flat-based, water pourer" as the definition of a tree, it would then suffer the fault of not being present in its instances, since that definition would not apply to a willow tree, for example, which is an instance of a tree. In these ways one understands that each of these examples cannot be the definition of a tree.

In general, with respect to positing a definition, some are presented from the point of view of function and others from the point of view of both function and essential nature of the relevant entity. The examples of a definition from the point of view of function are: a pillar is "that which is capable of functioning to support a beam," or a cause is "that which produces its effect." The examples of a definition from the point of view of both function and essential nature are: fire is "that which is hot and burning" and consciousness is "that which is clear and cognizing." Here "hot" refers to the essential nature of fire and "burning" to its function, and "clear" refers to the essential nature of consciousness and "cognizing" to its function.

The definition of "definiendum" is "that which possesses all three properties of an imputed existent (the designation)." These three properties are: (1) it is an object defined, (2) it is established with respect to its basis, and (3) it does not serve as the definiendum of any other definition. To illustrate these three by taking, say, *thing* as an example: First, "thing" is itself an object defined. Second, it is established with respect to its basis, such as a vase, for example. Third, other than being the definiendum of its definition "that which is capable of function," it does not serve as the definiendum of any other definition. Hence all three properties of a designation are complete.

The definition of "definiens" is "that which possesses [278] the three properties of a referent existent (the designatum)." The three properties are: (1) it is a definition, (2) it is established with respect to its basis, and (3) it does not serve as a definiens of any other object defined. To illustrate these three properties by taking *that which is capable of function* as an example: First, "that which is capable of function" is itself a definition. Second, it is established with respect to its basis, such as a vase, for example. Third, other than being the definition of its object-defined "thing," it does not serve as the definiens of any other object defined. In brief, "that which is capable of function" is that which characterizes or defines a thing. *Thing* is the object defined or that which is characterized by "capable of function." *Vase* is an instance: it is a basis characterizing a phenomenon that is a thing, since it is "that which is capable of function."

In the relationship between "that which is capable of function" and "thing," there exist eight modes of pervasion: (1–2) two modes of entailment in terms of being, (3–4) two modes of entailment in terms of not being, (5–6) two modes of entailment in terms of existing, and (7–8) two modes of entailment in terms of not existing.[209]

Further, since it is the case that in general the ascertainment of "the object defined" by valid cognition is contingent on a prior ascertainment of the definition by valid cognition, there is a logical sequence in which the two—the definiendum and its definiens—are ascertained by a valid cognition. [279]

## One and Many[210]

To present the topic of "one and many" (or identity and difference), in general all knowable objects are either one or many. The definition of being *one* is "a phenomenon that is not diverse or multiple." "Not diverse" refers to its mode of appearance to conception and its mode of delineation in language, which in itself "lacks diversity." For example, the word "pillar" does not express anything else, such as a vase, that is distinct from a pillar. So too no other objects, such as a vase that is distinct from a pillar, appear to the conception apprehending a pillar. Hence pillar is posited as being one as opposed to many. The same is true of the way in which things such as earth, water, fire, wind, and so on are posited as being *one*. As for something like "the horn of a rabbit," although it is expressed individually in words and it appears individually to conception, it is not "one" since it is not a phenomenon.

*Many* is defined as "phenomena that are diverse or multiple." "Diverse" phenomena refers to their mode of appearance to conception and their mode of delineation in language, in the sense of "possessing diversity." Examples of this include, for instance, the pair earth and water, and the pair being produced and being impermanent. Although the pair being produced and being impermanent are the same (or one) in terms of their essential nature or reality, they are posited as "diverse" since the object-universals of both appear separately to conception and they are understood separately through their different proper nouns. Therefore in general, since *one* (or the same) and *many* (or distinct) need to be defined from the perspective of thought and language, they must be understood as referring to one and many in terms of their conceptual isolates. [280]

Buddhist epistemological texts speak of different senses of being one or many. For instance, there is the sense of being one or many in terms of their entity or essential nature. Similarly, there is the sense of being one or many in terms of their substance. Then there is the sense of being one or many with respect to their conceptual isolates (namely, their conceptual identities).

In a similar manner, phenomena that possess the same entity are defined as "phenomena whose essential natures are not separable"; for example, the pair "a knowable object and existent," the pair "produced and impermanent," and the pair "a pillar and a knowable object." The definition of a phenomenon with the *same entity* as another phenomenon is "a phenomenon whose essential nature is not separable from another phenomenon." The same entity, the same nature, and the same intrinsic nature are equivalent and mutually inclusive. The reason why a pillar itself has the same entity as a knowable object is because merely establishing a pillar establishes the natural essence of a knowable object. Equivalent (or mutually inclusive) phenomena must have the same essential nature, and even particular instances of a phenomenon must have the same essential nature as that phenomenon. But the principle of one and many cannot be applied to the nonexistent, since they lack any intrinsic nature.

The meaning of *distinct entities* is "phenomena whose essential natures are separable"; for example, earth and water, or a pillar and a vase. Also, there are those who do not assert that unconditioned phenomena have distinct essential natures, for distinct entities must appear separately to conception and they must be delineated separately in words. Not only that, but each phenomenon must not be established as the essence of the other. Therefore those phenomena that are distinct entities must be distinct, but those that are distinct are not necessarily distinct in entity or essential nature; for example, produced and impermanent.

The definition of *same substance* is "phenomena that arise without separate essential natures." *Same conditioned entity* and *same substance* are equivalent; for [281] example, a pillar and a golden pillar, or produced and impermanent. These are the same substance since they are conditioned phenomena that do not appear separately to the perception to which they appear. Though unconditioned phenomena are posited as the same entity, they are not posited as the same substance, for if something is the same entity it is not necessarily the same substance, but if something is the same substance it must be the same entity.

In general the *same substance*, the *same coexisting substance*, the *same indifferentiable coexisting substance*, and *same coexisting entity* are different. The meaning of the *same substance* is as before. The meaning of the *same coexisting substance* is "those phenomena that have the same substance that arise, cease, and endure at the same time"; for example, a vase and the eight basic particles comprising the vase. Therefore, if they are the same substance, they do not need to be the same coexisting substance, for though both today's vase and a vase are the same substance, they are not the same coexisting substance, since they do not arise, cease, and endure at the same time. The meaning of the *same indifferentiable coexisting substance* is stated in *Exposition of Valid Cognition*:

> Because of that by seeing things
> one sees all qualities.[211]

If they have three characteristics—the same time of arising, ceasing, and enduring—and if one appears to perception the other must appear, then they are the same substance; for example, a pillar and an impermanent pillar. An impermanent pillar must appear to the nonconceptual perception to which a pillar appears, [282] and since these two share three characteristics—the same time of arising, ceasing, and enduring—and since they share the same substance, they are the same indifferentiable coexisting substance.

That which is the same coexisting substance but not the same indifferentiable coexisting substance is, for example, a vase and the eight basic particles assembled in a vase. Though they are not divisible in substance, they are not the same indifferentiable coexisting substance since the wind particle in the assembly of particles comprising a vase does not necessarily appear to the perception to which a vase appears.

*Same coexisting entity* refers to phenomena that possess three characteristics: the same arising, ceasing, and endurance. Those that are coexisting entities but not the same coexisting substance are, for example, the eight basic particles existing in a composite vase.

The definition of *same substantial type* is "distinct conditioned phenomena arisen from the same substantial cause"; for example, both a large and small barley seed arisen from a single seed of barley that is their substantial cause. Substance that is not the same substantial type is taken to be the substantial cause or earlier substantial continuum. *Exposition of Valid Cognition* states:

> Distinct states are from distinct causes.[212]

In conformity with that, Master Śākyamati states in his *Commentary on the Exposition of Valid Cognition:*

> Therefore the categories of distinct continua of similar and dissimilar type arise owing to substantial causes.[213] [283]

Those phenomena that are the same substantial type are not necessarily the same substance. For example, though both a large and small barley seed arisen from a single barley seed are the same substantial type, not only are they not the same, since they have distinct proper names and are thus understood separately, but they are also not the same substance, since they do not appear in the same place to eye consciousness.

The definition of *distinct substances* is "phenomena that arise with separate essential natures." *Distinct conditioned entities* and *distinct substances* are equivalent (for example, a pillar and vase). When these two appear to nonconceptual cognition they are posited as distinct substances, since they are conditioned phenomena that appear with separate essential natures. Thus the ways in which things can be one and many, identical or different in reality as entities, and identical in their essential nature yet have conceptually distinct identities or isolates are indicated in *Exposition of Valid Cognition* by the consummate logician Dharmakīrti:

> Because all things naturally exist with their
> own individual essential natures

they are dependent on conceptual isolates
of similar and dissimilar class.

For that reason anything that possesses a cause
is differentiated through its conceptual isolates.
Owing to differentiating such isolates
comprehension arises, so they are applied.

Therefore understanding arises
owing to distinction and attributes,
but other means lack this capacity.
Thus they stand as distinct objects.[214]

Since things exist in reality with their own individual essential natures unmixed with others, [284] they exist separately from all other phenomena, both those of similar class and those of dissimilar class. Therefore a single basis such as sound may be differentiated in terms of many conceptual isolates of similar class such as "differentiated from a vase" and conceptual isolates of dissimilar class such as "differentiated from space" and so on. The basis on which the production and impermanence of sound are differentiated as distinct conceptual isolates is sound itself, and the reason for differentiating production and impermanence as distinct conceptual isolates with regard to sound is that sound is established as arising from causes through refuting distinct counterpositions, such as sound is unproduced and permanent.

Then, taking that as the reason, "production of sound" and "impermanence of sound" are differentiated as possessing the same essential nature or entity yet distinct in terms of conceptual isolate. Since sound itself exists separately from all those of dissimilar class such as nonsound, unproduced, permanent, and so on, sound has as many conceptual isolates as it has counterpositions. Given the difference of such conceptual isolates, a word like "produced" lacks the capacity to give rise to the specific understanding of impermanence. So too, since that is true for other abstract phenomena, it is said that phenomena such as the production of sound and the impermanence of sound possess the same essential nature but they stand as distinct objects in terms of language and concept. [285]

### *Explaining the Negation of Being and the Negation of Not Being*[215]

Furthermore, the *negation of not being* and *being* are posited as mutually inclusive, or equivalent, and so too the *negation of being* and *not being* are posited as equivalent. Although the *negation of not being* exists for possible phenomena, it does not exist for impossible phenomena.

If applied to a specific instance, both the *negation of being a thing* and *not being a thing* are equivalent and both the *negation of not being a thing* and *being a thing* are equivalent. So too, both the *negation of not being a vase* and *being a vase* are equivalent and both the *negation of being a vase* and *not being a vase* are equivalent. The negation of the negation of not being a vase, in even multiples, and not being a vase, are equivalent. The negation of the negation of being a vase, in any multiple (odd or even), and a single multiple of the negation of not being the basic vase are equivalent.

In brief twice, three, four, five times *the negation of the negation of being a phenomenon*, in even multiples, is just that phenomenon. But that phenomenon, in odd multiples, is posited as not that phenomenon.

In general this is closely related to the presentation of negation and affirmation. *Exposition of Valid Cognition* states:

> Also in that regard, reversing the other and opposing the other.[216]

Many such statements occur in the context of presenting the theory of exclusion of others (*anyāpoha*). [286]

## UNIVERSALS AND PARTICULARS[217]

In general the question of whether universals exist separately from particulars, what might be the essential nature of these universals, and what is the essential nature of the relation between universals and the particulars they pervade and so on have been matters of detailed investigation by classical Indian logicians. These questions shall be addressed in the presentation on philosophies, where the *apoha* (exclusion) theory as well as the topic of how language engages objects according to Dignaga's and Dharmakīrti's views will be discussed. Here, with a specific focus on how to assist beginners to train their minds on the path of reasoning, we shall briefly explain the topic of universals and particulars as they are presented in the summary texts of Chapa Chökyi Sengé.

The definition of a universal is "a phenomenon encompassing its specific instantiations" where "its specific instantiations" are necessarily understood to be its particulars. Its instances include such things as a pillar, a thing, a knowable object, existence, and what appears as a vase to the conception apprehending a vase. The manner in which, say, a pillar encompasses its particulars is this: the sandalwood pillar, juniper pillar, stone pillar, and so on are all specific instantiations or particulars of the kind of pillar, and the type pillar pervades all those particulars. Hence the pillar itself is said to encompass its instantiations.

So if one were to define "universal" with respect to a specific example, the definition of a universal of pillar would be "a phenomenon that encompasses pillars as its instances." Its examples include impermanence, material form, knowable objects, and so on. [287]

If universals are classified, on the basis of how the term "universal" is applied there are three categories:[218] type-universal, composite-universal, object-universal.

A type-universal is defined as "something that encompasses its own kinds." Its examples include such things as a pillar, a human being, existence, a thing, and so on. Since the meaning of "encompassing its particulars" and "encompassing its own kinds" remains the same, universal and type-universal are equivalent.

A composite-universal is defined as "a coarse form that is the collection of its multiple parts." The two—composite-universal and coarse material form—are equivalent. Its examples include such things as a vase, a tree, a stream, and so on. Since a vase is a physical entity composed from the collection of its numerous parts, such as its spout, base, belly, and the numerous subtle particles, it is a composite-universal.

An object-universal (or generic image) may be defined by the example of a tree, "that which appears to the conception apprehending a tree to be identical to a tree, even though it is not identical to it." It is called an "object-universal" in that it is merely a generalized concept of "tree" grasped by thought. This is similar to, for example, the generalized idea one has of one's dwelling even when one is elsewhere.

The definition of a particular is "a phenomenon that has a type that operates as its pervader." Its examples include permanent phenomena, impermanent phenomena, earth, water, fire, wind, and so on. We may illustrate the relationship of pervader and pervaded in the context of, say, water. There exists something that operates as the pervader of water, such as "thing," and since water is necessarily a thing, "thing" is the pervader while water is that pervaded. Thus thing is the *type* and [288] water is a token of that type, namely, thing.

Furthermore, there are three conditions that must be present for something to be a particular pervaded by a specific universal. In the example of positing a pillar as a particular token of a knowable object: (1) A pillar is a knowable object. (2) A pillar is intrinsically related to knowable objects. (3) There exist many things that share a common locus between that which is not a pillar and a knowable object.

The same can be extended to all similar examples of particulars. That which is a universal but not a particular is, for example, a *knowable object*. That which is a particular but not a universal is, for example, a *vase* and a *pillar*.[219] That which is neither is, for example, the *horn of a rabbit*. Since a vase is the universal of a gold vase and, at the same time, a particular (or subset) of a thing, it is both a universal and a particular. Similarly, one can speak of material form as being the universal of a *vase*, a thing as the universal of *material form*, and *existence* as the universal of a *thing*. What has a greater field of pervasion is characterized as the universal, whereas that which has a smaller field of pervasion is its particular.

## Substantial Phenomena and Abstract Phenomena[220]

In general in Indian Buddhist epistemological texts "substantial phenomena" refers primarily to unique particulars that have real existence. "Abstract phenomena," in contrast, are defined in terms of differentiations based on the elimination of that which is not it (i.e., others), brought about by thought and language. Other similar usages of these terms also exist. "Substance" is understood to connote substantially existent, and "abstract phenomena," in contrast, is understood to mean nominally existent.

In Chapa's *Collected Topics*, [289] however, the terms "substantial phenomena" and "abstract phenomena" must be understood in terms of his presentation of what are known as the *eightfold substantial and abstract phenomena*. We may illustrate some of these briefly by taking a vase as an example. Since a vase is a knowable object, and a vase is a vase, and non-vase is not a vase, it is said to possess all four properties of a substantial phenomenon: (1) it is an established base, (2) it is itself, (3) non-it is not it, and (4) its isolate (i.e., it itself) is not mutually exclusive with substantial phenomena.[221]

"Definition" is an example of something that does not possess all four properties of substantial phenomena. In general, though a *definition* is a knowable object (or an established base), since the definition of *definition* exists, in itself *definition* is something that is defined and not a definition. Also nondefinition is not a definition. Therefore something like *definition* possesses the following four properties of an abstract phenomenon that is not itself: (1) it is an established base, (2) it is not itself, (3) non-it is not it, and (4) its isolate (i.e., it itself) is not mutually exclusive with nonabstract phenomena.

We see Chapa's presentation on substantial phenomena and abstract phenomena as an innovative logical instrument to help clarify in the mind the distinctions between things themselves (self-isolates) and their instantiations (instance-isolates). Therefore we have mentioned this topic in a brief introduction. Those who desire to comprehend these points in detail should consult the texts of the *Collected Topics*.[222] [290]

## Explaining Contradiction

The definition of contradictory is "distinct phenomena with no possible common locus between them." If differentiated, it is twofold: (1) contradictory through incompatibility, and (2) contradictory through mutual exclusion. For example, Kamalaśīla's *Light of the Middle Way* says:

> Thus contradiction among things is to be posited as being of two types: those characterized as incompatible through not abiding together and those characterized as mutually exclusive through eliminating each other.[223]

The characteristics of these two are then explained:

> When the presence of something constitutes the incompleteness of a specific effect, then the presence of that confirms the absence of its opposite. Such examples are posited as the first type of contradiction; for example, the tactile sensation of hot and cold.[224]

Thus Kamalaśīla explains how certain types of phenomena are posited as contradictory through not abiding together because a potent agent, which undermines the arising of its opposite when none of its causes is incomplete, prevents the continued presence of that phenomenon. The same text then continues:

> When the positive determination of one is not possible without the exclusion of the other, then two things are posited as the second type of contradiction;[225] for example, [291] with respect to a specific basis (*dharmin*), where one exists and the other does not exist.[226]

This statement clearly explains "direct contradiction" or "contradiction through incompatibility." First, *contradictory through incompatibility* refers to things that cannot assist each other in a supportive capacity. For those that are incompatible, an instance of direct contradiction is posited as the pair hot and cold, and an instance of indirect contradiction is posited as the pair strongly billowing smoke and cold contact. In the second case, though strongly billowing smoke cannot directly harm cold tactile sensation, it can indirectly harm it. Furthermore, wherever strongly billowing smoke exists, there fire necessarily exists, and since fire is capable of directly excluding the sensation of cold, then smoke (that is its effect) excludes cold tactile sensation and cannot abide together with it.

Question: Exactly in what way do the contact of hot and cold act directly as the agent that harms and the object harmed for each other?

Reply: It is stated that the two—the heat that possesses numerous temporal moments and the cold that possesses numerous temporal moments—act mutually and directly as the agent that harms and the object harmed. For example, Dharmottara says in his *Extensive Commentary on the Drop of Reasoning*:

> Given that both hot and cold are phenomena that are produced by their causes and are the producers of their effects, it is their two continua that exclude each other. The individual moments of each are not mutually exclusive or contradictory.[227] [292]

The way in which the pair light and darkness and the pair hot and cold act as agents that harm and objects that are harmed are different. If we take the example of light and darkness, they each act as the agent that harms and the object harmed, since the production of light and the cessation of darkness are simultaneous. And that is because at a specific place the production of light as a manifest state and the cessation of darkness that eliminates light as a manifest state are simultaneous. Furthermore, it is because of the simultaneity of the light *being already produced* and the darkness that counters that light *having already ceased* at a specific site that neither the continuum of light or darkness is able to assist the other in a commensurate supportive capacity in that location.

If we take the example of hot and cold, when a hot sensation counters a cold sensation, the object harmed by it, it counters it rapidly in the third moment. In the first moment both hot and cold make contact, in the second moment the potency of coldness declines, in the third moment the continuum of coldness is stopped. And though this is true regarding the rate of interaction, still the greater the potency of the countering agent in the course of a series of moments, the smaller the corresponding potency of the object countered. Dharmottara's *Explanation of Ascertainment of Valid Cognition* states:

> Therefore with respect to that which removes the other [its opposite], if it is done rapidly, the other will be eliminated by the third moment.[228]

The moment referred to here is not the shortest moment but a moment required for completing an action that is composed of many such shortest moments of time. *Explanation of Ascertainment of Valid Cognition* [293] states:

> Given this, mutual exclusions here are not of single moments; they are phenomena possessing continua.[229]

Furthermore, when contradiction through incompatibility is differentiated, there are two types: (1) contradictory material forms that are incompatible, and (2) contradictory cognitions that are incompatible.

Examples of the first include hot and cold or light and darkness, while an example of the second is love and hatred. If it is impossible that two things abide together it is not necessarily the case that they are contradictory through incompatibility. For example, it is impossible that both the seed and the sprout that is its effect abide together, since they are cause and effect, but they are not contradictory through being incompatible, since these two exist in the relation of the agent that benefits and the object benefited, and not the agent that harms and the object harmed. Therefore something such as hot and cold do not abide together because they do not abide together continuously without one acting to harm the other. *Twenty-Five Thousand Verse Perfection of Wisdom Sūtra* states:

> Subhūti, it is like this . . . when sun shines no possibility remains for darkness to exist.[230]

The pair light and darkness are an example of contradiction through incompatibility. The way in which contradiction through incompatibility extends to how certain types of subjective mind act as opposing agents for other types of cognition that they exclude will be explained later, for it is extremely important to understand this key point. [294]

With regard to contradiction through incompatibility, when a specific basis is fully established as an object of cognition, its opposite is necessarily excluded; for example, existent and nonexistent, permanent and impermanent, produced and not produced, one and many, a vase and a nonvase, fire existing and fire not existing.

Further, if illustrated by the pair existence and nonexistence, when you know that a basis such as a vase exists, its nonexistence is negated, and when you know that a vase is not nonexistent you indirectly understand that it exists. Therefore the existence and nonexistence of a basis such as a vase are directly contradictory.

So too if illustrated by the pair permanent and impermanent, the cognition that positively determines or fully comprehends that something such as space is permanent excludes space being impermanent or can comprehend it is not impermanent. The cognition that excludes space being impermanent or comprehends it is not impermanent positively determines or comprehends it to be permanent. Therefore contradiction is labeled in dependence on the processes of determination and exclusion.

Furthermore, contradiction is twofold: direct contradiction and indirect contradiction. An example of the first is the pair permanent and impermanent, and an example of the second is the pair a vase and a pillar.

The pair permanent and thing are not directly contradictory because they do not fulfill the criteria for direct mutual exclusion. That is because doubt may arise in the minds of some beings that a third alternative exists, and that cannot be eradicated by means of the cognition that determines one alternative by the power of refutation excluding another alternative, even though there is no third alternative other than permanent and thing, since there is no phenomena that is neither. [295]

## Explaining Relation

In general there can be different types of relation: like the one between a composite entity and its parts, such as a vase and the components of that vase; or like the one between a juniper and a tree, which is a relationship of that pervaded (particular) and its pervader (universal); or like the one between being produced and being impermanent—things that are mutually inclusive—which is an intrinsic relationship. Among these, there must exist an intrinsic relationship between mutually inclusive phenomena, between a general set and its subsets, between the parts and the whole, and between a composite entity and its constituents. The relationship that exists between any phenomena that are cause and effect, on the other hand, is that of causal relation.

The definition of relation is this: "*It* is different from *that* and the absence of *that* necessarily entails the absence of it." If differentiated, relation consists of two types: intrinsic relationship and causal relationship.

The definition of something intrinsically related to a phenomenon is: "Something that is distinct from a phenomenon because of being intrinsically the same as that phenomenon, and if that phenomenon does not exist then it necessarily does not exist." *Exposition of Valid Cognition* states:

> That which is contingent on the nature of another
> is related through its very existence.

If the other doesn't exist, then its existence ceases,
because they are not distinct in nature.[231] [296]

Because the entity of impermanence that is innately related to a product's very existence is negated, the entity of the product is negated. Therefore "product" is intrinsically related to "impermanence."

The definition of something causally related to a phenomenon is "Something that is related to a phenomenon through being intrinsically distinct from that phenomenon, and if that phenomenon does not exist it necessarily does not exist." The *Rice Seedling Sūtra* states:

When this exists, that comes to be;
with the arising of this, that arises.[232]

*Exposition of Valid Cognition* states:

From presence and absence
the presence of the effect is observed.
Its nature is that of possessing a cause,
therefore it does not arise from a dissimilar class.[233]

The process therefore is that if an earlier phenomenon acts to benefit another thing, there is the presence of its arising; if it does not act to benefit it, then there is the absence of its arising. In dependence on this evidence the latter phenomenon is seen to follow the earlier, and that which has such a nature possesses a cause or is an effect of the earlier phenomenon. Furthermore, smoke, for example, is ascertained to follow fire in just that manner. Therefore, since it does not arise from a dissimilar class, it is ascertained to arise from fire. [297]

## Negation and Affirmation[234]

The presentation on negation and affirmation is closely related to grasping the theory of *anyāpoha* (exclusion of others) or *apoha* (exclusion), which is essential for understanding the nature of reality as well as how the mind engages reality. Also, when one engages in negation and affirmation it becomes crucial to recognize the scope of negation through logical reasoning. In all such matters, understanding of the nature of negation and affirmation remains of vital importance.

In general whatever the given phenomenon, so long as it is something that has a defined identity, one must define its identity in terms of negating what it is not. This said, phenomena such as a vase, a pillar, a tree, or a flower, which are objects of perception by sensory consciousness, by themselves must be posited principally as positive phenomena. However, the negation of nonvase or of nonpillar, which are defined by the elimination of their dissimilar classes, since they are defined by the negation of their opposites, must be understood as forms of negation. Thus when one states, "that which is not a nonvase" or "that which is not a nonpillar," the reference is to something that is a vase or a pillar. As the saying goes, "two negatives make a positive." This said, since the two—not a nonvase and not a nonpillar—are defined by the negation of nonvase and nonpillar, they are forms of negation, not affirmation. [298]

Similarly, in the case of the *absence of a vase*, except for defining it in terms of the negation of its opposite, no identity can be posited by way of its essential nature. In contrast, something like a *vase* can be defined by way of its own identity without depending on explicitly negating its opposite (nonvase). Therefore the "absence of a vase" is posited as a negation, whereas "vase" is posited as an affirmation.

First, in defining "negations" and "affirmations," negation is "a phenomenon that must be cognized by the mind that directly comprehends it, directly precluding its negandum (object of negation)." Negation, exclusion (*apoha*),[235] and exclusion of other (*anyāpoha*) are equivalent; for example, not a vase, not a pillar, and a house without humans.

If the means of comprehending an object were that it appeared to cognition owing to cognition directly precluding its negandum, then the negandum would appear as an aspect that is precluded. For example, the statement "there is no vase" must be ascertained by means of the negandum—the vase—appearing as an object of cognition, and that negandum merely not existing or it being refuted.

Affirmation (or a positive phenomenon) is "a phenomenon that is not comprehended by the mind that directly comprehends it, directly precluding its negandum"; for example, a pillar, form, consciousness, or produced. If illustrated by a pillar, then it is not comprehended by the mind that directly comprehends it, directly precluding its negandum, for a pillar is posited as an affirmation, since it is perceived through being established as the perceived object of that cognition.

*Detailed explanation through briefly elaborating the classifications of negation*

If negation is classified, there are two types: nonimplicative negation and implicative negation.

The difference between these two [299] is whether another phenomenon isn't or is implied in the wake of negating its negandum. The former is called a *nonimplicative negation* and the latter is called an *implicative negation*. Further, the verses of *Lokaparīkṣa* referred to by Nāgārjuna and quoted in Bhāvaviveka's *Lamp of Wisdom* state:

> The negation of existence
> is not the apprehension of nonexistence,
> just as stating "It is not black"
> does not mean "It is white."[236]

Thus saying "That object is not black" indicates merely that it is not black. It does not indicate it is some other color such as white. So too, saying "That object lacks inherent existence" indicates merely the negation of inherent existence or merely an absence of inherent existence. It does not establish that noninherent existence exists and so on.

If such statements indicated something other than the mere negation of the negandum, then *lacking inherent existence* would become an implicative negation, and through that the difference between the two negations can be understood. Thus all negations are subsumed as either implicative or nonimplicative negations. Śāntarakṣita states in his *Compendium on Reality*:

> Thus there are two types of exclusion (*apoha*),
> implicative and nonimplicative negation.[237] [300]

Therefore the definition of a nonimplicative negation is "that comprehended by directly precluding its negandum, where no other phenomenon is implied to awareness in the wake of negating its negandum"; for example, the statement: "There is no elephant here." For if it is asked, "Is there an elephant here or not?" then stating "There is no elephant," negates there is an elephant here, and since that statement does not imply to awareness anything other than an elephant, such as a tiger and so on, it is posited as a nonimplicative negation. So too, space is the mere negation of obstructive contact, and even though its proper name does not contain a term of negation, it is posited as a nonimplicative negation since nothing other than the mere negation of obstructive contact appears when space appears to mind. Further, *Blaze of Reasoning* states:

> A nonimplicative negation is taken to be the mere cessation of the essential nature of a thing. It does not establish another thing that is similar to it but not it. For example, the statement "Brahmans should not drink alcohol" merely prohibits that act, and it does not recommend "Drink something else" or "Do not drink something else."[238]

If nonimplicative negations are classified, there are two types: (1) a nonimplicative negation whose negandum (object of negation) exists, and (2) a nonimplicative negation whose negandum does not exist. An example of the first is *there is no pillar* or *there is no house*. The negandum is *there is a pillar* and *there is a house*, and both are possible phenomena. [301] A nonimplicative negation whose negandum does not exist is, for example, *there are no horns of a rabbit*, because it is impossible that its negandum (i.e., the horns of a rabbit) exists.

Therefore the definition of an implicative negation is "that comprehended by means of directly precluding its negandum, where another phenomenon is implied in the wake of negating its negandum"; for example, the opposite of not a vase. That is posited as an implicative negation since a vase is implied or comprehended in the wake of the mind that directly comprehends a vase, directly precluding that which is not a vase. *Blaze of Reasoning* states an example:

> Implicative negations negate the nature of a thing and establish the nature of another thing similar to it but different from it. For example, the negative statement "This is not a brahman" establishes a person to be a śūdra [or lower caste] who is inferior

> in terms of austerity, study, and so on, one who is not a brahman
> but similar to a brahman, something different.[239]

If implicative negations are classified according to how other phenomena are implied there are four: (1) negation directly implying another phenomenon, (2) negation indirectly implying another phenomenon, (3) negation both directly and indirectly implying another phenomenon, and (4) negation implying another phenomenon owing to context.

For example, the statement "A mountainless plain" directly implies another phenomenon. The statement "Fat Devadatta does not eat food at night" indirectly implies another phenomenon. The statement "Fat Devadatta who does not eat during the day is not thin" both directly and indirectly implies another phenomenon. And the statement "That person is either of warrior or brahman [302] caste but not of brahman caste" implies another phenomenon due to context.

The way this occurs is:

1. The statement "A mountainless plain" directly indicates *a plain* through literally (or directly) negating *mountains*, its negandum.

2. The statement "Fat Devadatta does not eat food at night" literally negates *eating at night*, the negandum, and by stating "fat" one understands "eats food." Since the time of eating is ascertained as either day or night, this statement indirectly indicates that he eats during the day, since "he doesn't eat at night."

3. The statement "Fat Devadatta who does not eat during the day is not thin" indirectly indicates that he eats at night, since it negates that he eats during the day and directly conveys that he is not thin.

4. When "That person is either of brahman or warrior caste, but he is not of brahman caste" is stated, it is understood that he is of warrior caste owing to context.

Furthermore, Avalokitavrata says in his *Explanation of the Lamp of Wisdom*:

> Negation revealed through the meaning,
> negation established by a single phrase,
> negation that possesses both, and negation not revealed
> by its words.
> The rest not included here are different.[240]

Further, a similar presentation is found in Nabidharma's *Summary of Negations*: [303]

> Negation understood through the meaning,
> negation established by a single phrase,
> negation that possesses both, and
> negation not revealed by its words.[241]

"Negation understood through the meaning" refers to an implicative negation that indirectly implies another positive phenomenon—or affirmation—in the wake of negating its negandum. "Negation established by a single phrase" refers to words that directly imply another positive phenomenon in the wake of negating its negandum. "Negation that possesses both" refers to words that both directly and indirectly imply another phenomenon in the wake of negating its negandum. "Negation not revealed through the words" refers to that implied owing to context in the wake of negating its negandum.[242] Thus both texts teach the fourfold classification of implicative negation.

Moreover, Vasubandhu states in his *Explanation on the First Dependent Origination and Its Divisions*: (1) the negation of existence, (2) another, (3) similarity, (4) inferiority, (5) a small amount, (6) absence, and (7) antidote.[243] The first of these seven classifications of negation is nonimplicative negation, and the remainder are implicative negations. From among those:

1. Nonimplicative *negation* is, for example: "There are no humans in a place without humans," where the term of negation negates the existence of humans in that place and does not indicate anything other than the mere absence of humans there. [304]

2. Negation indicating *another* is, for example: "The fat man does not eat during the day," where the term of negation not only negates eating during the day but also indirectly indicates that he eats at night, and that is different from eating during the day.

3. Negation indicating *similarity* is, for example: "Not a god (*asura*)," where the term of negation does not indicate a nonimplicative negation that merely negates "god" but also indicates a person who is similar to a god.

4. Negation indicating *inferiority* is, for example, when it is said: "A bad man is not a man," where the statement does not negate that he is a man but indicates he is an inferior man.

5. Negation indicating *a small amount* is, for example: "There is no hair on the head of that man," where the term of negation does not merely indicate that there is no hair on that man's head but also indicates he has a small amount of hair.

6. Negation indicating *absence* is, for example: "A person who lacks clothes," where the term of negation does not indicate merely "*a lack* of clothes" but it indicates a person *without clothes*. So too, a monastery without water is a similar example.

7. Negation indicating the *antidote* is, for example: "Nonattachment," where the term of negation does not indicate a mere lack of attachment but also indicates that the root virtue of nonattachment is the antidote acting as the excluding factor of attachment. So too, ignorance and lack of merit are similar in type.

Also, if extensively classified because of the different ways a term of negation can be applied to its proper name, there are fifteen classifications. *Summary of Negations* states: (1) nonexistence, (2) another, (3) similarity, (4) inferiority, (5) timidity, (6) weakness, (7) subtlety, (8) small amount, (9) swift passage, (10) smallness, (11) no extra, [305] (12) from some not all, (13) incompatibility, (14) antidote, and (15) absence.[244]

Thus there are fifteen terms of negation that express fifteen negations:

1. A term of negation indicating *nonexistence* is, for example, the statement "There are no humans in a place without humans." This statement merely negates the existence of humans in that place and indicates nothing else.

2. A term of negation indicating *another* is, for example, the statement "The fat man does not eat during the day." This statement not only negates eating during the day but also indirectly indicates that he eats at night, and that is different from eating during the day.

3. A term of negation indicating *similarity* is, for example, the statement "Not a god." This statement does not indicate a nonimplicative negation that is the mere negation of a god but rather indicates a being similar to a god.

4. A term of negation indicating *inferiority* is, for example, the statement "A bad man is not a man." This statement does not merely negate that he is a man, but it also indicates that this man is inferior.

5. A term of negation indicating *timidity* is, for example, "mindless sleep." This term indicates that the mind weakly engages its object as strong mental engagement ceases when asleep.

6. A term of negation indicating *weakness* is, for example, the statement "The horse doesn't go" for a horse that cannot go fast. That statement does not merely negate the horse going but also indicates that its going is weak.

7. A term of negation indicating *subtlety* is, for example, the statement "Meditative attainment of nondiscernment." That term does not indicate that discernment is completely nonexistent [306] but indicates that subtle discernment exists through negating coarse discernment.

8. A term of negation indicating *a small amount* is, for example, a being who is not very active being called "inactive." This term indicates a small amount of activity and does not just negate all activity.

9. A term of negation indicating *swift passage* is, for example, the term "He does not linger here" for a being who does not stay for a long time but quickly departs. This term does not merely indicate the negation of staying but also indicates staying for a short period.

10. The term of negation indicating *smallness* is, for example, the term "lacks a body" for a person with a small body. This term does not indicate that person has no body at all, but indicates a small body through negating having a large body.

11. The term of negation indicating *no extra* is, for example, the statement "I have no clothes" for a man who possesses just one suit of clothing and no extra clothing when others beg for clothes.

12. The term of negation indicating *from some not all* is, for example, the statement "Sons are not born from all women." This term indicates that though sons are born from some women, and that is not denied, it is not necessary to assert that sons are definitely born from all women.

13. The term of negation indicating *incompatibility* is, for example, the term "unfriendly." This term not merely negates friendliness but also indicates that it opposes friendliness.

14. The term of negation [307] indicating the *antidote* is, for example, the statement "It is not dark," for the purpose of indicating that light is the antidote of darkness. That term does not indicate that light is merely not darkness but also indicates light that acts as the antidote to darkness. So too the terms "nonattachment," "nonhatred," and "nondelusion" are terms of negation indicating the antidote.

15. The term of negation indicating *absence* is, for example, the phrase "a lack of clothes" for a person who lacks clothes. This term does not indicate a mere lack of clothes but indicates a person without clothes. So too

the terms "a monastery without water" and "a house without humans" are terms of negation indicating absence.

The first is a term of negation indicating a nonimplicative negation and the remaining fourteen are terms of negation indicating implicative negations. Further, how to subsume the fifteen negations in eight and in two negations is explained in *Summary of Negations*:

> (1) Timidity, (2) weakness, and (3) inferiority,
> and the six from subtlety and so on, are each combined
> as one. Thus the number of negations is eight.[245]

Both the negation indicating *timidity* and the negation indicating *weakness* are included in the negation indicating *inferiority*, and both timidity and weakness are taken together as the negation indicating inferiority. The six negations from *subtlety* and so on are included in the negation indicating subtlety, and those six together are called "negations indicating subtlety." Thus in total there are eight negations: negation indicating (1) nonexistence, (2) another, [308] (3) similarity, (4) inferiority, (5) subtlety, (6) dissimilar states, (7) antidotes, and (8) absence.[246]

Further, negation indicating timidity and negation indicating weakness are included in negation indicating inferiority because those three are similar in negating that which possesses good qualities. The reason why the six negations, such as subtlety, are included in the negation indicating subtlety is because those six are similar in negating what is coarse.

*Summary of Negations* states that there are seven negations:

> There are seven by combining incompatibility with antidotes.[247]

They are taken as seven negations by including incompatibility in antidotes from among those subsumed in eight negations.

*Summary of Negations* subsumes them in six negations:

> Inferiority, smallness, timidity,
> the number of negations is six.[248]

Of the seven negations, since both negation indicating inferiority and negation indicating subtlety are included in the negation indicating timidity, they are taken as six negations indicating: (1) nonexistence, (2) another, (3) similarity, (4) timidity, (5) antidotes, and (6) absence.

*Summary of Negations* subsumes them in five negations: [309]

> That understood through the meaning,
> that established through a single phrase.
> Negation that possesses both, and
> that not indicated through its words.[249]

Thus there are five negations: (1) nonimplicative negation, (2) that which directly implies another positive phenomenon in the wake of negating its negandum, (3) that which indirectly implies another phenomenon, (4) that which both directly and indirectly implies another phenomenon, and (5) that which implies another phenomenon by the power of context.

*Summary of Negations* subsumes them in four negations:

> Refuting imputed entities,
> timidity from one's own side,
> others, and exclusion makes four.[250]

Thus there are four: (1) negation refuting an imputed entity, (2) negation indicating timidity, (3) negation indicating another, and (4) negation indicating exclusion.

1. The first are nonimplicative negations. An example is "sound is not permanent," for "sound is permanent" is an imputed entity or a superimposition. "Sound is not permanent" is called *negation refuting an imputed entity* since it is a negation refuting "sound is permanent." All nonimplicative negations are similar in type.

2. Negation indicating timidity includes negations indicating inferiority up to negations indicating from some, not all. Since they are similar in negating great or coarse qualities and so on that are their specific negandum, they are called *negations indicating timidity.*

3. Negation indicating another is, for [310] example: "A fat person does not eat food during the day." Since this statement indicates that the person eats at night, and that is different from eating during the day, it is *negation indicating another.*

4. Negation indicating exclusion is, for example, ignorance, nonattachment, nonaversion, and nondelusion. Since the statements expressing each state indicate exclusion that constitutes contradiction through incompatibility, they are said to be *negations indicating exclusion*.

*Summary of Negations* subsumes them in three negations:

> There are three engaging the meaning of
> nonexistence, another, and the antidote.[251]

Thus there are three: (1) nonimplicative negation, (2) negation indicating another, and (3) negation indicating the antidote.

*Summary of Negations* subsumes them in two negations:

> The entire enumeration of negation
> is included in both nonimplicative and implicative negation.[252]

Thus there are two: nonimplicative negation and implicative negation. As for their essential nature and illustrations, these have already been explained before.

To conclude, the principles of negation and affirmation are especially emphasized in the Buddhist texts in general and in the writings on the Middle Way philosophy and epistemology in particular because they are connected with correct understanding of how the mind engages with reality as a whole. More specifically, they are connected with understanding clearly the manner in which language and concepts operate by way of negation or affirmation with their relevant objects. The principles of negation are also connected with other issues. For example, when engaged in negation or affirmation through reasoning, one can correctly ascertain the types of negation and affirmation involved as well as discern, through logic, the boundaries of what is negated. For these reasons the presentation on negation is accorded such critical importance in the treatises. [311]

## The Three Classes of Ascertainable Object

To explain the three types of ascertainable object, if one were to categorize all types of ascertainable objects, they would be summarized in two types: (1) evident phenomena (*abhimukhi*), which can be ascertained by direct perception, and (2) hidden phenomena (*parokṣa*), which cannot be ascertained by direct perception and must be ascertained through reasoning.

Within hidden phenomena, there are two kinds: slightly hidden phenomena, which can be established through logical reason based on objective facts, and extremely hidden phenomena, which must be ascertained in dependence on valid testimony. Thus the threefold classification with respect to the division of the types of ascertainable object is exhaustive. These points are discussed in Buddhist texts from an extremely early period. For example, the *Unraveling the Intention Sūtra* states:

> That all conditioned things are momentary, that there are world systems beyond, and that virtuous and nonvirtuous karma endure without losing potency, all of these are correctly deemed impermanent, and they are observed by direct perception. Similarly, that different types of sentient beings engage in different types of karmic action is observed by direct perception. And that sentient beings who experience happiness and suffering engage in virtuous or nonvirtuous karmic action is also observed by direct perception. But for those [312] who cannot directly perceive such facts, they need to be inferred. . . .[253]

Thus characteristics such as impermanence, which reflects the momentary change pertaining to phenomena such as material form, can be directly observed by ordinary beings through their experience. When these attributes cannot be ascertained by way of direct experience, there is the need to rely on inference. Vasubandhu's *Autocommentary* similarly states the following:

> Any phenomena that exist may be observed with direct perception when there is no obstructing agent; for example, the six objects and the mental faculty.[254]

Thus when there exists no impediment for an object being evident, it is then capable of being observed with direct perception.

Therefore those objects of knowledge that are perceived directly by the senses are labeled "evident phenomena." Those types of phenomena that cannot be perceived directly but can be ascertained through inference grounded in facts established by direct perception are labeled "slightly hidden phenomena." And those types of phenomena that for the time being cannot be ascertained even through inference based on objective facts and must therefore be ascertained by the testimony of others alone are labeled "extremely hidden phenomena." Dharmakīrti's *Exposition of Valid Cognition* says, for example:

> Other than perceptible and hidden,
> no other ascertainable objects exist.[255] [313]

Thus Dharmakīrti speaks of ascertainable objects being confined to the two categories of directly perceivable evident phenomena and hidden (or obscure) phenomena. The second category consists of slightly hidden and extremely hidden phenomena. Thus Dharmakīrti states ascertainable objects to be definitively subsumed within this threefold classification.

First, to explain directly perceivable evident phenomena, the definition of an evident phenomenon is "a phenomenon that can be cognized by direct perception without relying on the use of a correct reason." Examples include such external objects as visible form, sound, aroma, taste, tactility, and so on, which can be cognized by direct perceptual experience. It also includes internal objects such as the feeling of happiness or suffering, or coarse levels of ideation and so on.

The definition of a slightly hidden phenomenon is "an object that is necessarily ascertained in dependence on reasoning based on objective facts." To give an example, even though phenomena such as a visible form share the nature of being momentary, they are established by direct perception. Nevertheless their existence characterized by momentariness can be inferred by reasoning such as they are dependent on their causes and they do not depend on any additional third factor for their disintegration. This characteristic of momentariness cannot be established by direct perception that cognizes the entities themselves, such as a visible form.

Similarly, when a person who is walking on a trail sees smoke rising over the summit of a distant mountain, he or she can infer the presence of a fire there. [314] This inference is possible because the person had previously ascertained the relationship between fire and smoke, such that if there is no fire no smoke will arise, having directly perceived smoke at that time. In contrast, if fire were to exist at night it would appear to direct perception from a distance, whereas smoke is not likely to appear even if it were present. At such a time, therefore, one can infer the absence of smoke by reason of no fire. In those two instances the person cannot ascertain the presence of fire and the absence of smoke through the power of direct experience because these two facts remain hidden.

The definition of an extremely hidden phenomenon is "an object that cannot be ascertained by direct perception or inference grounded in objective facts and must be established in dependence on testimony validated through a threefold analysis." The phrase "validated through a threefold analysis" refers to the following: (1) scriptural statements or testimony pertaining to evident facts are not controverted by valid direct perception, (2) statements pertaining to slightly hidden facts are not controverted by valid inferential cognition based on objective facts, and (3) statements pertaining to extremely hidden facts have no internal contradictions between earlier and later passages as well as between earlier and subsequent propositions.

To give an example conforming with our own Buddhist view: In general the fact that karmic effects only arise from the accumulation of consonant karmic causes can be ascertained by inference based on objective facts affirming the general principle of causation that specific types of cause give rise to specific types of effect commensurate with them. [315] However, as to what extremely subtle aspects of the effects are produced by specific subtle aspects of the cause, these remain for the time being extremely hidden since they pertain to the extremely subtle functions of karma. Such obscure issues must be ascertained in dependence on scriptural testimony, which is itself validated through the threefold analysis.

With regard to this third category of ascertainable object, extremely hidden phenomena, even from the point of view of everyday worldly convention there are instances where the relevant facts need to be comprehended on the basis of testimony. For example, many past events must be understood in dependence on historical accounts. Similarly, the date of one's own birth must be understood on the basis of reliable statements of others, such as one's parents. That there are special contexts in which certain facts need to be established by taking the Buddha's scripture as authoritative is stated in Dharmakīrti's *Exposition of Valid Cognition*:

> It was said that inference itself
> does not depend on scripture regarding its referent object.

What is established by it is properly established
at that time and does not rely on treatises.[256]

Thus Dharmakīrti states extremely clearly that if a fact falls within the domain of inference based on objective facts, since it is established by an inferential cognition grounded in objective facts, there is no need for reliance on scripture. If, however, the relevant fact belongs to the third category and is extremely hidden, then:

When proceeding to the third type of object [316]
it is logical to reference treatises.[257]

So when engaging with a category of facts that are extremely hidden, it is the way of a logician to rely on scriptural testimony.

The manner in which one should take scriptural testimony as valid is the following. For example, when scriptural testimony is used to infer that wealth comes from the act of giving, it is not done on the basis of citing the scriptural statement itself as the proof, such as "Wealth comes from the act of giving because that is stated in scripture." Rather, one establishes that a scriptural statement is nondeceptive with regard to its subject matter by citing the fact that the particular scriptural statement has been validated through the threefold analysis. On this point, *Exposition of Valid Cognition* says:

Related to what is beneficial, the appropriate means,
and the purpose of beings, through meaningful speech.[258]

This passage expresses the appropriate subject of a syllogism. Further:

When the meaning of seen or unseen entities
is not harmed by perception
or either type of inference
such statements are nondeceptive.[259]

This passage identifies the reason for such a syllogism.

In terms of other proofs, the Buddha formulated examples to establish that scripture teaching *evident perceptible entities*, which are [317] the first type of ascertainable object, as well as *slightly hidden objects* such as the four noble truths, are nondeceptive regarding their subject matter because they are authenticated by the three examinations. The Buddha also stated the means of indirectly establishing that scriptural statements teaching *extremely hidden entities* are nondeceptive regarding their subject matter. *Exposition of Valid Cognition* states:

> Words of conviction are nondeceptive
> and inferred through universals.[260]

Moreover, it is stated in Āryadeva's *Four Hundred Stanzas*:

> Whoever has doubt about what the Buddha said
> regarding that which is obscure
> should rely on emptiness
> and have conviction in him alone.[261]

And *Exposition of Valid Cognition* states:

> The Protector has seen the path and taught it;
> he does not teach falsely, for that lacks any [good] result.
> Because he possesses compassion, everything he does
> is undertaken for the sake of others.[262]

So the words of the Buddha are proven authoritative on the basis that the speaker Gautama Buddha is an embodiment of genuine compassion for all sentient beings and there exists no reason whatsoever for him to speak falsely about statements pertaining to extremely hidden facts. Therefore, although the Buddhist tradition does acknowledge specific instances where scriptural testimony can be taken as valid, the manner in which such testimonies are authoritative involve the convergence of numerous conditions, including the contents of the scriptural testimony are not controverted by direct perception or reason, no internal contradictions exist regarding the words of the scripture itself, and the speaker has no other special goal or reason for making the statements. [318]

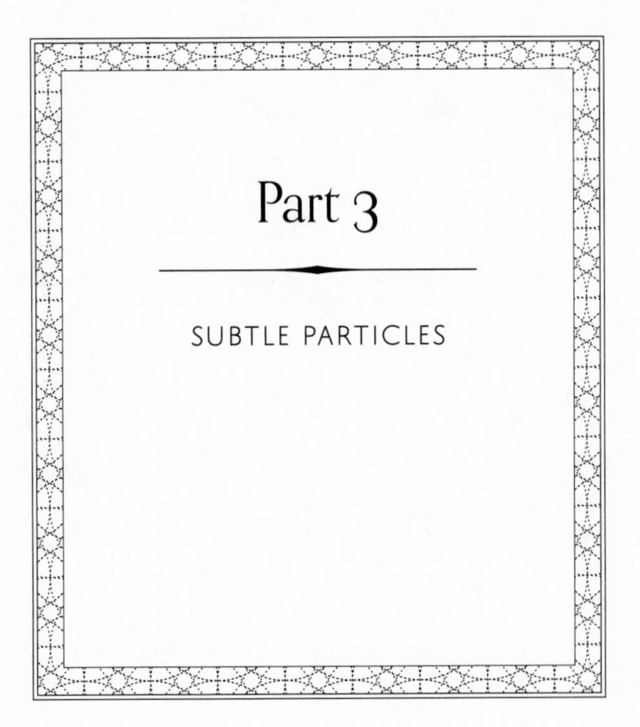

# Part 3

## SUBTLE PARTICLES

## THE SEARCH FOR THE ULTIMATE CONSTITUENTS

PART 3 OF OUR volume is about the theory of atoms in early Buddhist thought and the critique of atomism developed by other influential Buddhist schools, especially the Yogācāra and Madhyamaka. On the thinking behind the view of the Buddhist atomists, Vasubandhu writes in his *Treasury of Knowledge*:

> An atom (*paramaṇu*), a syllable, and a single moment (*kṣaṇa*)
> are the smallest unit of matter, of words, and of time. (3.85bc)

In these two lines Vasubandhu succinctly captures the reductionist impulse of some of the early Buddhist theorists, especially those belonging to the Sarvāstivāda school. The logic is simple. Just as a body of text can be dissected into its smaller units of sentences, words, and syllables, so too can a length of time, such as a day, be reduced to its smaller units of hours, minutes, and seconds. This reductionist principle, the logic goes, must equally apply to the world of matter as well, which means that the everyday objects of the physical world that we experience with our senses must be composed of and reducible to their smaller units, ultimately to their most fundamental irreducible constituents.

Earlier we discussed the Abhidharma's *dharma* theory, the view that there are existents that are irreducible units of reality with their own intrinsic nature. In the earliest sources, such as the sūtras as well as the seven books of Abhidharma, however, there does not seem to be any mention of atoms (*paramaṇu*). Rather, the ultimate constituents of matter are spoken of in terms of the four "great elements" (*mahābhāta*)—earth, water, fire, and wind—and the macroscopic world of our everyday experience is understood to be composed of and is reducible to these fundamental elements. These four great elements are understood to be inseparable, and they act as each other's coexistent causes. The question is if there are only four basic elements that are always inseparable, how does one account for the diversity of forms that exist in the phenomenal world? The answer suggested is that this diversity is a function of the predominance of the substance of specific elements in any given composition. The assertion is that, for example, smoothness arises from a predominance of water and fire elements, coarseness from a predominance of earth and wind elements, lightness from a predominance of fire and wind elements, heaviness from a predominance of earth and water elements, and so on. So the inseparability of the four elements and their interdependence do not entail that they always exist in equal proportion in any given material composition.

Vasubandhu's *Treasury of Knowledge Autocommentary* lists five ways in which the great elements cause or compose the derived material objects: (1) the great elements *produce* the derived matter as a child is produced by her parents, (2) they *influence* the derived matter as a pupil is under the influence of his teacher, (3) they *support* the derived matter in that their transformations are functions of the modifications of the great elements, (4) they *sustain* the derived matter for they ensure its continued existence, and (5) they *nourish* the derived matter in that they give rise to its growth (see page 132). Although the early Abhidharma texts are not explicit on whether these great elements are best conceived as material elements or as forces, judging by the way they are characterized in terms of their functions, they may be better understood as forces.

## The Buddhist Theory of Atoms

It is generally accepted that the earliest proponent of atomic theory in Indian thought is the non-Buddhist Vaiśeṣika school of Kaṇāda, whose date is estimated to be sometime between the sixth to the second century BCE. According to Kaṇāda's *Vaiśeṣika Sūtra*, atoms are conceived to be material substances that are small, eternal, uncaused, and indivisible. Vaiśeṣikas maintain that the macro objects of the everyday world, which are subject to change, are composed of atoms that are themselves changeless. This unchanging nature of the ultimate constituents accounts for the identity of a substance through time, for it was presumed that the substance's essence must lie in something unchanging.

Sometime around the beginning of the Common Era, or possibly even earlier, Sarvāstivāda Buddhists also began speaking of matter in terms of an aggregation of atoms. Unlike Vaiśeṣikas, Buddhist theorists did not see atoms as eternal and unchanging; "they spring into being from time to time and then are destroyed, lapsing seemingly into nothingness" (McGovern 1923, 127). Nonetheless, these Buddhist atomists agree with the Vaiśeṣika view of atoms being indivisible and imperceptible to the senses. They are, to use a Buddhist expression, mental objects inferable only through their effects.

The *Great Treatise on Differentiation*, one of the earliest Buddhist sources explicitly referring to atoms, describes an atom as being the smallest constituent of matter. An atom cannot be split apart, nor can it be seen, heard, smelled, or touched. Thus an atom (*paramāṇu*), rendered in this volume by the translator as "subtle particle," is said to be the finest or smallest of all matter (*rūpa*). This description of an atom echoes the etymology of the Greek word *atom*, which literally means "not divisible."

As explained in our volume, Vasubandhu speaks of two types of atom, the *substantial atom* (*dravya paramāṇu*) and the *aggregate atom* or molecule (*saṃghāta paramāṇu*). A substantial atom is indivisible, devoid of any parts, has no spatial dimension, and is imperceptible. A single substantial atom can never exist on its own but only in combination with other atoms. In contrast to a substantial atom, an aggregate atom is composed of at least eight substances—the atoms of the four elements as well as those of color, smell, taste, and tangibility. Specific material aggregations have different balances of intensity among these. Thus when, say, the color atom is predominant, we perceive color rather than some other derived matter, such as smell or taste.

Sarvāstivāda sources, including the *Great Treatise* as well as the *Treasury of Knowledge*, present a principle of atomic agglomeration that proceeds by sevenfold increments. For example, seven atoms are said to equal a minute particle, seven minute particles an iron particle, and so on. Although some of the details in the progressive list of sevenfold increments differ in various sources, as noted in our volume, the basic doctrine of this increment seems to be shared widely among Buddhist atomists. The number seven derives from a single atom having a nucleus with one atom in each of the four cardinal directions as well as one above and one below. It is this sevenfold progressive agglomeration that is understood to account for the emergence of perceptible objects of our everyday world from indivisible atoms. To my knowledge, the Sarvāstivāda sources do not tell us explicitly which of the two atoms—the substantial atom or the aggregate atom—is the basic unit in this sevenfold agglomeration. Since the Sarvāstivāda maintains that no single substantial atom can occur by itself and that in the world atoms always exist as part of an aggregate of eight substances, the basic unit in this sevenfold increment appears to be the aggregate atom not the substantial one.

There is no doubt that the introduction of the theory of atoms, possibly borrowed from Vaiśeṣika, brought a level of sophistication and detail to the Abhidharma account of the material world. However, it also gave rise to conceptual problems, especially in harmonizing atomic theory with the earlier account of the four great elements. It is also unclear how the view of eight substances relates to the sevenfold progressive increment. Even within the basic molecular unit, the atoms of color, smell, taste, and tangibility are each themselves supported by the four fundamental elements. This means that even the smallest aggregate atom or molecule is composed of in fact twenty individual atoms—$(4 \times 4) + 4 = 20$. It is a sevenfold increase of these twentyfold individual atoms within a single molecule that gives rise to the large number of 823,543 atoms within a single particle of sunlight (see page 219). Perhaps as a reaction to such unresolved tensions with the Sarvāstivāda theory of atoms, the Sautrāntikas opted for a simpler theory. In their view, the four fundamental elements are understood not in terms of atoms but as forces or "seeds" (*bijatas*) within a given composite, where atoms of color, smell, taste, and touch coexist to form a molecule.

One important debate between the Sarvāstivāda and Sautrāntika about atoms concerns the actual structure of the molecule. The Sarvāstivāda maintains that, even within the composition of a single molecule consisting of eight substances, the individual atoms do not touch each other. For if the individual atoms touch each other, they assert, these atoms will come to fuse with each other. To the question "If there is no contact between the atoms and if there is space in between the atoms, how is it that the molecule does not collapse?" the response given is that the force of the wind element protects the integrity of the molecular structure. The Sautrāntikas agree with the Sarvāstivāda that if the individual atoms touch each other, then these atoms would not qualify as being partless. However, Sautrāntikas reject the idea that there is space between the atoms but maintain that the structural relationship between the atoms is such that they form a tight unit.

In exploring the Buddhist theory of atoms, one historical fact to bear in mind is this: Although we have extensive resources on the Sarvāstivāda views, thanks especially to the *Great Treatise* (extant in Chinese translation) and Vasubandhu's *Treasury* and its commentarial literature, hardly any primary sources extant reflect the Sautrāntika view. The two sources that are available and used in this volume on Sautrāntika are Vasubandhu's *Treasury of Knowledge Autocommentary*, where he frequently upholds the Sautrāntika standpoint, and Śubhagupta's verse work *Proof of External Objects*. Beyond these two works, unfortunately, no substantial Sautrāntika work survives to elucidate the views of this school.

In Western thought, although the theory of atoms was first proposed by Democritus before the Common Era, it was only with the work of John Dalton in the eighteenth century that a systematic theory of the atom emerged. In the Buddhist world, however, the systematic theory of the atom emerged quite early in history. Broadly similar to Dalton's theory, the Buddhist theory of the atom states that (1) all matter is made up of tiny indivisible atoms, (2) atoms of the same element are identical in nature and properties, (3) compounds are formed from different kinds of atoms, and (4) there is something—wind element for Sarvāstivāda and the four elements as forces for Sautrāntika—that ensures the integrity of molecules.

## The Critique of Atomism in Buddhist Sources

The broad contours of the above theory of atoms came to be accepted by most Buddhist schools with the exception of the Cittamātra, or Yogācāra, which rejected the reality of external objects. This said, the existence of indivisible unitary atoms became a major object of critique by Madhyamaka and Yogācāra thinkers. Although an earlier variation of this critique can be found in Āryadeva's *Four Hundred Stanzas*, the most well-known formulation is the one found in Vasubandhu's *Twenty Verses* and its autocommentary. There Vasubandhu points out that if an atom is indivisible and is devoid of parts or dimensions, as asserted by the atomists, it would be impossible to account for the macro world of everyday objects through the aggregation of atoms. If there is such a thing as an atom that is the minutest unit of matter, he reasons, one must still admit it as having dimensions or parts. Say, for example, that a single atom is surrounded by six other atoms: the atom at the center will come to have six parts. This is because the spatial locus of one atom cannot be that of the other, for if it did, the aggregation of seven atoms would in fact become a single atom! If, on the other hand, these seven atoms have distinct locations, this means that the atom at the center will have directional surfaces that face some surrounding atoms and not others. The atomists are confronted with the following choice: either admit that what they postulate as an atom is not without parts or be saddled with an unbridgeable explanatory gap on how to account for the macro world through the aggregation of partless atoms. From this argument, Vasubandhu as a Yogācāra draws a radical conclusion, rejecting not only the theory of the atom itself but even the reality of any matter that is external to the perceiving mind.

Unlike Vasubandhu's *Twenty Verses*, Dignāga's *Analysis of Objective Conditions* focuses on the untenability of atomically composed matter as being objects of our perception. The heart of his critique is that, given that atoms are conceived to be identical in size and shape, they cannot account for the diversity of forms we perceive, such as vases and bowls. Atoms can neither be perceived individually nor constitute the everyday objects of our senses. Dharmakīrti develops Dignāga's idealism further and formulates the well-known argument of "constant co-cognition" (*sahopalambaniyama*) that perception and its object are always experienced simultaneously, which implies what we think to be an external reality is in fact constituted by our own perception, echoing the quantum mechanics view of the role of the observer.

The critique of atomism in Āryadeva's *Four Hundred Stanzas* is directed more specifically at the Vaiśeṣika version, with a special focus on refuting its claim that atoms are eternal and unchanging. Āryadeva's arguments, originally presented in a terse form in verse, are explained in detail by Candrakīrti and cited at length in this volume (see pages 232–34). Like their Yogācāra counterparts, the Mādhyamikas too reject the idea of substantial atoms proposed by the Buddhist atomists. In fact, the Mādhyamika view questions the very project of grounding reality, be it matter or mind, in any kind of ultimate indivisible elements. What are said to be indivisible atoms by others are nothing but smallest conceivable units of matter that cannot be physically divided further but that still can be characterized as possessing spatial locus and sides. These atoms always occur in complexes, combined with other atoms, and never in isolation, so indivisibility and being without parts are incompatible even for these minute material elements. Nāgārjuna and his Mādhyamika followers such as Āryadeva and Candrakīrti appear to suggest that a more useful way of understanding reality, including the material world, is through the principle of dependent origination and appreciation of complex relationships rather than through the reductionist method of seeking matter's ultimate constituents.

In language and arguments strikingly similar to quantum physics, these Mādhyamika thinkers challenge any attempt to seek a mechanistic, reductionist, and essentialist account of reality. Today, any serious attempt to give a scientific account of reality has to deal with challenges posed by mind-boggling quantum phenomena like wave/particle duality, non-locality, and superposition. Nāgārjuna and his school question our common-sense object-property based perception of the world, which assumes the things we perceive to have some kind of self-defining essence that give their identity. And these thinkers demonstrate the utterly contingent and composite nature of all our concepts. Every dichotomy—one and many, self and other, subject and object, inner and outer—we take for granted, as well as the very bivalent logic that underpins these dichotomies, collapses when we seek its grounding in objective reality. What we are left with is a paradox, a picture of the world constituted by complex relationships with no objectively real relata. The only description we can have of reality is on the conventional level, a perspective that is limited within the framework shaped by our perception, expectations, and needs. Thus these thinkers employ the framework of two levels of reality, the ultimate level and the conventional level, to offer a more coherent understanding of reality. On the ultimate level, they assert, we can say nothing in the way of affirmation at all, and any attempt to develop an account of reality at this objective level is doomed to failure. It is on the conventional level, then, that we can speak justifiably of things and events and their reality, and here too, we need to be vigilant so that our account of reality does not succumb to the temptation to isolate some kind of objective grounding. Hence the preference for the language, logic, and perspective of relationality and dependent origination.

## Further Reading in English

For a succinct explanation of the Buddhist theory of atoms and how it fits within the larger Abhidharma project of developing a comprehensive understanding of the material world, see part 2 of William Montgomery McGovern, *A Manual of Buddhist Philosophy* (London: Kegan Paul, 1923).

On the presentation of atoms in the *Great Treatise on Differentiation* and Saṅghabhadra's views, including especially his critique of Vasubandhu's positions found in the *Treasury of Knowledge Autocommentary*, see chapter 8 of Bhikkhu K. L. Dhammajoti, *Sarvāstivāda Abhidharma* (Hong Kong: The Buddha-Dharma Centre, reprinted 2015), especially 226–35.

For a brief but succinct introduction to Kaṇāda's Vaiśeṣika theory of atoms and its subsequent developments, see chapter 5 of Karl Potter, *Encyclopedia of Indian Philosophies*, vol. 2 (Princeton, NJ: Princeton University Press, 1977; reprinted by Motilal Banarsidass, 1995).

For a full text of Vasubandhu's *Twenty Verses* and its autocommentary in English, see Stefan Anacker, *Seven Works of Vasubandhu* (Delhi: Motilal Banarsidass, 1984; corrected edition, 1998). For a translation and study of Dignāga's *Analysis of Objective Conditions* in English, see Douglas Duckworth et al., *Dignāga's "Investigation of the Percept": A Philosophical Legacy in India and Tibet* (New York: Oxford University Press, 2016).

For an engaging discussion on similarities between some aspects of Nāgārjuna's Madhyamaka philosophy and quantum mechanics, as well as the epistemological challenges of understanding reality raised by both, see chapter 3 of the Dalai Lama, *The Universe in a Single Atom* (New York: Morgan Road Books, 2005); and chapters 1–4 in Arthur Zajonc, ed., *The New Physics and Cosmology: Dialogues with the Dalai Lama* (New York: Oxford University Press, 2004).

For a quantum physicist's exploration of Madhyamaka philosophy to help clarify some important aspects of the philosophical foundation of quantum physics, see Victor Mansfield, "Madhyamika Buddhism and Quantum Mechanics: Beginning a Dialogue," *International Philosophical Quarterly* 29.4 (1989): 371–91.

# 14

# How Subtle Particles Are Posited

In general the subtlety or coarseness of material form is distinguished by mass, size, and the way it appears to the sense faculties. It is not classified according to the ease or difficulty by which it is comprehended by the mind, as in the case of the coarse or subtle level of impermanence. Such physical form exists in a broad range of dimensions from coarse matter that is the basis of the external cosmos and subtle physical form such as a particle of sunlight. Among subtle material forms too there are numerous kinds, among which subtle particles are the smallest.

What exactly is the ultimate nature of the subtle particles that are the constituent elements of coarse material form? How do gross material phenomena of the macroscopic world come to be formed from subtle particles? Are subtle particles perceived by the senses or not? And are these subtle particles partless or do they possess constituent parts? It is with regard to these questions that numerous divergent opinions have arisen among classical Indian thinkers.[263]

The view that gross levels of matter emerge from the accumulation of subtle particles is shared by thinkers of both Buddhist and non-Buddhist schools. Nevertheless, from among classical Indian philosophical traditions, historically [320] it is the Vaiśeṣikas and the Naiyāyikas—two influential philosophical schools—that are reputed to be the earliest and most prolific expositors of atomic theory.

## Vaiśeṣika and Naiyāyika Views

The Vaiśeṣikas and the Naiyāyikas classify all knowable objects into six categories: (1) substance (*dravya*), (2) qualities (*guṇa*), (3) action (*karma*), (4) universals (*sāmānya*), (5) particulars (*viśeṣa*), and (6) inherence (*samavāya*).

*Substance* is so called because it is self-sufficient and acts as the basis of qualities that are different from substance itself. Substance possesses three attributes: it has activity, it has qualities, and it is the cause of inherence. *Possessing activity* means it has movement, *possessing qualities* means it has multiple qualities, being the *cause of inherence* means it is the cause of entities interacting with one another.

Substance is of two types: substance not permeating all and substance permeating all. Substance not permeating all refers to the four great elements and mind (*manas*). Substance permeating all refers to the remaining four: self (*ātman*), time (*kāla*), spatial dimension (*dik*), and space (*ākaśa*). Alternatively, substance is classified in two types: permanent substance and impermanent substance.

The four substances permeating all—self, time, spatial dimension, and space—are permanent substances in their entirety. As for self, time, spatial dimension, and space, there are two levels—coarse form and subtle particles. The elements pertaining to coarse form are impermanent, whereas those belonging to subtle particles are permanent. For if subtle particles were impermanent [321] then necessarily atoms composed of them would emerge, which is not the case. As for coarse material form composed of subtle particles, Vaiśeṣikas and the Naiyāyikas assert these to be impermanent. They assert that in the past when the world was in the phase of voidness, subtle particles of the four primary elements necessarily endured without destruction. If at that time subtle particles did not exist, just as coarse material form did not exist, then there would be no cause at all for the regeneration of the coarse material world. So during that phase of voidness, even though subtle particles exist they are permanent since they lack any cause. Thus when external world systems and their inhabitants first come to be formed, it is on the basis of the permanent and indivisible subtle particles of the four primary elements pervading every direction, and the intention of Maheśvara to create beings, that external world systems and their inhabitants gradually came into being. These two schools therefore assert that the ultimate cause of the universe is partless permanent atoms.

This Vaiśeṣika and Naiyāyika viewpoint is described with clarity by Candrakīrti in his *Commentary on the Four Hundred Stanzas*, where he writes:

> Subtle particles (*paramāṇu*) are solely permanent because they exist in the nature of seeds as the source of coarse material form.

> For if these permanent particles did not exist then coarse form would come to arise in the absence of any seed state. Thus it is from just this source that coarse material form arises in the first eon. At the time of destruction, however, all composite material substance disintegrates and only subtle particles remain, and there is no possibility that coarse material form exists.
>
> If, in contrast, [322] subtle particles were not to exist just as coarse material form did not exist, then at that time coarse material form would lack any cause. Therefore subtle particles that are the cause of the substance of composite matter surely exist, and they are also permanent because they exist but lack any cause.[264]

In Kamalaśīla's *Commentary on the Compendium on Reality* it is said:

> The intrinsic nature of subtle particles such as earth and so on is permanent because subtle particles are permanent. But what is composed of them and so on is impermanent owing to the logic "that which possesses a cause is impermanent."[265]

The Vaiśeṣika and Naiyāyika philosophical systems accept the view that the four primary elements, including their subtle particles, possess, respectively, four qualities, three qualities, two qualities, and one quality. For instance, earth and the earth particle possess the four qualities of material form, smell, taste, and tactility. Water and the water particle possess three qualities, since they lack the quality of smell. Fire and the fire particle possess two, since they lack the qualities of smell and taste. Wind and the wind particle possess one, since they merely have tactility. Bhāvaviveka's *Blaze of Reasoning* states:

> The subtle particles of earth, water, fire, and wind are themselves permanent, [323] and they possess, in their respective order, four qualities, three qualities, two qualities, or one quality.[266]

Also:

> That which possesses material form, taste, smell, and tactility is the substance earth. That which possesses form, taste, and

> tactility is water, for it is wet and moist. Fire possesses material form and tactility, and wind possesses tactility.[267]

## The View of the Lower Abhidharma System

As for the presentations in the Buddhist texts, in the Seven Treatises of Abhidharma it is said that material phenomena are composed of the four great elements, which act as their cause. It is, however, in the *Essence of Abhidharma* by Dharmaśrī, a second-century Abhidharma master, and in the *Great Treatise on Differentiation*, a treatise that extensively establishes the views of Sarvāstivāda Abhidharma, that detailed discussions on atoms are found. Compiling the essential points of the *Great Treatise* and presenting them succinctly in a systematic manner, Vasubandhu composed the *Treasury of Knowledge* and its *Autocommentary*. It is in these texts that the tradition of the lower Abhidharma system presents its view on subtle particles.

On this view, physical entities are formed from the combination of eight substantial particles: the four primary elements of earth, water, fire and wind, and the four element derivatives of form, smell, taste, and tactility. And [324] it is in terms of these constituents that subtle and coarse physical entities are posited. It is further maintained that if we were to deconstruct composite material entities, they are reducible to the level of subtle particles, and that it is through the aggregation of many atoms that progressive levels of coarse physical entities emerge.

The way in which the eight substantial particles assemble and form physical entities can be explained by taking water as an example. Since water supports wood and so on by keeping it afloat, and since water promotes the growth of grass and trees, and since water moves, it arises together with earth particle, fire particle, and wind particle, respectively. Being water, the water particle is already present. And given that eye consciousness perceives the color of water, ear consciousness the sound of water, nose consciousness the smell of water, tongue consciousness the taste of water, and body consciousness the tactility of water, the particles of material form—of smell, taste, and tactility—arise simultaneously with water. This is also the case with other physical entities.

Now the subtle particle is the smallest among particles and the basic constituent of coarse physical form. But it is also said in early Abhidharma treatises that the subtle particle is partless and that it is surrounded by many other particles, and those particles do not mutually touch. They are understood in such terms because: (1) they cannot be divided into spatial dimensions or components while still retaining their identity; (2) one single subtle particle at the center is surrounded by other subtle particles; and (3) if one particle were to touch another, and if they made contact on all sides, they would merge in a single particle, while if one particle were to make contact on a specific side, then it would possess parts. [325]

In general such texts speak of two types of subtle particle—the unitary substantial particle (*dravyaparamāṇu*) and the aggregated atom (*saṃgrahaparamāṇu*), and of these the first is the smallest discrete unit of matter and the second represents the smallest compound unit of matter.

The unitary substantial particle is partless, whereas aggregated atoms are endowed with parts.

The unitary substantial particle is defined in these terms: It is a partless particle that resides as a constituent of material form that possesses obstructive resistance, and it cannot be destroyed by other physical entities. It cannot be mentally deconstructed into smaller parts and it constitutes the smallest discrete unit of matter. In contrast, the aggregated atom is defined thus: it is an atom, and although it can be mentally deconstructed into its constituent parts, it cannot be destroyed by other physical entities, and no other aggregated form exists that is smaller than it. For example, there are the four indivisible particles of earth, water, fire, and wind, and together with the four indivisible derivative particles of form, smell, taste, and tactility, they constitute the eight indivisible substantial particles, and each of these separately constitute a unitary substantial particle. The smallest unit that constitutes an aggregation of eight such particles represents the smallest compound unit of matter. That substantial particles consist of both elements and derivatives is evident from the fact of there being eight unitary substantial particles. On these points, the *Great Treatise on Differentiation* says:

> Subtle particles should be recognized as extremely subtle material form. They cannot be dissected, destroyed, or pierced. They cannot be taken up or relinquished, transported, compressed, held, or stretched. They are not long or short, square or round, even or uneven, high or low. They are without [326] parts, cannot be annihilated, seen, heard, seized, experienced, or touched. Therefore subtle particles are called *extremely subtle form*. Seven

> subtle particles constitute one atom. Among the forms that are perceived by the eye or eye consciousness, it is extremely subtle.[268]

In *Investigating Characteristics* it says:

> The subtle particle has two types: the unitary substantial particle and the aggregated atom. Unitary substantial particles are components of form that are subtler than all other material form that possess obstructive resistance and cannot be mentally deconstructed into other more subtle types of form. Thus they are called the "smallest limit of form," since they do not possess parts, just as a moment is the smallest limit of time, since a moment cannot be divided into smaller moments. The aggregated atom is a composite component, there is no subtler compound anywhere, and it cannot be destroyed by other form.[269] [327]

## The View of the Upper Abhidharma System

The presentation on subtle and coarse material phenomena as stated in the primary and commentarial texts of the upper Abhidharma system, such as those of Asaṅga, is as follows. In this view "subtle particle" refers to the smallest unit of matter that can still be conceived by the mind as a material entity, even though there is no ultimate limit of divisibility when subjected to mental deconstruction. Asaṅga's *Compendium of Abhidharma* states, for example:

> It is said that the aggregation of subtle particles is composite form, but what is referred to as the subtle particle (*paramāṇu*) should be recognized as lacking physical structure. They are determined to be subtle particles by the ultimate analysis of the intellect, which dispels the idea that they constitute a solid mass and denies they substantially exist as material form.[270]

*Elucidation of the Compendium of Abhidharma* says:

> No limit is reached in the process of deconstruction. So the statement that the subtle particle is posited on the basis of dissection by the mind means this: to the extent that the mind is able to delineate the limit of material form through the process of deconstruction, to that extent a subtle particle is posited.[271]

On the question of whether these subtle particles have shape, it is said that although the individual particles in a composite entity like a vase are mutually distinct substances, they are equally spherical with regard to their shape. This is stated in Dignāga's *Autocommentary on the Analysis of Objective Conditions*:

> Although the subtle particles are different substances, all are the same in [328] being spherical.[272]

Commenting on this section, Vinītadeva says in his *Commentary on the Analysis of Objective Conditions*:

> Although the subtle particles of a vase, a clay bowl, and so on are held as different substances, because subtle particles are not differentiated in terms of their sphericality, they are not classified separately.[273]

On the differentiation of the types of subtle particle, Asaṅga's *Compendium of Ascertainment* states:

> In brief, subtle particles are of fifteen types: five types of subtle particle consisting of the five sense faculties, five types of subtle particle consisting of the objects of those faculties, four types of subtle particle consisting of the four primary elements, and one type of subtle particle consisting of a substantially existent mental-object form.[274]

Thus the fifteen subtle particles—from the subtle particle of the eye sense faculty to that of the mental-object form—are said to belong to the category of "mental-object form."

### *Subtle Particles in the Madhyamaka Texts*

Bhāvaviveka not only accepts that physical entities are composed of the eight substantial particles in combination—the four primary elements of earth, water, wind, and fire plus the four element derivatives of form, smell, taste, and tactility—but he also says that the basic constituents themselves do not exist in isolation but as combinations of eight substances and as substantially existent entities. He asserts, for example, in his *Blaze of Reasoning*: [329]

> Thus subtle particles themselves are combinations of the eight basic substances, therefore they are asserted to be substances themselves.[275]

Of all the particles, subtle particles are said to be the subtlest. For example, Avalokitavrata's *Explanation of the Lamp of Wisdom* says:

> What is referred to as *subtle particles* are extremely subtle particles, such that nothing subtler than these exists anywhere. Hence they are called "subtle particles."[276]

In the Madhyamaka writings of thinkers such as Āryadeva and Candrakīrti, subtle particles are described as being resistant since they impede other subtle particles from occupying their space. They also maintain that any phenomenon that is resistant cannot be accepted as permanent, just as everyday coarse material objects like a pot are not permanent. Subtle particles are imputed existents merely designated in dependence on other factors and are composites of the eight substantial particles. Āryadeva's *Four Hundred Stanzas*, for example, states:

> A resistant entity that is permanent
> is nowhere to be observed.
> Therefore never do the buddhas speak of
> atoms that are permanent.[277]

In expounding these lines, Candrakīrti's *Commentary on the Four Hundred Stanzas* asserts:

Here too, it states that subtle particles are not permanent. Why? Because "A resistant entity that is permanent is nowhere to be observed." [330] This means that since a subtle particle cannot be comprehensively penetrated by another particle, subtle particles possess the characteristic of obstructive resistance. So, just like a vase that is resistant, it would be illogical for these particles to be permanent. Therefore no such thing as a permanent subtle particle exists. "Therefore never do the buddhas speak of atoms that are permanent." Those who see reality without distortion ascertain the facts just as they are. However, the subtle particles of the eight basic substances do exist as impermanent entities, and so too as imputed existents. For they are designated on the basis of the eight substances, just like, for example, a vase.[278]

# 15

## How Coarse Matter Is Formed

THERE IS THE following explanation in the texts of the lower Abhidharma system on how coarse material form is established from the accumulation of subtle particles. The smallest unit of matter is the subtle particle, the smallest unit of words is the letter, and the smallest unit of time is the shortest moment of time, and it is these that initiate composition. The accumulation of seven subtle particles represents the measure of one atom, seven atoms equal one iron particle, seven iron particles equal one water particle, seven water particles equal one rabbit particle, seven rabbit particles equal one sheep particle, seven [331] sheep particles equal one ox particle, and seven ox particles equal one sun-ray particle. It is said that a sun-ray particle can be seen by the eyes but those particles that precede it cannot be seen directly by the eyes.

In brief Vaibhāṣika and Sautrāntika thinkers assert that physical entities are established from the aggregation or assembly of indivisible particles, which are their basic constituents. Śubhagupta, for example, states in his *Proof of External Objects*:

> Not touching one another,
> they exist devoid of parts,
> therefore the earth and so on
> come into being from accumulation.[279]

Similarly, in Vasubandhu's *Treasury of Knowledge Autocommentary*, it is written:

> "The limit of form is a particle; names, a letter; time, a moment" means the limit of form is a subtle particle, the limit of time is a

moment, and the limit of names is a letter. In that regard, "Seven subtle particles equal an atom. Similarly, iron, water, rabbit, sheep, ox, sun-ray particle, egg of a louse, and the louse derived from that; and similarly, one finger joint, the latter seven times the former" indicates these subtle particles and so on should be recognized as increasing by a factor of seven.[280] [332]

The *Extensive Display Sūtra* also mentions:

> The bodhisattva states that seven particles that are subtle particles make one minute particle, seven minute particles make one small particle, seven small particles make one sun-ray particle, seven sun-ray particles make one rabbit particle, seven rabbit particles make one sheep particle, seven sheep particles make one ox particle, seven ox particles make one louse egg, seven lice eggs make one mustard seed, seven mustard seeds make one barley seed, seven barley seeds make a finger joint, twelve finger joints make one handspan, two handspans make one cubit, and four cubits make a bow.[281]

In Maudgalyāyana's *Presentation on the World* one reads:

> One forty-ninth of a sun-ray particle is a subtle particle. Seven subtle particles equal one minute particle. Seven minute particles equal one sun-ray particle. Seven sun-ray particles equal one rabbit particle. Seven rabbit particles equal one sheep particle. Seven sheep particles equal one ox particle. Seven ox particles equal the size of one louse egg. Seven lice eggs equal the size of one avagaṇa.[282] Seven avagaṇa equal the size of one vaṭikā.[283] Seven vaṭikā equal the size of one barley seed. Seven barley seeds equal the size of one finger joint. Twenty-four finger joints equal the size of one cubit.[284] [333]

*Detailed Explanations of Discipline* (*Vinayavibhaṅga*) states:

> Six subtle particles make one minute particle, six minute particles make one water particle, six water particles make one iron

> particle, six iron particles make one rabbit particle, six rabbit particles make one sheep particle, six sheep particles make one ox particle, six ox particles make one sun-ray particle, six sun-ray particles make one louse egg, six lice eggs make one louse, six lice make one barley seed, six barley seeds make one finger joint, six finger joints make one cubit.[285]

The *Vinaya Sūtra* states:

> Subtle particles, minute particles, water, iron, rabbit, sheep, ox, and sun-ray particles, louse egg, louse, barley, finger, where six of the former equal one of the latter.[286]

To compare these calculations, the sun-ray particle as defined in the *Treasury of Knowledge* and its commentaries equals 823,543 subtle particles. According to *Vinayavibhaṅga*, however, since these calculations increase by a factor of six, a single sun-ray particle equals 279,936 subtle particles. In contrast, a single sun-ray particle as defined in the *Extensive Display Sūtra* would be equivalent to 343 subtle particles, while a single sun-ray particle described in Maudgalyāyana's *Presentation on the World* would be equivalent to 49 subtle particles.

That these divergent explanations do not represent contradictions is argued in some of the commentaries on the *Treasury of Knowledge*. They explain how these terms were used as labels for specific levels of aggregation of subtle particles and so on, and [334] that the calculations should not be understood literally as the size of the tip of the hair of the specific animals whose names are used in the calculation. Some commentaries do mention, however, that the dust covering the tip of the hair of a rabbit is one rabbit particle, the dust covering the tip of the hair of a sheep is one sheep particle, and the dust covering the tip of the hair of an ox is one ox particle. Nonetheless, since the hair of a rabbit, the hair of a sheep, and the hair of an ox are differentiated by their coarseness, they assert that the dust abiding on the tip of the hairs is different in size.

In the texts of the Kālacakra system, an aggregated array of eight subtle particles is the measure of one atom, an eightfold array of atoms is the measure of a spherical hair-tip, an eightfold array of spherical hair-tips is the measure of a black mustard seed, one eightfold array of black mustard seeds is the measure of a louse, one eightfold array of lice is the measure of a barley seed, one eightfold array of barley seeds is the measure of a finger, twenty-four fingers are the measure of a cubit, four cubits are the measure of one bowlength, two thousand bowlengths are one krośa, four krośa are the measure of one yojana.[287]

In general, according to the lower Abhidharma system, when composite matter such as that of a vase is formed, the eight substantial particles consisting of earth, water, fire, wind, and form, smell, taste, and tactility [335] must assemble, and if any one of these is absent, the entity cannot come into being. For example, *Investigating Characteristics* states:

> It is said that the eight basic substances arise. They will not arise if any of the eight substances is absent.[288]

As for the sense faculties, given that they are substances that are different from the substance of external particles, the formation of the sense faculty requires the presence of the particle of the sense faculty in addition to the eight substantial particles. Therefore when the body sense faculty is established, for example, there are nine substances through the addition of the particles of the body sense organ. Similarly, when the eye faculty is formed, ten substances arise simultaneously through the addition of the particles of the eye sense organ. This is true as well for the remaining sense faculties of the ears, nose, and tongue, where particles of the specific sense organ are added. *Treasury of Knowledge* states:

> The eight substances are subtle particles arising in the
> desire realm not associated with the sense faculties or with
> sound.
> Those that possess the body sense faculty possess nine
> substances.
> Those that possess the other faculties possess ten substances.[289]

To illustrate these in their respective order, the aggregated atoms within a composite entity like a vase, which does not possess sound, represent the accumulation of eight substantial particles. The aggregated atoms of the body sense faculty of a fetus in the womb prior to the establishment of the link of the eye *sense faculty* and so on but after the establishment of the link of *name and form* are a collection of [336] nine substantial particles. The aggregated atoms of our eye sense faculty, ear sense faculty, nose sense faculty, and tongue sense faculty are aggregated atoms that are a collection of ten substantial particles. So too, it is said that if aggregated atoms are associated with sound but not associated with the sense faculties, there are nine, since sound is added to the eight substantial particles, and if associated with the body faculty there are ten. If the tongue faculty and so on were added to those, then eleven arise simultaneously. For example, *Investigating Characteristics* states:

> This is explained extensively, such as in the following passage: "Those subtle particles are associated with sound. . . ." Here, those that are not associated with the sense faculties have nine substances; those possessing the body sense faculty have ten substances; those possessing other sense faculties have eleven substances. These occur, on occasions, without the presence of the sound particle. Because sound is produced through the primary elements colliding, it is not present at all times.[290]

To illustrate these in their respective order, then, the aggregated atom associated with the sound of water and wind is a collection of nine substantial particles, and the aggregated atom associated with the sound of a person's hands clapping is a collection of ten by adding the substance of the particles of the body sense faculty. And the aggregated atom associated with the sound of a person clicking his tongue [337] would possess eleven substantial particles by adding both those of the body sense faculty and the tongue sense faculty.

Question: If the eight substantial particles that are said to arise simultaneously are asserted to be substances pertaining to the sense bases, then eight substances would be too many since the four primary elements would be included within the substance of the tactility base. Thus they would be enumerated as just four substances. But if they are viewed as mere substance in a general sense, then eight would be too few since multiple substances such as the substance of shape, weight, and so on exist.

In response it may be said that if such doubts arise, then in this context unitary substantial particles are posited as eight by omitting the substantial particles of sound and those of the sense faculties. And this is from the perspective of differentiating the substance of the primary elements that support and the nature of the sense bases that are supported by them. In this context Vasubandhu surmises in his *Autocommentary* that the extant substantial particles of all physical entities cannot be counted:

> Here some that are mere substances are held to be substance, for they support. Here some that are the substances of the sense bases are held as substance, for they are supported. In light of this, the substances of the primary elements are multiple because forms derived from the elements [338] are diverse and they are dependent on the fourfold division of the primary elements. In this context, and in terms of substantial type, they are held as substance, and other instances of the four primary elements do not surpass this substantial type.[291]

According to the texts of the upper Abhidharma system, the assertions of the Vaibhāṣikas and Sautrāntikas about the existence of indivisible particles and external physical entities being formed through the aggregation of such elementary particles are rejected. Furthermore, since the formation of composite material entities begins only when many particles accumulate, there is no difference between *accumulation* and *composition*. For example, Asaṅga's *Compendium of Abhidharma* says:

> The aggregation of subtle particles is composite form.[292]

When differentiated, fourteen such composite material forms are mentioned, these being the four primary elements of earth, water, fire, and wind, the five objects of form, sound, smell, taste, and tactility, and the five physical sense organs from eye organ up to the body sense organ. The latter group includes all the forms except for mental-object forms. Asaṅga's *Facts of the Grounds*, for example, states:

> In brief, such composite form possesses fourteen types of substance. They are earth, water, fire, wind, form, sound, aroma, taste, tactility, and the five physical faculties such as the eyes,

> except for the form that is the experiential object of the mental faculty.[293] [339]

Bhāvaviveka, for example, appears to maintain a distinction between accumulation and composite. He defines "accumulation" in terms of the collection of substantial particles of similar type within a single entity. In contrast, he defines "composite" in terms of the collection of numerous particles of substances of dissimilar type on the basis of diverse entities. For example, a vase is an *accumulation* in that it is a collection of many subtle particles of similar type, wherein these subtle particles that are its constituent elements exist as massed together. In contrast, a *composite* is posited as a collection of diverse things, such as an army that consists of elephants, horses, and chariots, and a "forest" that is a collection of many types of different trees, such as barberry trees, khadira trees, and so on. Thus *Blaze of Reasoning* states:

> Question: What is the difference between the pair accumulation and composite?
>
> Response: Subtle particles of similar type supported on the basis of a single entity are called *an accumulation*. The assembly of substances of dissimilar type on the basis of diverse entities, such as elephants and horses, or barberry trees and khadira trees, which are designated an army or forest and so on, is called *a composite*.[294]

Furthermore, since all coarse physical entities are established from subtle particles that are their basic constituents, and since such constituent subtle particles do not arise without a cause and their continua is sourced in their beginningless substantial streams, [340] regardless of whatever the resultant entity, one can never point to something and assert that this is its first cause. *Four Hundred Stanzas* affirms:

> When the original cause of
> even a single effect is not seen . . .[295]

To summarize, early Indian Buddhist thinkers appear to converge on the following key points: Physical entities come into being through the aggregation and the transition from subtler to coarser levels of form. The basic constituents of matter consist of subtle particles of the four primary elements. Atoms are formed through the accumulation of substantial particles and the diverse forms of physical entities emerge through the accretion of atoms.

# 16

## Analyzing Whether Indivisible Particles Exist

THE VIEW OF THE non-Buddhist schools, such as the Vaiśeṣikas and so on, is that subtle particles are partless or indivisible, that they are the cause of the material substances that constitute a composite whole, that they are spherical in shape, and that when the partless particles form coarse physical entities, they do so without coming into contact with one another. The view of these schools is summarized in Candrakīrti's *Commentary on the Four Hundred Stanzas*:

> "That which is the cause, is spherical, and lacks directional parts"—these are the characteristics of the substances that represent subtle particles.[296]

Bodhibhadra's *Discourse on the Compendium of the Essence of Wisdom* says: [341]

> "Furthermore, these non-Buddhist Vaiśeṣikas assert such that . . ." means they assert that when this world is destroyed, only permanent subtle particles of earth, water, fire, and wind, which lack spatial location, remain on their own.[297]

*Blaze of Reasoning* also notes that in the Vaibhāṣika system it is maintained that when indivisible subtle particles create coarse physical entities they do so without coming into contact with one another:

> Upon the aggregation of particles of a similar type that do not contact one another, these subtle particles initiate activities that are concordant and discordant with themselves.[298]

As for the Buddhist Vaibhāṣikas, they maintain that since these subtle particles themselves lack parts they do not come into contact with one another. And when a single particle is surrounded by other particles, an intervening gap remains between them. For if such particles were to touch one another from all spatial directions, they would merge in one single particle. And if such contact were to take place in specific spatial directions, there would be the erroneous consequence that such particles possess parts.

Question: If subtle particles within accumulated composite matter do not contact one another, how is it that they do not disintegrate?

In response, it is stated that such particles are held together by wind and that it is also through the connection sustained by the power of their substances that they remain without disintegrating. For example, Śubhagupta's *Proof of External Objects* states:

> Therefore they are described as surrounded
> by many specific directional parts.
> But such particles
> are not states that possess parts.[299] [342]

And:

> Like individual beads strung together,
> which are of mutual benefit to one another,
> atoms cannot be dissected
> just like a vajra and so on.
> Just as spirits or snakes are captured
> by the force of mantra,
> some say it is possible that atoms
> are held together by the force of substance.
> Others say this force is too weak.[300]

Vasubandhu writes in his *Treasury of Knowledge Autocommentary*:

> "Do subtle particles make contact with each other or not?" The Kaśmiri Sarvāstivada school says, "They do not make contact." Why? Because first if they were to make contact in all spatial

> directions their substances would merge. But if they made contact in a single spatial direction, it would follow they had parts, but subtle particles do not have parts.

And:

> Question: If they construct coarse matter owing to their accumulation, why should they not be destroyed?
> Response: Because they are supported by the wind element.[301]

According to the *Treasury of Knowledge Autocommentary*, the Sautrāntika master Bhadanta asserted that if subtle particles were to contact one another they would possess parts. Yet if there were gaps, then particles of light would penetrate during the day and particles of darkness would penetrate during the night. So although there exists no such gaps, particles exist without touching one another. The *Autocommentary* states:

> The Bhadanta says: They do not make contact, but they are discriminated as making contact [343] since their surfaces approach.[302]

Kamalaśīla's *Commentary on the Ornament of Madhyamaka* also states:

> Some say that they do not touch one another yet neither are there intervening gaps; they coexist solely in the manner of a structured array. For example, Bhadanta says: Even though the particles do not touch one another and have no gaps they are conceived as if in contact with one another.[303]

One might wonder—if subtle particles did not make contact with one another, then in consequence heat would not undermine cold and light would not eliminate darkness. But in response it is said that when particles of heat or light come into close proximity with their opposites, then heat particles prevent the near approach of cold particles, light particles prevent the near approach of darkness particles, and it is this activity that is characterized as "undermining" cold and darkness.

Sautrāntikas do not admit that the particles within composite entities come into contact with one another. This is stated, for example, in *Proof of External Objects*:

> Through some particles that exist nearby
> preventing the movement of mobile particles,
> they are described as impeding them.
> This does not occur because of other factors.[304]

Therefore although both Vaibhāṣikas and Sautrāntikas are similar in asserting that subtle particles are partless and do not come into contact with one another, an apparent distinction emerges between the two. The Vaibhāṣikas say that there are gaps, whereas Sautrāntikas maintain no such intervening gaps. [344]

The thinking of those who assert indivisible particles is as follows. Take coarse physical entities such as earth and rocks, for example, and break them into pieces or smaller parts. Through a process of reducing the entity into progressively smaller units, one must eventually reach a point beyond which no further dissection can take place. Now if that subtle particle possesses parts, it could still be divided further, which would mean that it is not the final smallest discrete unit. So either one must assert it to be devoid of parts or admit an infinity of divisible parts. If one were to accept such an infinity of parts, it would be impossible to reach the limit of divisibility even for a single drop of water. There is therefore no choice but to accept that subtle particles are indivisible or devoid of parts. Thus the Indian philosophical schools from the non-Buddhist Vaiśeṣika to the Buddhist Sautrāntika equally assert that if one were to reduce coarse physical entities to their smallest discrete units one would arrive at the indivisible subtle particles that are the basic constituents of coarse physical entities. The difference is that non-Buddhist schools such as the Vaiśeṣikas assert that these particles are permanent, whereas the two Buddhist schools, the Vaibhāṣikas and Sautrāntikas, assert them to be impermanent.

As for Madhyamaka and Cittamātra masters, they do not accept indivisible partless particles. In fact in their writings numerous logical arguments are presented to refute the concept of indivisible subtle particles. For example, the following refutation of indivisible particles occurs in Vasubandhu's *Twenty Verses*. If one were to analyze the smallest discrete particle that is asserted to be partless one must admit that it possesses parts, for when a single particle alleged to be the smallest elementary particle is surrounded on six sides—the four cardinal directions and above and below—one must accept that the particle at the center has parts since [345] the side of the central particle facing east does not face south or west and so on. If this were not the case, then the particles residing in the remaining five spatial directions would exist at the spatial locus of the subtle particle in the east. In that case, all six particles would, erroneously, merge into one. And if their spatial locations merged, then in consequence no matter how many subtle particles accumulated they could never compose a coarse physical mass, for their physical dimension would remain the same as that of a single particle. These arguments are presented in *Twenty Verses*:

> If six converge at the same time,
> an atom would have six parts.
> If these six occupied the same spatial location,
> a composite mass would be merely the size of a particle.[305]

*The Twenty Verses Autocommentary* states:

> If each subtle particle had no distinct spatial parts then when the sun shone how could a shadow fall on its other sides? This would mean there would be no other spatial directions where the sun's rays would not fall. If you do not assert there are different spatial directions, how could subtle particles be shaded by other subtle particles?
>
> If subtle particles had no other parts, then in the spatial direction where light shines, shade would be obstructed in the others. If there were no obstruction, then since all sides would merge at the same spatial position, all accumulated form would be merely the size of a subtle particle.[306]

Thus Vasubandhu presents the argument that if subtle particles lack obstructive resistance, then however many were to accumulate they would merely retain the size of a single particle. [346] Also, if such particles do not possess spatial dimension corresponding to specific directions, then consequently when sunlight strikes one face of a material entity, shade would not fall on any of its other sides. In this way, Vasubandhu establishes that so-called subtle particles must possess spatial dimensions and obstructive resistance.

Asaṅga, for instance, states in *Facts of the Grounds* that although subtle particles are endowed with parts they do not constitute composite wholes:

> Composite entities possess spatial dimensions. The subtle particles too have such spatial dimensions. Composite entities are, however, wholes, whereas subtle particles are not. Why? Because subtle particles are constituents and they are present within composite entities. However, subtle particles do not exist in other subtle particles. Therefore subtle particles are not composite wholes.[307]

Thus both subtle particles and composite form possess parts since they have spatial dimensions. Form that is an assembly of particles possesses components since it has subtle particles that constitute its parts. Subtle particles are indeed components since they are parts of form that is an assembly of particles, but they are not composite wholes.

The refutation of indivisible particles in the texts of the Madhyamaka Yogācāra masters Śāntarakṣita and his student Kamalaśīla can be understood from Śāntarakṣita's *Autocommentary on the Ornament of Madhyamaka*:

> Thus the followers of Kaṇāda say: The world is necessarily created through particles meeting and combining; [347] furthermore the unattached natural essence has intervening gaps and is encircled since it is maintained by common mutual force.
>
> Also, in conformity with that one says: The indescribable particle is not a state that possesses parts through merely some particles being fully encircled by many other particles from different spatial directions.
>
> Again, many have no intervening gaps, for it is said: How could subtle particles that do not make contact be discriminated as making contact, since there are no intervening gaps between them?
>
> But such statements are implausible since all these ideas are mere speculation and without merit.

> If particles converge on all sides then their natural essence would merge because the convergence of particles on one side would also be their convergence on other sides. If they converged on one side they would have parts, and particles with their own distinct essential nature would merge with other particles.
>
> Also, where there are intervening gaps between particles there would be the opportunity for both subtle particles of light and darkness to exist in those intervening gaps. Because the interstice is a state where darkness or light exist they would converge with them. Even those spatial directions without intervening gaps would not be different from the spatial directions that connect, and in lacking intervening gaps they would lack interstice. Meeting and connecting are not different in meaning, for the great chariots already point toward the great highway.[308]

Some assert that indivisible particles have no intervening gap between them, some [348] assert that particles are encircled by many particles with intervening gaps between them, while others say that particles merely appear to make contact since there are no intervening gaps between them, but they don't actually make contact.

But all these views merely serve to reify the particle. In accordance with the assertion that particles have no intervening gaps, even if it is said that particles do not touch, then they must meet, and if particles meet on all sides they would merge through interpenetration. And if they meet on one side and not another, they must be asserted to have parts. And even if it were accepted that there were intervening gaps between them, this would demonstrate the counterargument that "another particle other than the subtle particles of darkness at night and light during the day would intervene between them" and refute the existence of an indivisible particle. Also, for those who advocate subtle particles, spatial directional parts such as east, south, and so on must exist. Therefore this line of reasoning clearly establishes that particles have parts. So too, Kamalaśīla states in his *Proof of the Absence of Inherent Existence*:

> Given that the subtle particles are material, one will have to admit that they possess spatial dimensions. If that were not so,

> they would not have existence defined in terms of their spatial dimensions corresponding to east, north, and so on, which means aggregations such as mountains and so on would not exist.[309]

In the works of the Madhyamaka master Āryadeva as well, such as in his *Four Hundred Stanzas*, arguments are presented to refute the concept of indivisible partless particles. For example, he argues that if composite entities are formed through subtle particles coming into contact with one another, in that case if all their corresponding parts were to make contact, it would be impossible for them to increase in mass. If, on the other hand, they made contact with some parts and not others, those parts that came into contact would become the cause of composite matter while other parts that do not make contact would not be such a cause. [349] This would mean that subtle particles are divisible. In that case, subtle particles would become entities that have diverse aspects, with the implication that it would be illogical to assert that they are permanent. Furthermore, if subtle particles were devoid of spatial dimensions, there would be the flawed consequence that no particles could exist on each of a particle's four sides. If, on the other hand, they do possess spatial dimensions, such as east and so on, they would necessarily possess sides facing the east and so on. As such the assertion of indivisible particles would be untenable. Also, if subtle particles were devoid of spatial dimensions, they would then be devoid of the movement from one spatial location to another, which means that they could never link with other particles. This would signify that they could not be a cause for creating composite physical entities. On the other hand, if they were agents of movement from one spatial location to another, they would necessarily possess at least two spatial dimensions—one facing forward from the space they occupy and a second facing the rear space that has been left. Still, one cannot attribute such dimensions to particles that are indivisible. Thus it is through a variety of reasons that the notion of "indivisible particle" is refuted. *Four Hundred Stanzas* asserts:

> Some of its sides are causes
> and some sides are not causes,
> therefore it is something that is diverse.
> But it is illogical that diverse entities are permanent.

The spherical nature of the cause
is not present in its effect,
therefore the complete convergence
of atoms is not tenable.

That which is the position of one atom
cannot be asserted to be that of another,
therefore neither the cause nor effect
are asserted to possess the same dimension.

Whatever has an eastern side
also has an eastern part.
For any particle with sides
they must admit is not a partless particle. [350]

Whatever lacks both
the front advancing
and the back departing
is not an agent that moves.[310]

Candrakīrti also states in the *Four Hundred Stanzas*:

> When subtle particles do not completely converge, then any parts of one that converge with the parts of another subtle particle are causes, and any that do not converge are not causes. As such, some sides are causes and some sides are not causes, and thus they are diverse because they have multiple aspects. Therefore, because this demonstrates they are not permanent, like a picture it is said, "It is illogical that diverse entities are permanent."
>
> But if you think that because subtle particles have no parts their sides never converge but they do completely converge, [and that] therefore they possess fully qualified contact though they lack sides, then in response it is said: "The spherical nature of the cause is not present in its effect. Therefore the complete convergence of atoms is untenable."

You say that *cause*, *spherical*, and *without sides* are the defining characteristics of the substance of subtle particles. But if subtle particles completely converge with other subtle particles, though not in terms of their sides, then [351] it follows that any sphericality existing in the subtle causal particle would manifest as an effect in the second atom and so on. Therefore all composite wholes would exist beyond the range of the sense faculties because they would be merely the size of a subtle particle, but they are not merely the size of a subtle particle. Therefore subtle particles that converge on all sides are untenable.

Further, when subtle particles do not converge on all sides with others then: "That which is the position of one atom cannot be asserted to be that of another. Therefore neither the cause nor effect are asserted to be the same physical dimension." Therefore indeed the substance of the composite whole does not exist beyond the range of the sense faculties because subtle particles do not converge on all sides with other subtle particles.

The assertion that subtle particles possess sides is unshakable, and they are also diverse because they possess sides. Therefore it is said they are not permanent.

You say that this fault arises at the time when the substance of the effect is produced, but there is no erroneous consequence in stating that subtle particles have no parts prior to the production of the effect.

But even at this time subtle particles are divided into sides such as an eastern part and so on. At this time it is certain that "whatever has an eastern side also has an eastern part." Therefore you say that like a vase and so on, they do not become a subtle particle because they have sides. "For any particle with sides they must admit [352] is not a partless particle."

If subtle particles lacked parts they would not converge with other subtle particles since they would not go anywhere, nor would they produce the substance of the composite whole. Thus when in the process of going due to possessing a body: "Whatever lacks both the front advancing and the back departing is not an agent that moves."[311]

In summary, with regard to the subtle particles that are the basic constituents of coarse physical entities, there does exist a divergence of views in the texts of the lower and upper Abhidharma systems as well as those of the Madhyamaka as to whether such particles are indivisible and also whether the material forms composed of the accumulation of such particles possess real external existence. This said, broadly speaking there exists no difference in how subtle and coarse levels of material form are posited. When examined by the Cittamātra and the Madhyamaka schools, even the smallest unit of matter, the subtle particles that the two realist Buddhist schools assert to be indivisible, turn out to possess spatial dimensions. Furthermore, so long as something is posited to be a material entity characterized by obstructive resistance, it will have to be admitted to possess dimensions. Even though one may analyze the dimensions of the particles and the dimensions of those dimensions for an eon, for as long as something is defined as possessing the identity of a particle, there can be no end or limit to the dimensionality or division of such particles. So a particle can never transcend its identity as a material entity. It is, however, important to recognize that the kind of deconstruction referred to here is really from the perspective of thought experiments and not through physically deconstructing such particles. [353]

In brief, Madhyamaka and Cittamātra thinkers seemed to view the "subtle particle" to be a material form conceived by the mind that represents the smallest point arrived at through the process of mentally dividing material entities into their constituent elements. Since it is thought to have shape and impedes other subtle particles from occupying its place in space, it is characterized by obstructive resistance. Furthermore, it does not exist in isolation but as part of a collection of the particles of the four primary elements. Also, given that it possesses spatial dimensions and is part of a composite of other particles, it is a component but not a composite whole. In itself it is understood to be a phenomenon emerging from a process of deconstruction and belongs to the category of "derivative form."

Despite the fact that Madhyamaka and Cittamātra thinkers are similar in rejecting the notion of "partless indivisible particles," the final grounds on which they reject indivisible particles are different. For the Cittamatrins, there is no such thing as an atom that has an objective external reality, so what is called "an atom" represents a subtle state defined in terms of a progressive division undertaken on the basis of what is conceived in the mind. Therefore composite coarse physical entities formed through the aggregation of such particles are but perceptions of the mind and thus not objective external reality. This is a standpoint shared similarly by the Svatantrika-Madhyamaka masters Śāntarakṣita and his heir Kamalaśīla. As for the great Madhyamaka masters of the Prāsaṅgika Madhyamaka school, such as Candrakīrti, whether it is the subtle atoms or coarse physical entities formed by them, apart from their reality as mere imputations—designated by the mind in terms of "this" or "that"—[354] they admit nothing that can be found objectively as being "this" or "that" phenomenon. Therefore, as it emerges extremely clearly from their writings, these thinkers do not accept any notion of "indivisible particles" or "indivisible moments of consciousness."

The justification for the non-Buddhist schools of the Vaiśeṣikas and Naiyāyikas as well as the two essentialist Buddhist schools [Vaibhāṣika and Sautrāntika] to postulate the notion of indivisible particles as the ultimate constituents of matter is the following: That the objects of external reality do have an existence is something affirmed by our direct perceptual experience. When one examines the existence of these physical entities it is established that coarse physical entities are composed from the accumulation of their subtler constituent elements. And when the process of reduction from coarse to subtle is pursued, an endpoint has to be admitted where no further deconstruction can take place. If no such endpoint exists there would be no explanation at all for how subtle coarse physical entities come to exist through aggregation. This is the fundamental idea behind their views.

They uphold the basic tenet according to which, at least with respect to causally efficacious entities, when the referent object is sought, a real objective basis is found. However, Nāgārjuna and his students, especially those who are known as Prāsaṅgika Madhyamaka, such as Candrakīrti and Śāntideva, explain that such a view is held because of confusion about the mode of being of phenomena, which is the absence of inherent existence. Thus these masters refute any notion of "objective inherent existence."

In their own system such Madhyamaka masters assert that for any phenomena, whether external or internal, their mode of existence can only be posited in a nominal conventional sense. [355] In an ultimate sense, moreover, notions such as "identity and difference," "parts and the whole," "components and the composite entity," and so on cannot be posited at all. Hence they maintain that one cannot speak of the essential nature or characteristics of something like indivisible particles. But on a conventional level, however, even subtle particles that others view as indivisible are dependent originations that are solely reliant on other phenomena. Hence nothing can exist that possesses real, independent existence, and even the mind cannot establish such existence. Thus they declare no phenomenon, whatever it may be, transcends the nature of dependent origination. [356]

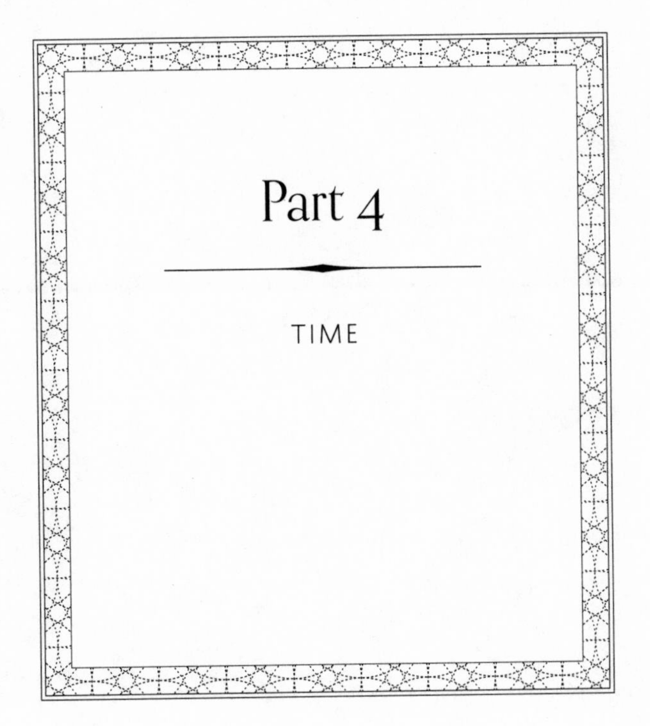

# Part 4

## TIME

## The Nature of Time

In part 2, we observed that, in Buddhist thought, time belongs to the category of nonassociated conditioning factors. This is to say that time, though belonging to the world of the conditioned, is neither material nor part of the category of mind. For all Buddhist schools, including the Sarvāstivāda Abhidharma, time is a relational concept and inseparable from the temporal phenomena of conditioned things. No Buddhist school accepts the notion of time as an independent eternal receptacle within which temporal phenomena unfold in a natural sequence. Rather, time is understood as a construct, albeit an indispensable one, that is extrapolated from the occurrence of events in the world. As cited in our volume, Asaṅga expresses this basic Buddhist view succinctly in his *Compendium of Abhidharma*: "What is time? Time is imputed to the continuous occurrence of causes and effects." Besides conditioned things, material or mental, which come into being from their causes, their momentary existence, and their cessation, say the Buddhists, there is no independent entity called time. Nor is there an overarching container like time that regulates events taking place.

This Buddhist view of time is very different from the non-Buddhist Vaiśeṣika view, which asserts time to be one of four independent eternal substances alongside the spatial direction, space, and selves. In this view, time is conceived, like space, to be a ubiquitous substance that is uniformly present everywhere. Even in this view of time as an eternal entity, time does not appear to be understood as a kind of receptacle. Rather, time is conceived as that which connects an object to a specific temporal measure, such as a day. So, according to this view, "to say that *A* is older than *B* is to say that there is an entity, time, that connects the sun and *A*, as well as the sun and *B*, every day that *A* and *B* have existed, and that the number of such connections in the case of *A* has been greater than in the case of *B*" (Potter 1977, 93).

Whereas the Vaiśeṣika school conceived time as the temporal correlate of spatial dimension—nearness and remoteness in time and space—few, if any, Buddhist sources correlate time and space in this manner. Units of measurement for space and time are almost always presented together, but for Buddhists, time is more closely connected to consciousness than space. The *Atthasālinī*, a text in the Theravāda canon, describes a mutual relationship between time and consciousness: "By time the Sage described the mind and by the mind described . . . time" (Nyanaponika 1998, 93). Whereas the state of consciousness is limited as well as described by time, mental factors determine time "by furnishing the measure of the time unit, which consists only in the duration of that temporary combination of factors" (94). This intimate connection between time and consciousness becomes all the more apparent when we consider the perception of time experienced by different subjects below.

For Sarvāstivādins, time is a nonassociated conditioning factor with a less substantial reality than matter and consciousness, yet they still maintain that time is real and possesses some substantial reality. Not only is time substantially real for them, they maintain that all three temporal phases—past, present, and future—possess substantial existence, hence the name Sarvāstivāda, "those who propound existence of all [three] times."

## The Three Temporal Phases and the Relativity of Time

The nature of the three periods of past, present, and future is at the heart of one of the earliest debates concerning time in the Buddhist world. While the Sarvāstivāda posits that all three phases have real existence, the Sautrāntika asserts the reality of only the present. In contrast, Nāgārjuna and Āryadeva argue not only for the relativity of all three temporal phases but also for the absence of intrinsic existence of time itself.

The challenge for the Sarvāstivāda, by asserting the real existence of all three periods, is explaining what remains constant and what is changed in time. Both the *Great Treatise on Differentiation* and the *Treasury of Knowledge* (5.26) list four distinct views on this. When a thing endures temporally although its substance remains constant, there must be a change in (1) its mode of being, (2) its characteristic, (3) its state, or (4) its position relative to time (see pages 259–62). Of these four positions, Vasubandhu prefers the third, asserting it is the state that is changed. In this view, the periods are established through activity. That is, when a conditioned thing has not accomplished its activity, that is the future, when it is accomplishing its activity, that is the present, and when its activity has ended, that is the past. Sarvāstivāda proponents of each of these four positions agree that conditioned things exist in all three time periods, and that just as the present thing that is directly observable exists, so do the past and future entities.

Sautrāntikas critique this Sarvāstivāda assertion about the reality of all three periods, but the most devastating refutation comes from Nāgārjuna and his disciple Āryadeva. In his *Fundamental Treatise* Nāgārjuna devotes a chapter of six verses to the analysis of time. If time has substantial existence, he says, then each of the three periods of past, present, and future must also possess intrinsic reality. This means either that each of the three time periods exists independently, which is absurd, or that all three exist at the same time, which is also absurd. In the final verse, Nāgārjuna rejects the position held by some fellow Buddhists that although time is a conceptual fiction, its bases, conditioned things, possess intrinsic existence and are thus ultimately real. Apart from temporal phenomena and their relations, Nāgārjuna seems to suggest, there is in fact no time at all. Time is nothing but a useful fiction that helps us conceive the temporal relationships among events. It is not an entity in its own right, nor is it an eternal vessel in which things and events occur or a regulator of events.

Āryadeva's *Four Hundred Stanzas* elaborates on Nāgārjuna's critique by refuting two specific proofs of the reality of time. In the first proof, time is said to be real because its basis, the conditioned things, are real. The second proof asserts that because we can have valid recollection of past events, these past events must be real. Āryadeva refutes the premise of the first proof by arguing that nothing endures, since all conditioned things are momentary. For if things endure—if there is such a thing as real abidance—then there can be no decay and cessation. Things would remain frozen in some kind of eternal present (9.17). If it is claimed that although conditioned things are impermanent, when they are in their abiding state, it is the abiding character that is predominant and not their impermanence, then how can the impermanent character assert itself later? Thus the consequence of a frozen present still follows (9.21–22). The second proof, related to memory, is based upon the seemingly straightforward notion that what distinguishes recollection from actual perception is its object, which is a past event. Just as the present object of perception exists, so too do the objects of our recollection. Āryadeva refutes this memory argument on the ground that what is once perceived cannot be perceived again, for the momentary nature of conditioned things is such that one and the same object cannot be perceived by two distinct cognitions. For if what has already been perceived can be perceived again, this object would exist in the present, which is absurd (9.25ab). The past event *is* the object of recollection, Āryadeva says, but unlike perception, recollection is a form of construction that involves a conceptual confabulation (9.25cd).

In brief, what we find in Nāgārjuna and Āryadeva's works is a critique of time as an eternal substance or an ultimately real entity. Instead time is seen as utterly contingent and radically relative. Thanks to remarkable discoveries in physics today most of us take the idea of the relativity of time as normal, regardless of whether we individually understand the arguments. But this is a fairly recent development. For much of the history of science, people viewed time in Euclidean terms as one dimension, something independent of motion and proceeding at a fixed rate in all frames of reference. Given the paucity of the scientific theory of matter in the second century, great Buddhist thinkers like Nāgārjuna and Āryadeva never entertained any idea akin to the notion in modern physics of space-time as an interwoven continuum. This, however, should not prevent us from recognizing the great conceptual and philosophical leap represented by these Buddhist thinkers' insights on the relativity of time so early in the history of ideas.

## "THE SHORTEST MOMENT OF TIME"

In part 3, we were introduced to Vasubandhu's statement that just as an atom is the smallest unit of matter, a single moment (*kṣaṇa*) is the smallest unit of time. As to how short this shortest moment is, there seems to be quite a divergence of opinion. When analyzing these divergent views, two frames of reference may be helpful. The first may be called a conventional timeframe, where the perspective is that of the human mind and activity. The second, less conventional, framework found in Buddhist sources recognizes different planes of time.

In the first model, a "single moment of the completion of an action" is distinguished from a "single shortest moment of time." The examples for the first moment include the typical duration of such brief human acts as the snapping of fingers, the blink of an eye, the time it takes to say "ah." Early Abhidharma theorists, including the Sarvāstivāda, propose that a sixtieth of the first type of moment is the shortest moment of time—that is, 1/60 of the duration it takes to snap your fingers. In some Abhidharma sources, this shortest moment is said to be 1/90 of the duration of a thought. As discussed in our volume, others posit this shortest unit of time to be 1/365 of the snapping of fingers. Dharmakīrti also proposes an intriguing criterion when he states that the shortest unit of time is the time it takes for an atom to rotate (*Exposition of Valid Cognition*, 3.495). In any case, Vasubandhu's *Treasury*, after identifying the shortest unit of time as 1/60 of a finger snap, explains how 120 such moments make one gross moment, 60 of these make one long minute, and so on, culminating in the measures of day, month, year, and eons.

The second, what I call unconventional, timeframe is found mostly in Mahāyāna sources, especially when describing the cognitive powers of advanced meditative states. These discussions point to a highly relative notion of time such that what is a brief moment for one person could be experienced as a lengthy timespan for another. Our volume, for example, cites a passage from the *Flower Ornament Sūtra* that speaks of how within what is a single moment for one person, another with more advanced attainment could perform tasks as if that instance lasts longer. In a moment when Ānanda teaches one Dharma, Śāriputra can teach ten such discourses, and in the instance when Śāriputra teaches one such Dharma, Maudgalyāyana can teach ten such discourses, and so on (see page 269). Looking at similar notions of different planes of time in the Theravāda sources, the noted Theravāda scholar Nyanaponika Thera speaks of "time planes below and above the range of average human consciousness, which may likewise be either inferred by deductive methods or actually experienced in the 'experimental situation' of meditative practice, in which the range and sensitivity of average consciousness may be greatly expanded" (Nyanaponika 1998, 99). Some Buddhist Vajrayāna texts speak in mind-boggling terms of the Buddha's power to "stretch an instance into an eon" and "contract an eon into an instance." So when you take this idea of different planes of time, it appears that for the Buddhists, especially for thinkers like Nāgārjuna and Āryadeva, the relativity of time entails not only the notion that time is relative to events but also that time is relative to different planes of consciousness or frames of reference.

## Buddhist Theory of Momentariness

Perhaps one of the most controversial aspects of the Buddhist view of reality is what is referred to as the doctrine of momentariness, the assertion that everything within the world of conditioned things is in a state of constant flux and things remain for only a single moment. Although the impermanence of all conditioned things is accepted by all Buddhist schools, early Abhidharma schools such as the Sarvāstivāda did not interpret impermanence in terms of momentariness. Theirs is the standpoint that upholds the progression of four characteristics of a conditioned thing—arising, abiding, decay, and cessation. The Sautrāntika rejects this view and argues instead for three, not four, characteristics of conditioned things—arising, abiding, and ceasing—and asserts that all three characteristics should be understood as being simultaneous. So the arising of a conditioned thing, say a sprout, for example, is simultaneous to its abiding as well as cessation.

In other words, the moment a sprout comes into being, right there and then the process of its cessation has also begun. And as the sprout at time1 ceases, it gives rise to its next instance, time2, and in this way forms an uninterrupted flow of momentary entities of the same type that are causally connected to each other. What we perceive as a continuum is in fact a construct projected by the mind onto a series of instances, somewhat akin to the perception of continuous motion we experience when watching a film made up of a series of individual frames.

This theory of momentariness, which seems at such odds with our common sense and everyday experience, attracted sustained critique, especially from the non-Buddhist Indian schools. Buddhists in turn developed systematic arguments to defend and prove their thesis of universal momentariness. What might be called the traditional proof involves deducing momentariness on the grounds that things do not require an external cause for their cessation. Their cessation is built in, such that everything that arises also ceases spontaneously, with no need for an external agent, like a hammer, to bring about its cessation.

In the seventh century, Dharmakīrti formulated a new proof for the momentariness of things based on the idea that anything that has real existence must possess causal efficacy, and causal function is possible only for things that are momentary. In one of the formulations of this argument, Dharmakīrti asserts if a nonmomentary thing were to produce its effect in the first moment, given that it persists in time, it will have to produce the same effect again and again in all subsequent moments of its existence. If, on the other hand, the thing is not capable of producing its effect completely in the first instance, it should not be able to do so later, for to do so would entail a change in its nature. And the coexistence of two contradictory properties—producing and not producing its effect—is incompatible with the identity of the thing as a single entity. Thus Dharmakīrti concludes that causal efficacy necessarily implies momentariness.

In the eighth century, in response to objections against Dharmakīrti's "new" proof of momentariness, Dharmottara developed an innovative argument that combines the traditional proof of causelessness of destruction and Dharmakīrti's new proof. At the heart of Dharmottara's new argument is the refutation of what the opponent cites as a cause of destruction as being an example of a cooperative cause. Understanding this argument requires some technical explanation of the structure of typical classical Indian syllogism and the logical relations among the members of such a syllogism, something beyond the scope of this brief essay. Those who wish to engage with Dharmottara's highly technical arguments can do so by consulting the paper listed below in the further readings.

Buddhist theorists like Dharmakīrti are aware of how this theory of momentariness runs counter to our naïve common-sense perception of the world, which suggests constancy and continuity. This disparity is strikingly analogous to the disparity between our common-sense perceptions and the picture of reality presented in contemporary physics. Just as in this scientific view, everyday objects like tables and chairs are constructs of our mind and what you in fact have is a constant flux of particles and energy, in the Buddhist view too, our everyday perception of objects extended in space and time are nothing but convenient mental constructs. In reality, what exists are nothing but momentary entities, which are constantly in flux and have no capacity at all to remain static and unchanged for even a single instance. For the Buddhists the realization of this momentary nature of things—the insight into impermanence—opens the way to a more constructive approach to reality where we become less fixated and attached to things.

## Further Reading in English

On detailed analysis of the four viewpoints within the Sarvāstivāda Abhidharma on the three temporal phases, see Erich Frauwallner, *Studies in Abhidharma Literature and the Origins of Buddhist Philosophical Systems* (Albany: State University of New York Press, 1995).

For a succinct explanation of the Vaiśeṣika theory of time as an eternal substance, see Karl Potter, *Encyclopedia of Indian Philosophies*, vol. 2 (Princeton, NJ: Princeton University Press, 1977), 90–93.

For a lucid translation and succinct commentary on the chapter on time in Nāgārjuna's *Fundamental Treatise*, see Mark Siderits and Shōryū Katsura, *Nāgārjuna's Middle Way* (Boston: Wisdom Publications, 2013), 207–12.

For an English translation of the chapter on the analysis of time in Āryadeva's *Four Hundred Stanzas* based on Gyaltsap Jé's commentary, see chapter 11 of Geshe Sonam Rinchen, *Aryadeva's Four Hundred Stanzas on the Middle Way*, translated by Ruth Sonam (Ithaca, NY: Snow Lion Publications, 2008).

On the Abhidharma calculation of the various units of time, including the measure of what constitutes the shortest moment, see William Mongomery McGovern, *A Manual of Buddhist Philosophy* (London: Kegan Paul, 1923), 41–48.

For an informative overview of the notions of time as presented in the Pali Abhidhamma texts, see Nyanaponika Thera, *Abhidhamma Studies: Buddhist Explorations of Consciousness and Time* (Boston: Wisdom Publications, 1998), 93–114.

For a detailed contemporary analysis of the Buddhist theory of momentariness based on Dharmakīrti's philosophy, see Rita Gupta, "The Buddhist Doctrine of Momentariness and Its Presuppositions," *Journal of Indian Philosophy* 8 (1980): 47–68.

For an analysis of Dharmottara's argument for momentariness, see Sakai Masamichi, "Dharmottara's Interpretation of the Causelessness of Destruction," *Journal of Indian and Buddhist Studies* 58.3 (2010): 1241–45.

# 17

## The Definition of Time

TIME IS AN essential topic that needs to be understood as part of the category of "nonassociated formative factors," which has been discussed earlier. In general there is a shared recognition among classical Indian philosophers that when exploring the nature of reality it is vitally important to also present the nature of time. However, a divergence of views exists as to what is the actual nature of time. If we were to take the view of the non-Buddhist Vaiśeṣika school, for example, they classify objects of knowledge in terms of six categories (substance, quality, motion, universals, particulars, and inherence).[312] Substance, in turn, is differentiated by nonpermeating substance and permeating substance. Time, self, spatial direction, and space constitute permeating substances, and they view them as substances that are unborn and permanent. Furthermore, they say:

> Time matures the elements.
> Time assembles the people.
> Time awakens from sleep. [358]
> It is extremely difficult to transcend time.[313]

Thus they assert time to be distinct in nature from temporal entities, which is the agent of all activities, and a permanent substance. Time is revealed or illustrated by moments, by day and night, and by hours and so on.

From among classical Indian Buddhist schools, the Vaibhāṣikas assert that since the three times possess the characteristics of conditioned phenomena, time is not permanent. This said, they do maintain time to be substantially existent yet distinct from temporal substance. They assert that since all conditioned phenomena are characterized by arising, enduring, and disintegrating, the three times exist, and furthermore all conditioned phenomena that exist in the three times are substantially established.

As to how phenomena are said to possess substantial existence, in general Vaibhāṣikas assert that any phenomenon must be capable of bearing its own distinct identity, and since for them no phenomenon can be posited merely as a projected reality imputed to some other phenomenon, they assert that all phenomena are substantially established. In particular they assert that the three times exist as entities that retain their own distinct identities, not as something imputed upon another phenomenon. Furthermore, every entity, such as a sprout, for example, possesses the three times and also exists at the time of the future sprout as well as of the past sprout. In this way they speak of all three times as substantially established. [359]

Thus because they proclaim the three times to be subsets of substantially real entities, they are called Vaibhāṣikas.[314] The etymological explanation is that they assert the three times to be subsets of substantially real entities, on the basis of which the three times are posited. For example, there are three times in which a sprout exists as a sprout, and the sprout is the universal, whereas sprouts of the three times are its instantiations.

As to how a vase, for example, exists in the past and future, and how one posits a past vase and a future vase as vases, it is as follows. When a person is about to look at a vase, both the eye sense faculty within the person's mindstream as well as the vase that will be its object are posited as *future*, since these two are in the process of arising. In contrast, at the time of actually seeing the vase, both the eye sense faculty within the continuum of the person as well as the object being seen are posited as *present*. Immediately after it has been seen, both the eye sense faculty within the mindstream of the person who has seen the vase and the vase that was its object are posited as *past*. Therefore the Vaibhāṣikas assert the vase to be existent in both the past and future of the vase. This is not to say that they assert today's vase exists yesterday or tomorrow. Furthermore, given that the past vase and future vase already existed as a vase and will exist as a vase, and given that they belong to the same class as the present vase, the past and future vases are also posited as being vases. By analogy, even though some of the trees in a forest have not yet been turned into firewood, because they belong to the same class as firewood they too are referred to as "firewood." Similarly, because the milk in the udder of a milk cow belongs to the same class as that already extracted, it is referred to as "milk." These points are discussed in Vasubandhu's *Treasury of Knowledge Autocommentary*: [360]

> Do past and future entities substantially exist or not? If they exist substantially they would be permanent because they would exist in all three times. But if they do not exist substantially how do they exist? Are they permanent, or possessed by it, or free from it?
>
> Vaibhāṣikas do not assert that conditioned phenomena are permanent because conditioned phenomena are characterized as conditioned. But they clearly assert that "times exist in all three times" (5.23a). Why? It is said, "Because the Buddha said so" (5.23a). For the Bhagavan stated at length, "Monks, if a single past form did not exist, then the fact of learned ārya śrāvakas not considering past form would not arise. Because past form exists, therefore the fact of learned ārya śrāvakas not considering past form arises. If a single future form did not exist then the fact of learned ārya śrāvakas disliking future form would not arise. Because future form exists..." Also it is said, "Because it depends on both," (5.23b) for consciousness arises in dependence on both. What are these two? They are the eye sense base and the form base... up to... the mental sense base and the phenomena base, for if the past and future did not exist then consciousness observing objects would not be dependent on both sense bases and bases. For example, occasionally the past and future are taught to exist in scripture. They are also established through logic, for it is said, "Because the object exists" (5.23b), which means that if the object exists then it is not the case that [361] knowledge of them would not arise. But if the past and future did not exist they would not be observed. Owing to that, consciousness of them would not arise because they would not be observed. It is also said, "Because the result exists" (5.23b), which means if the past did not exist how would the effects of virtuous and nonvirtuous karma come to arise in a future life? When the effect arises, the current maturational cause ceases to exist, thus the Vaibhāṣikas say the past and future surely exist.
>
> It is well known that the Sarvāstivādin school and the

> rest necessarily assert this beyond any doubt. Thus it is said, "Because they advocate 'all exists' (*sarvāsti*) they are asserted to be 'those who proclaim all exists' (*sarvāstivāda*)," (5.23c–d) and whoever proclaims that *all* that arises in the past, present, and future *exists* are Sarvāstivādins. Whoever makes the distinction "some exist that have arisen in the present but pass without their effect being generated, but none whatsoever exist that have not yet arisen and pass with their effect already generated" are Vaibhājyavādins.[315]

Thus the following arguments are given in response to the objection that if the past and future were to have substantial existence conditioned things would then become permanent. Since conditioned things possess the four characteristics (of arising, enduring, decay, and disintegration) they do not become permanent, the three times are not permanent, [362] present consciousness observing the past or future has objects, consciousness arises in dependence on its object and sense faculties, and the effects of past karma can come to fruition in the present. In this way, except for the Vaibhājyavādins—one of the seven subschools of Sarvāstivāda—this school maintains that both past and future possess substantial existence. The Vaibhājyavādins, on the other hand, make the distinction that present time and those elements of the past that have not issued forth their effects possess substantial existence, whereas the future and those elements of the past that have already produced their effects do not possess substantial existence. *Investigating Characteristics* states:

> Since the term "all" is definitely applied to all three times, the term "all" expresses the three times. Other sects that proclaim the existence of unarisen results speak of distinctions (the Vaibhājyavādins). The Sautrāntikas, on the other hand, proclaim the existence of things that occur in the present alone.[316]

## Time as Dependent and Imputed

In general "time" is defined as a nonassociated formative factor that is imputed on the basis of aspects of a material entity or a consciousness and is characterized by the change that permeates all conditioned phenomena.

More specifically, time is posited merely on the basis of the three states of conditioned things: (1) not yet arisen, (2) arisen but not yet ceased, [363] and (3) arisen and ceased. Other than that no substantially real independent entity exists that can be identified as time. So the texts explain that "time," which is defined in terms of such states of existence, is so characterized because it relates to entities of cause and effect that have already come, are coming, and will come into being. Alternatively, it is so called because such conditioned things are consumed by impermanence alone. For example, Asaṅga's *Compendium of Abhidharma* states:

> What is time? Time is imputed to the continuous occurrence of causes and effects.[317]

So too it is stated in Maitreya's *Differentiating the Middle and Extremes*:

> Effects and causes already expended,
> so too those not expended are the others.[318]

Vasubandhu states in his *Commentary on Distinguishing the Middle from the Extremes*:

> What are the "others"? It refers to the three times. The term "time" is applied appropriately in the following manner. It is called *past* time since the cause and its effects have already been expended. It is called *future* time since the cause and its effects have not yet been expended. It is called *present* time because although the cause has been expended its effects have not yet been exhausted.[319]

Thus the three times of past, present, and future are posited in relation to the conditioned phenomena subsumed within the class of causes and effects in terms of having *already occurred*, *yet to occur*, and *occurring*, respectively. [364]

As for Nāgārjuna and his followers, in general they posit time as a phenomenon that is utterly contingent and exists as a dependent origination. They reject any notion of time that is a uniquely characterized real phenomenon not dependent on temporal phenomena. Even so, from the perspective of everyday worldly convention and in relation to a specific context, a shared common time can be posited. For example, temporal units such as a year, a month, a day, a minute, a second, a moment, and so on are posited on the basis of the accumulation of multiple temporal entities. One cannot posit such temporal units as a year, a month, and a day independently by themselves.

Nāgārjuna states that if each of the three times were to have an inherently existent nature, the three times would merge. *Fundamental Treatise on the Middle Way* states:

> If the present and future
> relied on the past
> the present and future
> would exist in the past.[320]

If time described as "the present" or "the future" existed inherently, would either present or future time be reliant on the past? If they were reliant on the past then it follows that both would exist at the time of the past. But inherently established phenomena that do not exist anywhere do not rely on the past, and for such phenomena, the basis relied upon and the phenomena reliant upon it must be posited as existing simultaneously. To assert that those two times exist in the past [365] is untenable because past time must be posited in terms of what has passed from the present, and the future must be posited as future in terms of what is not yet obtained in the present. Thus, if these two were to have an inherently existing nature, their dependence on the past would be untenable. The *Fundamental Treatise on the Middle Way* states that on the conventional level at least, if these two times were not dependent on the past, they could not exist:

> Without relying on the past
> neither the present nor the future could be established
> because the present and future
> do not exist in the past.[321]

If the present and future did not rely on the past, they could not be posited as existent. By this same logic, both the past and future are established as reliant on the present and both the past and present are established as reliant on the future, respectively. Thus Nāgārjuna clearly elucidates that all three times are phenomena that are utterly interdependent. So too Āryadeva states in the *Four Hundred Stanzas*:

> A future vase does not exist without
> a present vase or a past vase.
> Because those two have not appeared,
> therefore the future does not yet exist.[322]

Candrakīrti's *Commentary on the Four Hundred Stanzas* says:

> These three times exist interdependently because any of the three would not exist [366] without relying on the other two. Without reliance on the past and the present, a future time cannot be posited. For the future is that which has not yet come, and the nature of the future itself has not yet appeared at any time, but without doubt it will transform into the present and past. But if it does not yet exist, when will what is distinct from the future become the future? The presentation of the other two times should be understood by extending this explanation.[323]

# 18

## Positing the Three Times

It is on the basis of their causes and conditions that conditioned phenomena transform from the future to the present and from the present to the past, and when such phenomena transform in that way they are posited as future, present, and past, respectively. Take, for example, a tree. The phase when it has not yet arisen at the time of its cause is posited as *the future*, the phase when it has arisen but has not yet died is posited as *the present*, and the phase when it has died and its essential nature ceases is posited as *the past*.

In the Vaibhāṣika system the well-known way the four masters posit the three times [367] differs slightly. Moreover, the system of Bhadanta Vasumitra is [generally] regarded as best, and according to him the future, present, and past are posited as *action that has not yet arisen*, *action that has arisen but not ceased*, and *action that has been destroyed*, respectively. For example, when the seed has been planted in a field and before the sprout has arisen, it is posited as the *future sprout*; when it has arisen and until it is destroyed, it is posited as the *present sprout*; and when it has been destroyed it is posited as the *past sprout*. This Vasumitra asserted to be transformation from the future to the present, or transformation from the present to the past. The differing assertions of the four masters is stated in *Treasury of Knowledge*:

> They are four: the transformation of state, of characteristics, of context, and of mutual reliance.[324]

Thus there are four different presentations of the three times, advocating: (1) the transformation of state, (2) the transformation of characteristics, (3) the transformation of perspective, and (4) the transformation of mutual reliance.

1. Bhadanta Dharmatrāta is the advocate of the transformation of state. In his system the three times are distinguished as a result of the different ways they are engaged by thought and language. When, for example, a sprout enters the present from the future or enters the past from the present, its state transforms even though its substance does not transform. Thus the *Autocommentary* states: [368]

> The proponent of the transformation of state is Bhadanta Dharmatrāta. He says if a phenomenon arises in the [three] times, the thing transforms but its substance does not transform. For example, if a gold vessel transforms owing to being destroyed, its shape transforms but its color does not transform. If milk transforms into curd, then its taste, potency, and maturing agent are relinquished, but not its color. If a phenomenon passes from future time to present time, the future time of the entity is relinquished, but not the substance of the entity. So too, if that phenomenon passed from the present time to the past time, the present time of the entity is relinquished, but not the substance of the entity.[325]

2. Bhadanta Ghoṣaka is the advocate of the transformation of characteristics. In his system, although something, such as a sprout, possesses the three characteristics of past, present, and future, it is posited as past and so on by means of which characteristics are dominant. Thus *Autocommentary* states:

> The proponent of the transformation of characteristics is Bhadanta Ghoṣaka. He says if a phenomenon enters the three times, the past possesses the characteristics of the past, yet it also possesses the characteristics of the future and present. [369] The future possesses the characteristics of the future, yet it also possesses the characteristics of the past. So too the present also possesses the characteristics of the past and future. For example,

> it is like a male who has desire for one female but is not free from attachment to other females.[326]

3. Bhadanta Vasumitra is the advocate of the transformation of perspective, and his mode of assertion has already been briefly explained. For him the three times are labeled according to what characteristic is dominant from their specific perspectives. When a sprout, for example, passes from the future to the present and from the present to the past, the perspective of the action not yet performed is called the *future*, the perspective of the action being performed but not yet ceased is called the *present*, the perspective of the action having been performed and ceased is called the *past*. Therefore transformation is in name alone and there is no transformation of a uniquely characterized real substance. Thus *Autocommentary* also states:

> The proponent of the transformation of perspective is Bhadanta Vasumitra. He says if a phenomenon arises in the three times, it is described as *different from the other* by means of the difference in context through passing from one context to another, though it is [370] not different in substance. For example, if a bead is placed in the first slot of an abacus it is called *one*, if placed in the hundredth slot it is called *one hundred*, if placed in the thousandth slot it is called *one thousand*.[327]

4. Bhadanta Buddhadeva is the advocate of the transformation of mutual reliance. In his system the three times are labeled in reliance on what occurs earlier or later because when a single entity is located in the three times, it is future in reliance on the past and present that occurred earlier, it is past in reliance on the present and future that occur later, and it is present in reliance on the past that occurred earlier and the future that occurs later. Thus *Autocommentary* states:

> The proponent of the transformation of mutual reliance is Bhadanta Buddhadeva. It is well known that he says that if a phenomenon engages the three times, it is called *different from the other* in reliance on earlier and later times. For example, a single

> female is called a *mother* as well as a *daughter*. Thus these four are Sarvāstivādins.[328] [371]

In any case, here in the Vaibhāṣika system it is said that time and conditioned entities are equivalent or mutually inclusive. For instance, *Treasury of Knowledge Autocommentary* states:

> "They are time, the basis of language, definitely transcended, and possess a basis" (1.7c–d). "They" refers to conditioned phenomena. They constitute time because they have gone, are going, or will go, and they are consumed by impermanence.[329]

The Buddhist tenet systems of the Sautrāntika, Cittamātra, and Svātantrika Madhyamaka are similar in how they posit the three times. They assert that the past and future are necessarily permanent and nonimplicative negations, whereas the present and functional things are mutually inclusive or equivalent. When, for example, a thing such as a sprout is destroyed, every part of a sprout ceases and no other thing is obtained whatsoever. Since the past as such is the mere elimination of a negandum, it is not a functional thing. So too, the future sprout is nothing other than the fact of not having arisen at some point due to its causes and conditions being incomplete. And since except for that fact an instance of a functional thing is not found, it is not described as a functional thing. For example, Dharmakīrti's *Exposition of Valid Cognition* states:

> Stating "it is nonexistent"
> is also to state it is not a thing.[330] [372]

Dharmakīrti means that *the disintegrated state* and the past are not functional things. Since time is necessarily a nonassociated formative factor, something like a vase, for example, is *present* but it is not *present time.* Though the past and future are permanent, *past time* and *future time* are posited as functional things and nonassociated formative factors. Further, one must make the distinction that in general a *past vase* is *past,* but in relation to the vase itself it is *future*. Similarly, although *future vase* is *future,* in relation to the vase itself it is *past.*

In general in accordance with these systems, the definition of the past is "the disintegrated factor regarding the entity that is the negandum."[331] Past, ceased, and disintegrated states are mutually inclusive. For example, a tree does not exist at the time of its disintegrated state and a vase does not exist at the time of its effect.

The definition of the present is "arisen but not ceased." Present, functional thing, and conditioned phenomena are mutually inclusive. Examples include a vase, a pillar, and so on.

The definition of the future is "a thing that has not arisen owing to incomplete conditions even though the cause of its arising exists." For example, the nonarising of a sprout in a field in winter or the nonarising of a vase at the time of its cause.

If illustrated on the basis of a sprout, the time the sprout is destroyed is the *past time of the sprout*. The time when the sprout has not yet arisen owing to causes and conditions being incomplete at a specific place and time, such as an eastern field in winter, is the *future time of the sprout*. The time when the sprout exists is the *present time of the sprout*. Therefore both the time of its past and future is posited in reliance on the time of its present. [373] *Compendium of Abhidharma* states:

> In what way is something past? How many past states are there? Why is something considered past? Its defining characteristics are "arisen and ceased," where its cause and result have already been activated.
>
> In what way is something future? How many future states are there? Why is something considered future? Its defining characteristics are "its cause exists but it has not arisen," where its characteristics have not been obtained, and its cause and result have not been activated.
>
> In what way does something arise in the present? How many states arise in the present? Why is something considered to arise in the present? Its defining characteristics are "it has arisen but not ceased," where its cause has been activated but its result has not been activated.[332]

Prāsaṅgika Mādhyamikas view—as distinct from the three systems of the Sautrāntika, Cittamātra, and Svātantrika Madhyamaka—all three times of past, future, and present as functional entities. They do not assert that the past and future are nonimplicative negations, but rather that they are types of implicative negation. Nonetheless, even though they accept the three times to be functional things, they differ from the Vaibhāṣikas since they do not accept the existence of the sprout itself at the time of the future sprout or the past sprout.

According to this view, the past is defined as "a thing that has already disintegrated"; for example, a disintegrated sprout. The future is defined as "a thing that has not yet arisen owing to incomplete conditions even though its productive cause exists"; for example, an unarisen sprout at the time of the cause of the sprout. The present is defined as "that which has arisen but not yet ceased, and that is neither a disintegrated thing nor a [374] future thing"; for example, a sprout. To illustrate these definitions in relation to a sprout, the sprout itself is *present*, the disintegrated sprout is *past*, and the unarisen sprout is posited as *future*, since in general even though its cause exists, at this time its causes and conditions remain incomplete. Thus the past and future are defined on the basis of some other entity, in that past is defined on the basis of its cessation, whereas future is defined on the basis of the incompleteness of conditions even though its productive cause exists. In contrast, as for the present, it is not necessary to posit it on the basis of the cessation or nonarising of some other entity, since it is posited as the entity itself on the basis of it having arisen but not ceased. Therefore, from among the three times, the present is posited as *primary* and the other two as *secondary*. For example, when one speaks of the "future sprout," the "present sprout," and "the past sprout," the basis on which the three times are defined is the sprout itself. The future in the context of what is called *future sprout* is the future in relation to the time of the present sprout. Similarly, the past in the context of what is called *past sprout* is the past in relation to the present sprout itself. Therefore the present sprout is considered primary within the three times. *Questions of Subāhu Sūtra*, for example, states:

> You should well understand the three times. What are the three? They are the past time, future time, and present time, and these are called the *three times*. Any conditioned entity that

> has ceased is called *past time*, any conditioned entity that has not come and is unarisen is called *future time*, and any conditioned entity that has arisen in the present and exists [375] in the present is called *present occurring time*.[333]

Similarly, *Commentary on the Four Hundred Stanzas* states:

> Thus the future is something that has not reached the present time, the past is something that has passed from that state, and the present is something that has arisen but not ceased. The present is primary because it exists in the present. And it is on the basis of *not having arrived* and *having passed* that future and past are defined, so they are not primary.[334]

As to how the *disintegrated state* is established as a functional entity in the texts of the Prāsaṅgika Madhyamaka, this will be explained more extensively when presenting the views of the philosophical schools in a subsequent volume in the compendium series.

To summarize, according to the system of Nāgārjuna and his philosophical heirs, the three times are posited in the following way, taking as an example a sprout. A sprout that has arisen but not yet ceased is posited as one phase of that sprout that is called *present sprout*. The phase of that sprout when it is yet to occur is called *future*, that is, relative to the sprout that is the basis. The phase of the sprout having arisen and ceased is called *past*, that is, relative to the sprout. Thus it is on the basis of distinct phases of a single entity that the three times are posited. Therefore *past*, *present*, and *future*, due to their mutual reliance, are designated through conceptual imputation and established by popular convention. [376] None of the three times can be posited as inherently real. Also when one calculates a specific duration such as a year, a month, a day, an hour, a minute, or a second with respect to specific basis, one may say "up to this point is the past" and "from this point on is the future," but it is difficult to discern anything remaining that may be called the present.

# 19

## The Shortest Unit of Time

In general there are two types of moment found in the Buddhist texts: (1) the shortest moment of time, and (2) the moment required to complete an action.[335] In everyday language the shortest moment of time refers to a moment that is the briefest ultimate instant. However, in terms of its duration, Buddhist thinkers give divergent opinions. In the *Treasury of Knowledge* and its *Autocommentary*, Vasubandhu, for example, explains that the smallest material form is an indivisible subtle particle, the smallest unit of a word is a letter, and the briefest indivisible unit of time is the shortest moment. And this shortest moment is posited to be 1/65th of the time it takes a strong man to snap his fingers. He then goes on to explain how 120 shortest moments are one second, 60 such seconds are one minute, 30 minutes are one hour, 30 hours are one day, and 30 days are one month. For example, the *Treasury of Knowledge* states: [377]

> One hundred and twenty such moments
> are a second. Sixty
> of those are one minute.
> The three periods of hours, days and nights, and months
> are each progressively thirty times the former.[336]

The *Autocommentary* states in accordance with that analysis:

> The smallest limit of form is the subtle particle. The limit of time is a moment. The limit of names is a letter, as in the word "oxen." Again what is the measure of a moment? When conditions assemble it is the obtainment of a phenomenon itself, or when a phenomenon moves it is the time taken for it to move

> from one subtle particle to another subtle particle. Ābhidharmikas say: Just a single finger snap of a strong man lasts sixty-five moments.[337]

Dharmakīrti and Candrakīrti also understand "moment" in similar terms. For example, *Exposition of Valid Cognition* states:

> The time taken for a particle to turn or spin
> is asserted to constitute a moment that is the shortest discrete unit of time.[338] [378]

Thus the duration of a single subtle particle changing its locus or turning over is the shortest moment of time. Candrakīrti's *Commentary on the Four Hundred Stanzas* states that 1/65th of a moment required to complete an action is the shortest moment of time:

> What is called *a moment* is taken to be the briefest limit of time. A single finger snap of a strong man lasts sixty-five moments, and one such aspect perceived by consciousness in a moment is a single moment.[339]

In Dharmamitra's *Clarifying Words*, however, is it said:

> How should the meaning of *moment* be viewed in this context here? Some say that even within a single moment of a conditioned phenomenon, a moment that passes swiftly, such as the duration of the snapping of fingers, countless moments cease. This is taught in texts of the tenet systems. As for the duration of such a moment, the noble śrāvakas assert that in that period 368 moments pass.[340]

Thus the duration of the finger snap of a strong man is divided into 368 parts, and according to the śrāvaka system, one such part is the shortest moment. Other texts, for example, say that the aggregate of suffering passes without remaining for even a moment, like a water bubble or like a waterfall on a steep mountain, [379] and declare that one must cultivate recognition of their untrustworthy nature through contemplating how 365 moments pass in one finger snap. Thus the length of a finger snap is also said to have 365 parts.

The Mahāyāna philosophical system does not assert there is no moment shorter than 1/365th of a finger snap. There is, for example, the following calculation found in the *Flower Ornament Sūtra*. In the time it takes for a single wheel of a chariot driven more quickly than the flight of a garuḍa to turn once, a poisonous snake can circle a chariot seven times. And in the moment it takes to make one circle, the noble Ānanda can explain ten Dharmas and make them understood, Śāriputra can explain one thousand Dharmas and make them understood, Maudgalyāyana can transcend eighty thousand world systems, and the Tathāgata can present limitless displays of the three secret inconceivable states by means of the twelve deeds in every world system.[341] Nāgārjuna, for example, quotes in his *Compendium of Sūtras* the following from the *Flower Ornament Sūtra*:

> A human with an iron chariot with wheels of one thousand spokes drawn by a strong horse swifter than a garuḍa . . . up to . . . [380] the moment it takes a poisonous snake to circle that chariot once, Bhikṣu Ānanda explained ten Dharmas and their meaning was understood. In that moment Ānanda explained one Dharma and Bhikṣu Śāriputra explained one thousand Dharmas, and their meaning was understood. And in that moment Śāriputra explained one Dharma and Bhikṣu Maudgalyāyana visited eighty thousand world systems.[342]

## A Moment Required to Complete an Action

This refers to the moment it takes from the start of an action until it is completed, and its duration varies. For example, the period of a month from the first until the thirtieth, or the snapping of fingers by a person that lasts only briefly. As such different texts explain this point in different ways. For example, *Root Tantra of Mañjuśrī* speaks of the moment required for the completion of an action as blinking the eyes ten times:

> Blinking the eyes ten times
> is designated as a mere moment.[343]

In the *Vajraḍāka Tantra*, however: [381]

> A blink is closing the eyes.
> Six blinks are called one part.
> Six of those brief periods are a second.
> Six seconds are a minute.
> Six minutes are a short period.[344]

So six blinks are called *one part*, six brief periods are *a moment required to complete an action*, six of those are *one minute*, and six of those are posited as *a short period*. In Nāgārjuna's *Five Stages*, there is the following:

> A moment is said to be a finger snap.
> A brief period is the time taken to rotate a white mustard seed.
> A short period is just one exhalation and inhalation.[345]

The time taken for a strong person to snap his fingers is a moment required to complete an action, the time taken to rotate a mustard seed is a brief period, and the time required to inhale and exhale just once is a short period.

In general, no fixed length is assigned to a moment required to complete an action given the many varieties of actions and their durations. There is, however, a consensus among both higher and lower Buddhist schools that the shortest moment required to complete an action is the duration of snapping of fingers, blinking of eyes, uttering a short "a" syllable, or the shortest action of an ordinary person and so on. Similarly, there is consensus that the shortest moment of time is the time taken for the position of a subtle particle to change, or a single subtle particle to turn, or 1/65th of a finger snap. [382]

Thus, in general, Buddhist texts maintain that a moment to complete an action is the duration of a person's finger snap, and that finger snap divided into 1/65th or 1/365th or 1/368th part is the shortest moment of time. However, we need not understand the briefest limit of time solely in these terms, and this is evident from the previously cited passage from *Flower Ornament Sūtra* in Nāgārjuna's *Compendium of Sūtras*. Therefore, when examined carefully, it is apparent that what is referred to in Buddhist texts as "the shortest moment of time" need not represent the ultimate shortest unit of time.

Even so, just as the Buddhist śrāvaka schools assert that subtle particles that are the basic constituents of physical phenomena are partless, they also proclaim that with respect to consciousness that is a nonphysical conditioned phenomenon, the shortest moment of time arrived at when differentiated in terms of its earlier and later temporal increments (a moment that cannot be split further into subdivisions of earlier or later parts, such a temporal unit) is indivisible.

As for the proponents of Madhyamaka and Cittamātra tenets, since they do not accept any notion of an "indivisible entity," be it matter or consciousness, even though they do need to posit the "shortest single moment" as the shortest unit of time and the "subtle particle" as the smallest discrete material unit, a diverse range of responses emerge to the following question: If the *briefest discrete unit* and the *smallest discrete unit* are taken to mean that it cannot be divided into subtler parts, then how could such a thing not be partless? And if no ultimate limit can be reached when dividing them into their parts, what would be the meaning of positing something as the "briefest discrete unit" or "smallest discrete unit"? [383]

Nāgārjuna's *Precious Garland* states:

> Just as a moment possesses an end,
> so too consider it has a beginning and middle.
> Therefore, because it consists of three moments,
> the world does not endure for a moment.[346]

Since even the shortest moment of time possesses three parts: a beginning, middle, and end, such analysis establishes that it is not partless. But even if one were to ascertain the measure of the briefest unit of time and posit that as the identity of a moment of time, one cannot establish it as an inherently existing moment. Therefore, when one conceptually analyzes that moment of time, Nāgārjuna suggests one has no choice but to admit it definitely has three parts: a beginning, middle, and end. Therefore, just as one cannot posit an indivisible subtle particle, so too one cannot assert an ultimate unit of time that constitutes a partless moment. Madhyamaka and Cittamātra thinkers agree on this point.

# 20

## Positing Subtle Impermanence

IN GENERAL "impermanence" is posited on the basis of whether something undergoes change, and conditioned phenomena are subject to change owing to their causes and conditions.[347] So if something has causes and conditions, it is subject to change, and if it does not have causes and conditions, it is not subject to change. [384] The fact of such conditioned phenomena being subject to change is primarily a function of the productive causes that produce them. As such, all conditioned phenomena continuously undergo change without remaining static for even a single moment. For example, owing to the change of a tree's leaves, they fall to the ground with the arrival of cold in winter. It is not that they transform spontaneously, but rather they transform each day and week until in the end they fall to the ground. Those leaves that transform over many days do so naturally through merely being established. Therefore they transform moment by moment, and even though the eye does not see it, in reality they continuously transform. If subtle change did not exist moment by moment, then coarse transformation also could not arise.

With respect to how the four characteristics of conditioned phenomena are understood, the Vaibhāṣikas, for example, assert that when the three characteristics of conditioned phenomena—arising, enduring, and disintegrating—illustrate the conditioned nature of something, such as the form aggregate, they do not do so on the basis of something arising and so on. They do so by way of demonstrating that the given phenomenon possesses characteristics such as arising that are distinct from it. Therefore they don't assert these characteristics to be the action of arising and so on, but rather as substantially real, distinct entities that are the agents that generate, that endure, and that disintegrate. So although these four characteristics do exist simultaneously for conditioned phenomena, such as a material entity, when they occur on a specific basis, they maintain that first the action of arising occurs, next [385] enduring, then decaying, then the action of disintegration occurs in a sequential order. As for the four characteristics of arising and so on themselves, as mentioned earlier, they do not view these in terms of the *action of arising* and so forth. Rather they posit these in terms of substantially real, distinct entities that are the agent of generation, the agent of endurance, the agent of decay, and the agent of disintegration. Furthermore they interpret "conditioned phenomena" in terms of "produced through the aggregation of causes and conditions," that is to say, in terms of activities associated with active agents. For example, the *Treasury of Knowledge Autocommentary* states:

> Here owing to arising it generates that phenomena, owing to enduring it causes it to abide, owing to decaying it causes it to age, and owing to impermanence it causes it to disintegrate.[348]

*Explanation of the Treasury of Knowledge* states:

> Here they assert the characteristics of conditioned phenomena are by nature different substances.[349]

Therefore those characteristics are not posited in terms of extremely short moments. *Treasury of Knowledge Autocommentary* states:

> The Bhagavan teaches that "the continuum of conditioned phenomena is a conditioned phenomenon and a dependent origination" . . . up to . . . "these three are the characteristics of conditioned phenomena," but not for just a moment, for arising and so on do not manifest for just a moment.[350] [386]

For the Sautrāntikas, however, it is not the case that conditioned phenomena arise in the first moment, then endure in the intervening period, and then disintegrate through contact with the causes of disintegration. For the Sautrāntikas, the very instant the first moment of a given phenomenon came into being, it did so as something that does not remain still even for a single moment. And given that the first moment has the nature of something that does not remain static even for a single moment, the fact of it not remaining for a second moment takes place. Therefore they maintain that the specific character of the first moment not remaining even at its own time was created by the very same productive cause that produced the phenomenon in the first place.

So too, they view the characteristics of conditioned phenomena, such as arising and so on, to be the actions of arising, of endurance, and of disintegration of phenomena such as form and so on. To take a sprout as an example, *arising* is the new arising of a sprout that has never existed before, *enduring* is maintaining the continuum of its prior moment, and *disintegration* or *decaying* is the difference in characteristics between the earlier and later moments. The time when the sprout is at the point of arising is posited as the time of the sprout's cause, the time when the sprout arises is posited as the time of the first moment of the sprout, the time of the sprout's enduring is posited as the time of the three moments of a sprout—earlier, later, and middle—and the time when the sprout disintegrates is posited as the time of the first moment of the result of the sprout. The Sautrāntikas view these three characteristics to be simultaneous and not different in substance from conditioned phenomena. For example, *Blaze of Reasoning* states:

> For example, a Sautrāntika says: Arising is not arising just once but arising in a continuum derived from the power [387] of specifically defined causes and conditions. Enduring is enduring within the stream of earlier moments. Decaying is the factor of dissimilarity with earlier moments and disintegration is complete separation.[351]

So too, insofar as something is a conditioned phenomenon it exists in the nature of disintegration from the very instant of its coming into being. Dharmakīrti states in *Ascertainment of Valid Cognition:*

> There is no such thing called "impermanence only" that will come into being subsequently. The very fact that functional things endure only for a moment, this alone is what is meant by impermanence. This has been explained numerous times.[352]

Thus the Sautrāntikas are different from the Vaibhāṣikas in asserting that even a thing that does not remain for longer than the duration of a shortest moment possesses all three characteristics. *Explanation of the Treasury of Knowledge*, for example, says:

> Others say: Some measure of endurance exists even with regard to the final moment of a sound or a tongue of flame and so on. So too, since their transformation relies on earlier moments, hence they are posited as having the three characteristics.[353]

"Others" must be taken as referring to the Sautrāntikas. [388]

Just like the Sautrāntikas, the Cittamatrins and Mādhyamikas too view the three characteristics of conditioned phenomena—arising, enduring, and disintegrating—as actions and also as existing simultaneously. For example, *Compendium of Bases* states:

> The impermanence of conditioned phenomena will be understood by means of the three characteristics of conditioned phenomena as transformation through arising, enduring, and disintegrating. Also these three characteristics should be understood in dependence on the two continua of conditioned entities. As such there is (1) a continuum that is the continuation of one life to another and (2) a continuum that is the continuation of one moment to another.
>
> Regarding the first continuum, taking birth as a type of sentient being is *arising*, and dying at the end is *disintegration*. The phase of youth and so on that occurs between the first and last phase is change through enduring. It is *enduring* since it endures for the measure of this life however long it lasts. Since there exist specific later phases there is *change*.
>
> Then regarding the momentary nature of the second continuum, the newest arising of conditioned entities is *arising*. The moment of arising that does not abide beyond that is *disintegration*. Enduring in just the moment of arising is *enduring*.
>
> Transformation has two types: (1) transformation into the same entity and (2) transformation into distinct entities. Transformation into the same entity is the disintegration of the con-

> ditioned entity within the same continuum. Transformation into distinct entities [389] is the disintegration into different continua. But since a transformed body does not exist separately from its enduring state, these two merge as one and are designated a single characteristic.[354]

In brief, impermanence is defined as "that which is momentary." "Momentariness" in the context of the definition of "impermanence" refers to not enduring for a second moment beyond the time of its occurrence. Not remaining for a brief period is not the meaning of "momentariness." Thus for this moment there is a range of different lengths of duration. In the case of a year, for example, there are twelve months from the time of its occurrence, and when twelve months have passed the year no longer remains. Similarly, when not enduring for a second moment beyond its time of occurrence is applied to a day, the time of its occurrence lasts for the duration of twenty-four hours for a day and a night. Applied to a phenomenon that lasts only for the period of a minute, its duration becomes that of a minute.

One might ask: Does a century, for example, exist for one hundred human years or not? If it does not so exist, this would contradict the assertion that a year exists until the completion of twelve months. If, on the other hand, a century does exist until a hundred years are complete, it would contradict the assertion that it is impermanent in that it does not exist for more than one moment.

Response: This objection stems from the fault of not knowing how to posit a continuum, as stated in *Four Hundred Stanzas*: [390]

> If you view a continuum erroneously,
> you could be saying it is permanent.[355]

Thus Āryadeva states that if you view a continuum erroneously, that is, if you fail to understand that it is composed of parts, there is a danger of mistaking it to be permanent. Further, if illustrated by a year, the continuum possesses twelve months as its parts, and since the parts and the whole are the same entity, a year that is the whole disintegrates and so on owing to the first month that is part of that year disintegrating. All things are similar. However, the distinction must be made that when one month that is a part of that year disintegrates, the year that is the whole disintegrates but it is not yet destroyed, and though a year does not remain it is not without remainder.

In general there are two types of impermanence: impermanence in terms of a continuum and momentary impermanence. Prajñāvarman's *Exposition of the Collection of Aphorisms* states:

> Impermanence has two types: impermanence in terms of a continuum and momentary impermanence.[356]

Of these two, the first is called *coarse impermanence* and the second is referred to as *subtle impermanence*. To illustrate these two types of impermanence by taking Devadatta as an example, Devadatta not remaining after death is extremely coarse impermanence, since even a cowherd can ascertain this fact with sense perception. Compared to this, the fact of Devadatta not remaining from the time of his first moment to the second moment is subtler. Compared to this, the fact of Devadatta disintegrating even at the time of his first moment is even more subtle. Therefore, in order to cognize subtle impermanence, it must be preceded in general by comprehending the fact of coarse impermanence. [391]

Furthermore, if there were to be no subtle impermanence in the sense of moment-by-moment disintegration, one would not know how to posit impermanence in terms of a continuum. This fact of subtle impermanence in the sense of moment-by-moment disintegration has not been brought about by some other adventitious condition nor some subsequent factor. Rather, it is the productive cause of the entity itself that generates its momentary disintegration.

As for impermanence in terms of a continuum, this is explained in *Explication of the Five Aggregates*:

> A continuum commences from the birth of the five aggregates up until death, and that unbroken continuity of momentary characteristics is called *a continuum*. Disintegration refers to later moments not succeeding earlier moments owing to the momentary continuum of the five aggregates being severed. The arising of other moments of dissimilar type is called *disintegration*, and just that is designated *impermanence*.[357]

So a person who reaches the age of ninety is old and infirm owing to age, but he or she did not become old and infirm suddenly. First while abiding in the womb, then at birth, then in childhood, and so on up until becoming an elder, the person progressively changed until in the end he or she became old and infirm. Asaṅga's *Yogācāra Ground* states:

> What are the eight times? The time of abiding in the womb, the time of birth, the time of childhood, the time of youth, the time of a young adult, the time of an adult, the time of an elder, the time of frailty. [392] The time of abiding in the womb is the embryonic stage of arbuda and so on. The time of birth is the time beyond that up to and including the time of infirmity. The time of childhood is becoming mobile up to being able to engage in play. The time of youth is being able to engage in play. The time of a young adult is being able to engage in sex, up to and including the age of thirty. The time of an adult is up to and including the age of fifty. The time of an elder is up to and including the age of seventy. The time of infirmity occurs after that.[358]

In general when impermanence pertains to a coarse basis, such as the final moment of the flame of a butter lamp and something ascertained by direct perception, it constitutes coarse impermanence. In contrast, when it pertains to a subtle basis and needs to be ascertained through reasoning, it is subtle impermanence. Impermanence is analogous to what is meant by *not remaining for a second instant beyond the time of its establishment*. For when impermanence is applied to a day, it refers to the duration of thirty hours.[359] If, on the other hand, impermanence is applied to the briefest unit of time, it would endure for the period of a shortest moment. Therefore not remaining a second instant beyond the time it comes into being has both subtle and coarse dimensions.

The fact that all conditioned phenomena were created in the nature of disintegration by the very productive causes that brought them into being in the first place and that they do not remain in the second moment beyond the time they come into being are stated in *Sūtra on Gaganavarṇa's Patient Training*:

> These phenomena [393] arise in the first moment, disintegrate in just that moment, and do not exist in the second moment.[360]

Similarly, in *Exposition of Valid Cognition* it states:

> Because it has no other cause,
> disintegration is related to the entity itself.[361]

Here Dharmakīrti states that the disintegration of a product exists from the very moment of its existence and is related intrinsically to its very nature. The disintegration of a product is generated not by some other third factor but by the very cause that gave rise to its existence. As such the very cause that produced it also generates its disintegration.

To present this in a syllogism: Take a product; it is characterized by disintegration from the first moment of its existence because it arose from its own causes with the essential nature of disintegration without relying on some alternate cause of disintegration.

Again, *Exposition of Valid Cognition* states:

> If something were to disintegrate because of
> another cause of disintegration being generated . . .[362]

The opponent's position is that the disintegration of something, such as a vase, is not automatically established through the mere fact of the vase arising from its productive cause, for the vase disintegrates due to some other factor that causes its subsequent disintegration. To demonstrate that such a view is untenable, the same text then states:

> If it itself establishes the state of disintegration,
> how could a subsequent cause facilitate disintegration? [394]
> If it does not facilitate that state,
> how could it act on its dependent effect?
> How could there ever be an entity
> that lacked dependence? It does not exist.[363]

Thus Dharmakīrti states that since all conditioned phenomena are characterized by disintegration by the very productive causes that produced them, the subsequent factor that the opponent asserts to be the cause of disintegration neither facilitates the disintegration of form and so on nor harms it in any manner. And if this subsequent factor that is supposedly the cause of disintegration does facilitate the disintegration of form and so on, it is then illogical to maintain that this subsequent cause of disintegration and the disintegration of form and so on are related as the dependent effect and its dependent cause.

This logical establishment of the momentariness of conditioned phenomena is accorded great importance in Buddhist texts in general and epistemological texts in particular. A key point is the establishment of how disintegration does not depend on a subsequent separate factor or cause. We have presented such logical arguments only briefly here as an introduction. Those who desire to understand them in detail should consult the first chapter of Dharmakīrti's *Exposition of Valid Cognition*, such as the verses cited above, "Because it has no other cause, it disintegrates . . . ," and their related commentaries. *Ascertainment of Valid Cognition* states:

> Again, because it establishes production and impermanence [395] of the characteristics of aggregates, elements, and bases, there is no error.[364]

There is an extensive presentation after the above opening passage that addresses the following topics: (1) identification of what a conditioned phenomenon is, (2) logically establishing it to be impermanent, (3) rebutting objections to the view that disintegration does not require a subsequent factor as a cause, (4) how even so, disintegration does not become causeless, and (5) thus how the conclusion is derived that disintegration is an essential nature of functional things.

There are works that serve as supporting material for these two texts by Dharmakīrti, such as chapter 13 of Śāntarakṣita's *Compendium on Reality* and its commentary by Kamalaśīla, and Dharmottara's *Proof of Momentariness* and its commentary by Muktikalaśa. In them, Buddhist logicians present with refinement and in great detail the essential and subtle points of their reasoning to prove the momentary nature of conditioned phenomena. [396]

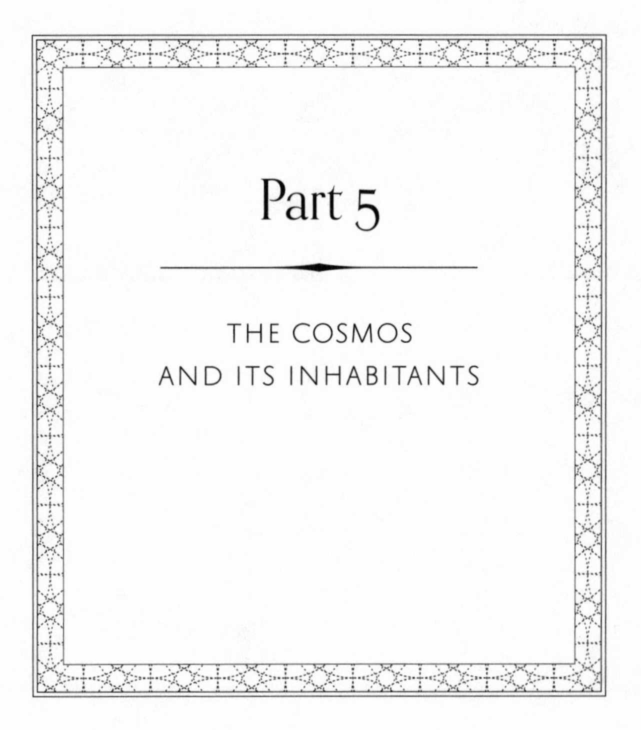

# Part 5

## THE COSMOS AND ITS INHABITANTS

## The Buddhist Rejection of Creationism

Part 4 of our volume deals with what in contemporary scientific terms we might call cosmology and evolutionary biology, namely the origin and formation of the world as well as the emergence of life and sentient creatures. In part 2, we discussed how, as in science, the law of causality is a fundamental explanatory principle in Buddhist understanding of reality and how the Buddha himself is credited with the statement "When this exists, that exists; from the arising of this, that arises." This is the famed principle of dependent origination (*pratīyasamutpāda*), which came to be embraced in the Buddhist world as offering an explanation not just of suffering and happiness but of the entire spectrum of reality, from the origin of the cosmos to the workings of each individual's mind. At the heart of this principle is the premise that nothing exists outside the complex network of causes and effects—neither a transcendent being nor a force—that is somehow responsible for nature. It is this principle that is enshrined in the slogan "this conditionality" or *idampratyaya*, sometimes rendered in English as "mere conditionedness." Archeological excavations of ancient Buddhist sites reveal that one of the most widely inscribed texts on the entrance and beams of Buddhist monasteries is the following verse on causality:

> All things arise from their causes,
> and the Tathāgata taught what those causes are.
> Also that which puts an end to these causes—
> this too was taught by the Great Monk.

A clear and systematic application of this principle to the question of the origin of the universe is found in Asaṅga's *Compendium of Knowledge*. Asaṅga describes three conditions that must be fulfilled for a satisfying explanation of how the universe came to be. First, the explanation cannot be based on the design of some creator, since such divine agency goes against the basic laws of cause and effect. Second, the cause must itself be impermanent, since nothing that produces effects is permanent. And third, the cause has to be something similar to the effect—not anything can arise from just anything.

In the opening section of chapter 2 of Dharmakīrti's *Exposition of Valid Cognition*, we find a remarkable refutation of a proof of theism that strikingly resembles Thomas Aquinas's well-known argument. In presenting the object of his critique, Dharmakīrti describes the proof of theism in this way:

> A creator must exist because of intermittence,
> particular shape, and efficiency. (2.10ab)

This argument for a creator is threefold: it says that there must exist prior intelligence because (1) we observe temporal regularity, how events unfold in a temporally ordered manner in nature, (2) things possess shapes or are configured in specific ways, and (3) things are efficient in performing their functions. Like the Indian proof, Aquinas's argument from design, too, proceeds from empirical observation of orderedness and the premise that the universe cannot be explained by pure chance but only on the basis of some kind of design or purpose, and design is a function of intelligence. Nature then must be a product of a grand designer, namely God.

Interestingly, in the Indian context, we do not see arguments similar to the so-called ontological argument attributed to Saint Anselmo (eleventh century), which asserts that because we can think of that which is perfect, this "that than which nothing greater can be thought," God, must exist. This is what is called in philosophical jargon *a priori* reasoning, or reasoning from deduction. In contrast, the argument from design is a form of *a posteriori* reasoning, an argument that proceeds from observation or experience.

Dharmakīrti refutes the above argument from design by questioning the key premise, the necessary existence of an eternal intelligence. He also points out that the argument is self-defeating because the question still remains as to who created that intelligence. If it is claimed that this intelligence is what Aquinas called a *necessary being*, which is permanent, Dharmakīrti points out the incompatibility of a permanent subject with impermanent objects of perception. Theism, in Dharmakīrti's view, undermines the very premise that it relies on—causality—for it is forced to admit that the designer or intelligence can neither have a cause nor be a cause. Those familiar with David Hume's *Dialogues Concerning Natural Religion* will recognize similar logical moves in its critique of the argument from design.

In brief, all Buddhist schools are united in their adherence to the motto "this conditionality" and seek to account for the origination of everything purely on the basis of the operation of causes and conditions. No force is outside the bounds of causality.

## The Formation and Structure of Our World

Buddhist sources converge on the basic intuition that there is no discernable beginning of the universe. The beginningless universe is, in a way, a logical consequence of the rejection of theism, for the only other option available would be origination *ex nihilo*, something arising from nothing, a standpoint anathema to all Buddhists given their commitment to the law of causality. The picture we obtain of cosmology from the Buddhist sources is a never-ending cycle of universes coming into being and ending through destruction. As explained in our volume, the texts speak of in fact four phases in this cycle—formation, abiding, destruction, and emptiness. When a new universe comes into being, it is said to begin with the emergence of the wind element, followed by the water element, the fire element, and finally the earth element. To the question what sets in motion this process for the formation of a new world, the Abhidharma texts point to the power of the karma of the sentient beings that will come to inhabit that world (see page 299). This said, as in science, the Abhidharma texts also understand that the emergence of sentient beings comes long after the formation of the external world. When the time comes for the end of a world, it is said to be destroyed mostly by fire, as the sun's power increases manifold. The Abhidharma sources speak in fact of a specific pattern in this destruction process—after every seven times destruction by fire, there will be one destruction by water, and after every seven destructions by water, there will be one destruction by wind (see pages 314–15).

One of the unresolved tensions in the above picture is the question of how the first elements of the new universe come into being. As observed in part 3, the Buddhists' Abhidharma rejects the Vaiśeṣika claim that atoms of the four elements, being permament, remain undestroyed at the end of a universe. This means that, for the Buddhists, when a particular universe comes to an end, all of its matter is destroyed, including even the atoms. Does this mean the Buddhist Abhidharma thinkers are forced to admit that origination of a new universe entails origination *ex nihilo*? I am not aware of any Abhidharma sources that explicitly address this tension. In the Kālacakra texts, an important Indian Vajrayāna tradition with extensive focus on cosmology and astronomy, there is the idea of "space particles," a kind of empty matter that is thought to remain when, in the aftermath of the destruction of a universe, there is a long period of emptiness. This suggests that when one universe comes to an end, through elemental destruction, all the matter collapses into these empty particles, a kind of nothingness but not entirely nothing, and it is from this the entire process begins, first with the emergence of wind elements, culminating in the emergence of an entire macroscopic world.

On the shape or structure of the universe, too, there seems to be some difference between the Abhidharma picture and the one we see described in the Kālacakra sources. At the heart of the Abhidharma model is Mount Meru, a massive mountain whose base extends downward into the depths of a cosmic ocean and whose tips touch the abodes of the celestial beings. It is said to be surrounded by seven concentric circles of mountains and oceans, with four continents in each of the four cardinal directions, each flanked by two smaller continents. Although Kālacakra too accepts this Meru-centric model, it differs from Abhidharma texts on the shape of Mount Meru as well as on the motion of the celestial bodies, such as the sun and moon. The technical details of the shapes and sizes, as well as the calculations pertaining to the motion of celestial bodies, according to these two Buddhist sources remain beyond the scope of this essay. Remarkably, in the fifth century the Indian mathematician Āryabhaṭa proposed an alternative cosmology and astronomy that accurately conceptualized the earth rotating on its axis on a daily basis, with apparent movement of the sun, moon, and stars being a function of the rotation of the earth (see page 323). Āryabhaṭa is thought to have been active at the ancient Nālandā University, availing himself of the astronomical observatory that existed at Nālandā. Although Āryabhaṭa's terse verse text the *Āryabhaṭayam* or *Āryabhaṭīya* survives in Sanskrit, as well as Bhāskara's sixth-century commentary, Āryabhaṭa's influence on scientific thinking seemed to have been felt more in the Arab world than in its homeland of India.

There is no denying that many of the details of the Abhidharma and Kālacakra picture of the size and distances between the celestial bodies are at odds with findings of contemporary science. That said, the Kālacakra system of astronomy does possess powerful predictive tools that accurately predict solar and lunar eclipses, explain the changing duration of daylight across seasons, and calculate the dates for the spring and fall equinoxes. My conjecture is that, unlike that of the Abhidharma, the Kālacakra astronomy was influenced and shaped in important ways by Āryabhaṭa's views, albeit by a circuitous route. That is, Āryabhaṭa's views influenced mathematics and astronomy in the Arab world, and the Kālacakra system was in turn influenced by the Arab system. There are frequent references in the Kālacakra sources to a rival astronomy, which the textual evidence strongly indicates to be the medieval Islamic astronomy.

Historically, both Abhidharma and the Kālacakra tantra remained influential in the Buddhist world. This is true particularly in the case of Tibet, leading some thinkers to even attempt reconciliation of these two different systems. Interestingly, the Tibetan tradition seemed to have settled on a middle way, whereby it takes the Abhidharma sources, especially the *Treasury of Knowledge*, as the primer on cosmology while relying on the Kālacakra system for astronomy and the lunar and solar calendars.

## The Emergence of Sentient Life

In both the Abhidharma and Kālacakra sources, the view of the origin of life we find can be described as that of a fall or descent. The first humans are described as possessing godlike characteristics, with bodies made of light and with no need for sustenance through food and so on. This comes from the pan-Indian view of historical evolution, or rather devolution, where humans pass through four *yugas*, or eons. First is the Satya Yuga, a golden age when everything was perfect and humans had godlike attributes. There was no crime or need for food, and human lifespans ran into thousands of years. In the second and third eras, the Treta Yuga and the Dvāpara Yuga, humans become progressively less god-like and shorter lived, until the final era, the Kali Yuga, the age of degeneration, when the human lifespan is reduced to around hundred years. This era, our own, is characterized primarily by discord and strife.

One might ask where these humans of the golden age came from in the first place. Here, we can recall another important aspect of Buddhist cosmology, the theory of the three spheres of existence—the desire realm, form realm, and formless realm. The last two realms belong to the gods alone, while the desire realm, or sense realm, is populated by beings of all six types of rebirth: humans, animals, hell beings, spirits, gods, and demigods. So when a new physical universe comes into being, it's the fall of beings in celestial realms that leads to the emergence of new life.

So for the Buddhists, the presence of life elsewhere, the object of so much fascination for modern science, goes without question. Given its massive cosmological scope, with multiple world systems, the presence of life elsewhere is taken for granted. This suggests an interesting paradox: while the physical world progresses from the primitive, namely from the aggregation of atoms and molecules to realization of its full potential in the emergence of the macro world, sentient life, on the other hand, degenerates over time. The Dalai Lama, in his *Universe in a Single Atom*, notes an important difference between Buddhist thought and contemporary science with respect to the emergence of life. Whereas science attempts to explain the emergence of living organisms from inanimate matter, Buddhists are more interested in understanding the emergence of sentient beings from what is essentially a non-sentient basis (Dalai Lama 2005, 111).

## Multiple Universes and the Complex Numerical Systems

One thing clear in the Buddhist sources is their embrace of the notion of many worlds. While some world systems are beginning to form, other world systems are entering into destruction. In speaking of these multiple world systems, Buddhist texts, especially the Mahāyāna sūtras, present mind-boggling numbers culminating in the "incalculable" or "unspeakable." It conjures the images one sees when one takes time-lapse photos from the Hubble telescope—every tiny segment of the sky reveals countless star systems. The *Flower Ornament Sūtra*, for example, speaks of these multiple universes in terms similar to the idea of the multiverse in modern string theory, which postulates the existence of multiple possible universes besides our own. Some texts even speak of the Buddha performing exactly the same twelve deeds—birth, enlightenment, giving the first public sermon, and so on—at exactly the same time in multiple world systems. It is, perhaps, due to this cosmological view of multiple universes that Buddhist sources, and the classical Indian tradition in general, uses such an elaborate numerical system. Western languages repeat the same word—"thousand," "million," and so on—for three place values each. In contrast, the Abhidharma system as presented in *Treasury of Knowledge*, as well as the Kālacakra system, uses a separate term for every place value all the way up to sixty places, with only a few instances where the same term is repeated as the "smaller *x*" and "larger *x*."

# Further Reading in English

For a contemporary exposition of Dharmakīrti's critique of Indian arguments for theism, see Roger Jackson, "Dharmakīrti's Refutation of Theism," *Philosophy East and West*, 36:4 (1986), 315–48.

For a pictorial description of the structure of the world with Mount Meru at its center as envisioned in Vasubandhu's *Treasury of Knowledge*, see Gelong Lodrö Sangpo, trans., *Abhidharmakośa-Bhāṣya by Vasubandhu* (Delhi: Motilal Banarsidass, 2012); for a depiction of the Kālacakra conception of Mount Meru, see Martin Brauen, *Mandala: The Sacred Circle in Tibetan Buddhism* (Boston: Shambhala, 1995), and appendix 9 in Khedrup Norsang Gyatso, *Ornament of Stainless Light: An Exposition of the Kālacakra Tantra*, translated by Gavin Kilty (Boston: Wisdom Publications, 2004).

For a detailed presentation of the Kālacakra system of cosmology and astronomy as understood by the Tibetan tradition, see Khedrup Norsang Gyatso, *Ornament of Stainless Light*, 80–144 and appendix 10.

For the Dalai Lama's views on the Abhidharma cosmology and the Kālacakra astromony and their convergence and divergence with contemporary scientific views, see chapter 4 of his *Universe in a Single Atom* (New York: Morgan Road Books, 2005).

For engaging dialogues between the Dalai Lama and the American astrophysicists Piet Hut and George Greenstein, see Arthur Zajonc, ed., *The New Physics and Cosmology: Dialogues with the Dalai Lama* (New York: Oxford University Press, 2004), 163–94.

For calculation of the mind-boggling numbers presented in the Mahāyāna sūtras, see *The Flower Scripture: A Translation of the Avatamsaka Sutra*, translated by Thomas Cleary (Boston: Shambhala Publications, 1993), 889–91.

# 21

## The Cosmos and Its Inhabitants in Abhidharma

GENERALLY SPEAKING, divergent views evolved among early Indian philosophers, who disagreed on the way external world systems and their inhabitants arose and were destroyed. Among them many non-Buddhist schools, such as Vaiśeṣika, asserted that before this present world system was formed, many subtle particles such as those of earth, water, fire, and wind existed separately without disintegration in empty vacant space. Then the god Maheśvara reflected, "I should create the cosmos within this empty space," and through the power of his intention and the complete maturation of the virtuous and nonvirtuous karma of sentient beings, two subtle particles of wind combined, then three subtle particles of wind, and in due course such wind particles created the resultant form of a great body of wind. Then water particles clustered together above it, forming a great aggregate [398] of water, then a great earth maṇḍala formed above it, and a great fire maṇḍala formed on top of that. Thus external world systems and their inhabitants came to arise in a graduated manner. The non-Buddhist Sāṃkhya school explains the emergence of these world systems on the basis of a "primal substance" and as the manifestations of one ultimate reality. There are also non-Buddhist schools that assert Brahmā, Viṣṇu, Śabdabrahmā, or a permanent self and so on to be the creator of the world. The Lokāyata (or Cārvāka) school speaks of the existence of this world in terms of a spontaneous emergence from material substance.

Among them, those who assert a creator of the cosmos propose the following logical proof for the existence of a creator:

*Subject*: The dwellings, bodies, and possessions of this world.
*Predicate*: They are preceded by the mind of a creator.

*Reason*: Because they operate in a temporally ordered manner, like a carpenter's adze; they possess specific shapes, like a clay vase; and they are capable of function, like a vase.

Although other types of logical proof exist, these constitute the principal arguments.

If the above reasons proposed by non-Buddhist schools merely establish in general the fact that dwellings, bodies, and possessions of the world arise owing to the existence [399] of a prior mind, then this point is already established among Buddhists, for Buddhists already accept that the world arises from karma, which is the mental factor of intention. So these arguments cannot constitute proofs that would establish something that has not yet been proven, that is, the creator, for Buddhists. If these proofs are supposed to demonstrate that things were created by a permanent mind, since the existence of such a permanent mind is impossible, the examples cited become untenable. Alternatively, if the claim is that these proofs establish the simple fact that a creator must precede the existence of these entities, one would then have to admit that the very creator that is said to precede all things must also have another prior creator. If this is admitted, it would contradict the assertion that the creator is permanent and self-arisen, and consequently the very reasons one has proposed would become inconclusive.[365] *Exposition of Valid Cognition* states:

> Things are temporally ordered, have particular shapes,
> and are capable of function, and so on (so a creator exists).
> Either the assertion is already proven, or the examples do
> not work,
> or the statement leads to doubt.[366]

Also, if a creator of the cosmos that is permanent and self-arisen is asserted, since there can be no change in the nature of something that is permanent, there could be no differentiation at all between its phases of producing or not producing effects. This implies that such an entity would be devoid of the characteristics of a cause since it would lack the following basic criterion: "When *it* is absent an effect cannot arise." Moreover, since such a creator of the cosmos is itself unarisen it is illogical that the entire cosmos, the external world system and its inhabitants, is created by such a being.

[400] In brief, given that things originate from the assembly of diverse causes and conditions, Buddhists demonstrate the untenability of things being the creation of a single creator. *Exposition of Valid Cognition* states:

> Without a change in its own nature
> it is not plausible as an agent of change.
> Because a permanent phenomenon is devoid of change,
> its capacity to act is hard to comprehend.[367]

Bhāvaviveka also states in *Heart of the Middle Way*:

> Because the self is unarisen,
> we do not accept that it produces effects.[368]

Its autocommentary, *Blaze of Reasoning*, states:

> The passage "There is no single creator anywhere, therefore my argument is not inconclusive" explains that the assertion of "a single creator" is untenable because all things originate from the assembly of causes and conditions. For if a single entity had the potent force of being a creator it would be pointless to investigate the multiplicity of causes and conditions. In that no single creator that is the cause of everything in the world can be observed anywhere, how could my reasoning be inconclusive?[369] [401]

Therefore Buddhists do not accept this cosmos of external worlds and their inhabitants to be established from the prior design of a creator, or established from a primal nature,[370] or causelessly, or through spontaneous origination and so on. Rather, they are established as "merely this conditionality," that is to say, they are established solely from their causes and conditions.[371]

There are numerous texts within the Indian Buddhist sources that contain presentations on the emergence and formation of the cosmos, such as the *Flower Ornament Sūtra*, the *Sūtra on Mindfulness*, *Presentation on the World*, the texts of the upper and lower Abhidharma systems, and the corpus of the Kālacakra tantra and so on. However, the principal sources are the Abhidharma texts and the Kālacakra tantras. In these two sources, although minor differences exist in their presentations on the formation and destruction of external world systems and their inhabitants, on the length of the lifespans of sentient beings, and on the shapes and sizes of external world systems, they are not incompatible in their basic outlook.

In brief, according to the viewpoints of Buddhist schools, common shared world systems come into being from subtle particles of earth, water, fire, and wind, which are the basis for the initial formation of the universe, together with the collective karma of sentient beings. And on the basis of the specific karma of individual beings their experiences of happiness and suffering also come to arise. [402] So Buddhists explain the evolution of the cosmos on the basis of the cycle of dependent origination alone, in terms of substantial causality and cooperative conditionality. With regard to our current cosmos, Buddhists explain that the external world was formed first and that sentient beings, the inhabitants, evolved later.

## The Formation of World Systems

With respect to this process of causal dependent origination, Asaṅga and others speak of "the three conditions": (1) the absence of prior design, (2) impermanence, and (3) potentiality. Asaṅga explains these conditions clearly on the basis of sūtras such as the *Rice Seedling Sūtra*. *Compendium of Abhidharma* states:

> What are the features of the conditions from which things arise? They arise from the condition of the absence of prior design. They arise from the condition of impermanence. And they arise from the condition of potentiality.[372]

Similarly, in the sūtra cited in Vasubandhu's *Explanation on the First Dependent Origination and Its Divisions* it states:

> Monks, the characteristics of dependent origination are threefold: the characteristic of arising from the condition of the absence of prior design, the characteristic of arising from the condition of impermanence, and the characteristic of arising from the condition of potentiality.[373]

Passages such as these provide an important summary of how this world system came into being through the power of the three conditions, and it is declared that Buddhists do not accept the view that this universe was created by an omnipotent being. [403] Texts of Asaṅga and Vasubandhu organize the meaning of the sūtras, such as the following from the *Rice Seedling Sūtra*, into three parts and explain their meaning in terms of these three parts. *Rice Seedling Sūtra* states:

> Dependent origination is like this: when this exists, that comes to be; with the arising of this, that arises; with ignorance as condition, formations come to be.[374]

"When this exists, that comes to be" presents the arising of things from the condition of the absence of prior design. "With the arising of this, that arises" presents arising from the condition of impermanence. "With ignorance as condition, formations come to be" presents the condition of potentiality.

1. *Condition of the absence of prior design.* The meaning of the first statement, that external and internal things originated from the absence of prior design, is just this: this cosmos, composed of external world systems and their inhabitants, did not come into being through the prior design of a creator, but rather through the power of their own productive causes and conditions. Hence the sūtra statement: "When this exists, that comes to be." Things come into being through the dependent origination of the causal process, not through the power of some alternate external agency not subsumed within the domain of causes and conditions. So too this world system and its inhabitants did not come into being spontaneously, devoid of causes and conditions.

2. *Condition of impermanence.* The meaning of the second statement, that things originated from the condition of impermanence, is this: Merely to assert that external and internal things originated from causes and conditions is inadequate, for the very causes from which they originated must themselves be conditioned impermanent phenomena. [404] It is illogical that effects that are impermanent originate from a permanent phenomenon, and this is the significance of the statement on the second condition. Since permanent phenomena are devoid of effective function, either sequentially or simultaneously, they are incapable of producing any

effects. Similarly, it is an established logical truth that if something cannot be impacted, helped, or harmed by any other phenomena, it too will have no capacity to help or harm any other thing. Therefore, so long as something is a cause from which external or internal phenomena arise, it must necessarily be a conditioned impermanent phenomenon. This necessity is indicated by the second condition. Hence in this specific context of the *Rice Seedling Sūtra* and the statement "With this arising, that arises," the meaning of "arising" is brought out with special focus.

3. *Condition of potentiality.* The meaning of the third statement, that external and internal things originated from the condition of potentiality, is this: It is not adequate for a cause, from which things arise, to be merely a conditioned impermanent entity, it must be commensurate with the effects it produces. Such a cause must have the potential to produce a specific effect. To underline this point, the phrase "the condition of potentiality" is applied. Therefore the sūtra states: "With ignorance as its condition, formation comes to be, . . ."

As a conclusion to how dependent origination is characterized by the three conditions, Asaṅga's *Compendium of Abhidharma* [405] states:

> What then is the meaning of dependent origination? The fact of there being no creator is the fact of dependent origination, the fact of arising from causes, the fact of no being, the fact of dependence on others, the fact of the absence of prior design, the fact of impermanence, the fact of momentariness, the fact of the unceasing continuity of cause and effect, the fact of cause and effect being commensurate, the fact of the diversity of causes and conditions, the fact of causes and effects being specifically determined, these are the facts of dependent arising.[375]

In brief, in the Buddhist system it is unacceptable that something arises without a cause, or that the cosmos is established from the prior design of a creator, or that a result arises from a permanent cause and so on. Instead Buddhists explain the unique path of causal dependent origination where the cause and result must be of similar type and diverse results arise from diverse causes and conditions.

As for how Buddhist Abhidharma texts explain the formation of world systems, a clear presentation exists in Vasubandhu's *Treasury of Knowledge Autocommentary*:

> Owing to the power of the karma of sentient beings, the wind maṇḍala that is supported in space comes into being. Its height [406] is 600,000,000,000,000 leagues (yojana), in breadth it is immeasurable, and it is hard such that even manifold great vajras could not destroy it. Above that, water forms to a height of 11,020,000 yojana. The term *maṇḍala* should be applied here to water as well. Owing to the karma of sentient beings, clouds gather in that wind maṇḍala and a continuous īṣādhāra rain[376] falls, forming a water maṇḍala 11,020,000 yojana in depth.
>
> "Why does that water not overflow?" one might ask. Some say this is due to the power of the karma of sentient beings. For example, it is like food and drink, for that which is consumed does not descend into the intestines when it is not yet digested. Others assert that water is held by wind just as a rice sack holds rice.
>
> Again, when the winds derived from the force of the karma of sentient beings completely churn the water, the upper level turns to gold in the manner of cream forming on boiled milk. From then on the water maṇḍala is 800,000 in depth, the remainder has turned to gold.
>
> Question: What is the remainder?
>
> Response: The upper layer of the water maṇḍala measuring 320,000 yojana becomes the foundation with the nature of gold. The height of the water and gold has already been explained.[377]

Thus first the wind maṇḍala is formed, then the water maṇḍala and the earth maṇḍala emerge in sequence. [407] Initially the lower foundation exists as empty vacant space, then a smooth rising wind that acts as a portent of the formation of world systems condenses over many years to form the lower foundation of the external environment. The name of this wind is *smooth one*, its shape is spherical, its color is variegated, and its essential nature is the eight basic substances such as the earth subtle particle, but since wind is the dominant element it is called the *wind maṇḍala*.

Above the wind maṇḍala clouds form in space and from them rain descends continuously for many years until the water maṇḍala forms. Then owing to the water maṇḍala being agitated by wind for a long time the earth maṇḍala progressively develops. As before, rain descends continuously on the water and earth maṇḍalas, and in dependence on that the great outer ocean forms. Then Mount Meru forms in dependence on the supreme elements of that great ocean being agitated by wind, the seven mountains such as Yugaṃdhara form from the intermediate elements being churned by wind, and the four continents and the subcontinents form through minor elements being churned by that wind.

Thus it is said that Mount Meru comes to be formed at the center, the four great continents in the four directions, and the eight subcontinents in the four subdirections with the sun, moon, and stars circling them. The great continent to the south is the continent of Jambudvīpa, our home. The texts explain that the great ocean fills the space between the continents, and all of this exists [408] supported or suspended in empty space by the wind maṇḍala.

One thousand arrays of the four continents, the sun, moon, Mount Meru, and so on is designated "one thousand world systems of a chiliocosm." One thousand such chiliocosms are called "one million world systems of a dichiliocosm." One thousand such dichiliocosms are called "one billion world systems of a trichiliocosm." In brief the one billion world systems, that is, from the four continents together with Mount Meru up to the Brahma realm, are called "one billion world systems of a trichiliocosm." That these world systems come to be ultimately destroyed together as well as formed together in the beginning is stated in *Treasury of Knowledge*:

> One thousand arrays of the four continents, the sun, moon,
> Meru, the desire-god abodes,
> and the mundane heaven of Brahma
> are asserted to be a chiliocosm.
>
> One thousand of those is the intermediate
> world system of a dichiliocosm.
> One thousand of those is a trichiliocosm.
> They cease and arise together.[378]

*Extensive Display Sūtra*, however, states that the one billion world systems from the four continents up to Akaniṣṭha are the one billion world systems of a trichiliocosm:

> In this world system of the four continents and so on [409] there are one billion world systems of four continents, one billion oceans, one billion surrounding regions and great surrounding regions, one billion heavens of Cāturmahārājakāyika (Four Great Kings), one billion heavens of Trāyastriṃśa (Thirty-Three Gods), one billion heavens of Yāma (Free of Conflict), one billion heavens of Tuṣita (Heaven of Joy), one billion heavens of Nirmāṇarati (Enjoying Emanations), one billion heavens of Paranirmitavaśavartin (Controlling Others' Emanations), one billion Brahmakāyikas (Brahma Lineage), one billion Brahmapurohita (Brahma Priests), one billion Mahābrahma (Great Brahma), one billion Parīttābha (Small Radiance), one billion Apramāṇābha (Immeasurable Radiance), one billion Ābhāsvara (Clear Radiance), one billion Parīttaśubha (Small Bliss), one billion Apramāṇaśubha (Immeasurable Bliss), one billion Śubhakṛtsna (Extensive Bliss), one billion Anabhraka (Free of Clouds), one billion Puṇyaprasava (Born of Merit), one billion Bṛhatphala (Great Result), one billion Avṛha (Lesser Measure), one billion Atapa (Free of Distress), one billion Sudṛśa (Sublime Appearance), one billion Sudarśana (Supreme Vision), and one billion Akaniṣṭha (Beneath None). This is called "one billion world systems of a trichiliocosm" that is broad and extensive.[379] [410]

In general Buddhists and others speak of the universe in terms of the three realms: the *desire realm*, the *form realm*, and the *formless realm*. These three are characterized, in their respective order, as principally dependent on external objects of sensual desire such as form, sound, and so on; on the internal bliss of absorption; and, being disenchanted even by the bliss of absorption, on tranquility permeated by the feeling of equanimity alone. Thus the cosmos is explained in terms of three distinct realms. If these three realms are explained further by means of their body, feelings, and resources, the desire realm is characterized as follows: The body is coarse, experience is predominantly a mixture of pleasure and pain, and beings depend mainly on coarse food. The form realm is characterized thus: beings there have bodies in the nature of light, their experience is permeated mostly by feelings of bliss, and they do not rely on coarse food. The formless realm is characterized as follows: beings there do not possess a coarse physical body that symbolizes their nature as sentient beings, they abide with feelings of equanimity transcending feelings of pain and pleasure, and they do not depend on coarse food but live their entire life in meditative concentration focused solely on objects such as limitless space, which resemble the cessation of all mental engagement in other objects. It is stated that beings who live in the higher levels of the desire realm and beings who live in both the form realm and formless realm are celestial beings. [411]

Further, in *Treasury* and its *Autocommentary*, presentations are found on the size of the sun, moon, and earth-sphere, the distance to the sun and moon, how night and day arises from the sun revolving about Mount Meru, as well as explanations of many phenomena that are directly observable, such as the length of the days in summer and winter and so on. Combining these Abhidharma explanations with the science of astronomy presented in the Kālacakra tantra, classical Buddhist scholars in both India and Tibet were able to produce calendars and almanacs as well as make accurate predictions about solar and lunar eclipses and so on.

An account similar to one presented in *Treasury of Knowledge*, how first the external world of Jambudvīpa (our southern continent) was formed and how then the inhabitant sentient beings emerge gradually, occurs also in the scriptures of the Vinaya (discipline) canon. *Detailed Explanations of Discipline*, for example, states:

> Monks, there was a time when this world first formed. After it formed, beings from the heaven of clear light died owing to the exhaustion of their lifespan, karma, and merit, and they arrived [412] in this world in accordance with their fortune to be human. Their physical form arose from the mind, lacking nothing and complete in faculties. They were in possession of all primary and secondary limbs, with pure features, a fine complexion, and innate radiance. Further, they were capable of flight, joyful, sustained by the food of joy, endowed with long life, and capable of living for a long time.

> At that time the sun and the moon had not formed, and even the stars had not yet appeared in this world system. Night and day did not exist, moments, minutes, and hours did not exist, there were no females, there were no males, and sentient beings were identified simply as *beings*. Then one voracious being took earth nectar on his fingertips and experienced its taste, and the more he tasted it the more he desired, and the more he desired the more he ate through treating it as coarse food. Other beings observed him experience the taste of earth nectar on his fingertips, and that the more he tasted the more he desired, and that the more he desired the more he ate through treating it as coarse food. After observing this, these beings took earth nectar on their fingertips and experienced its taste, and the more they tasted the more they desired, and the more they desired the more they ate through [413] treating it also as coarse food. Because such beings ate the earth nectar, treating it as coarse food, their bodies became solid and heavy, and their radiance disappeared, and darkness descended on the world.
>
> Monks, when darkness arose in the world, the sun and moon naturally arose, as well as the stars, night and day, moments, minutes, hours, months, half months, seasons, years.[380]

The sūtra goes on to state how the complexion of those who ate more food deteriorated, whereas the complexion of those who ate less remained clear, thus different complexions arose due to eating different amounts. Those with fine complexions developed strong pride and chided others by saying, "My complexion is excellent, but yours is bad," and owing to engaging in such nonvirtuous acts, the earth nectar deteriorated. After the earth nectar declined an earth essence arose with excellent color, taste, and aroma. Then after it declined, groves of sprouts arose with excellent color, taste, and aroma. Then owing to the sins of beings the groves of sprouts also declined. [414]

After that, uncultivated sālu rice grain grew with excellent color, excellent aroma, and excellent taste. When harvested in the morning it grew back in the evening, when harvested in the evening it grew back in the morning. Thus it was harvested again and again, harvesting when it grew back until it disappeared. The sālu rice grain that had not been cultivated by sentient beings was consumed regularly, and in turn beings produced feces and urine within their bodies, male and female organs developed, they engaged in sexual union, distinct moral and immoral conduct arose, and families began to emerge. *Detailed Explanations of Discipline* states:

> Through eating and possessing that food they had long life and lived long. Those who ate less food possessed good complexions, those who ate more food possessed poor complexions. Thus in dependence on those two foods, two types of complexion manifested. When these two complexions arose some sentient beings deprecated other sentient beings, stating: "Listen! I have an excellent complexion, but you have a bad complexion." When they developed manifest pride owing to their complexions, the earth nectar disappeared, caused by the arising of those immoral acts.[381] [415]

And:

> Monks, when the earth nectar of those sentient beings disappeared an earth essence arose with excellent color, excellent aroma, and excellent taste. Its color was just like the flower of a radish. Its taste was just like the unboiled honey of the bee. When they ate it and they had that food, they had long life and they lived long. Thus two types of complexion arose in dependence on how much of that food was eaten. When two types of complexion arose some sentient beings chided other sentient beings, stating: "Listen! I have an excellent complexion, but you have a bad complexion." When they developed manifest pride owing to their complexions, the earth essence disappeared, caused by the arising of those immoral acts.[382]

And:

> Monks, when the earth essence of those sentient beings disappeared groves of sprouts arose with excellent color, excellent aroma, and excellent taste. The color was just like the flower of a kadambuka.[383] The taste was just like the unboiled honey

> [416] of the bee. When they ate it and they had that food, they had long life and they lived long. Thus two types of complexion arose in dependence on how much of that food was eaten. When two types of complexion arose some sentient beings denigrated other sentient beings, stating: "Listen! I have an excellent complexion, but you have a bad complexion." When they developed manifest pride owing to their complexions, the groves of sprouts disappeared, caused by the arising of those immoral acts.[384]

*Detailed Explanations of Discipline* states more extensively:

> Monks, when the groves of sprouts of those beings disappeared, uncultivated sālu rice grain grew, without husks, without chaff, pure and clean, four inches long, and untangled. When harvested in the evening it flourished in the morning and grew back again. When harvested in the morning it flourished in the evening and grew back again. The entire harvest would grow back again, but then the harvest disappeared. When they ate it and they had that food, they had long life and they lived long. Because these beings treated the uncultivated sālu rice grain as coarse food, [417] therefore different sexual organs evolved. Some possessed male organs and some possessed female organs. Then some who possessed male organs and some who possessed female organs looked at each other, and in dependence on how much they looked at each other sexual desire arose, and through their desire they separated into couples, and having separated they engaged in immoral acts.
>
> When other sentient beings saw some sentient beings engaged in immoral acts with other sentient beings, they cast dirt, clods of earth, pebbles, gravel, and sand. They shouted, "You vile beings engaging in immoral deeds! You vile beings engaging in immoral deeds! O, why do you aggravate other sentient beings?" This is analogous to today's custom where when someone takes a new bride people sprinkle flour, as well as scents, flowers, and puffed rice. They then declare, "May you be happy." Similarly, when these and those sentient beings witnessed other beings engage in immoral acts with other sentient

> beings, they cast dirt, clods of earth, pebbles, gravel, and sand. They shouted, "You vile beings engaging in misdeeds! You vile beings [418] engaging in misdeeds. O, why do you aggravate other sentient beings?"
>
> Monks, thus what was previously regarded as nondharma is these days regarded as dharma, what was previously regarded as nondiscipline is these days regarded as discipline, and what was previously disparaged is these days praised. But these came to be abandoned just for one day, or abandoned for just two or three days or even seven days. Because these beings engaging in sinful conduct were treated with such intense anger, in order to engage in such acts these people constructed walls and said, "It's here we will engage in such inappropriate acts, here we will engage in such inappropriate acts," and the name "home" came to be. Monks, this was the first act of all the acts of householders that arose in this world.[385]

A similar presentation is found in *Presentation on the World*.[386] [419]

To summarize, Buddhist texts are in broad agreement with the general outlook of Indian schools on cosmology at that time. For example, explanations are found that when humans first became inhabitants in this world there emerged in sequence the era of perfect precepts (*kṛitayuga*), the era of keeping three precepts (*tretāyuga*), the era of keeping two precepts (*dvāparayuga*), then the era of conflict (*kaliyuga*). The era of conflict refers to the present time when humans have a lifespan of one hundred years. It is also called the time of the spread of "the five degenerations,"[387] and it is counted as an inferior age. The era of perfect precepts is when humans naturally practice Dharma and do not engage in any immoral acts. The era when the first immoral act arises in the world is the era of keeping three precepts and no longer the era of keeping all precepts. When two types of nonvirtuous action arise it is the era of keeping two precepts. When every type of nonvirtuous immoral act arises it is called the era of conflict. On these explanations there is general agreement among all Indian schools. [420]

# 22

# The Development of the Cosmos in Kālacakra Texts

THE FORMATION OF the world systems according to the explanations found in the Kālacakra texts is as follows. Prior to the development of our world system an empty vacuous state prevailed when all material phenomena existed as empty particles. Then after a certain period, wind particles started to accumulate together and the world began to form. Moreover, after the earlier cosmos was destroyed subtle particles called "empty particles" that existed as discrete material units in vacant space accumulated and the wind maṇḍala gradually formed. Then from among the subtle particles called empty particles, first the subtle particles of wind that resided invisibly in space and had not combined came to coalesce with one another. Previously, before coalescing they lacked the capacity to move even though they possessed the nature of lightness and mobility. But after coalescing they were able to move owing to their lightness and mobility, and they are referred to as "the wind that came to form the cosmos." This is said to be the first cause of the wind maṇḍala, which is the formative basis of the cosmos but not the actual wind maṇḍala. [421]

Then the subtle particles of wind and fire that previously existed separately gradually combined, and owing to making contact a fire-like electrical state associated with wind energy appeared called "fire that first created the fire maṇḍala." Then the subtle particles of fire and water accumulated and fully combined, and owing to making contact a rainlike state possessing wind and fire called "water that first created the water maṇḍala" appeared. Then the subtle particles of water and earth progressively combined, and owing to this contact a rainbow-like state appeared in space called "the earth that first created the earth maṇḍala."

Through the progressive increase of the four particles such as earth, water, fire, and wind, the circular maṇḍalas of earth, water, wind, and fire that constitute the lower foundation were established. Then the structure of Mount Meru came to be formed, round in shape, narrow at the base, and broad at the summit.[388] Then the seven mountains, seven seas, and seven continents surrounding Meru, and the planets and stars above, came to be formed in progressive stages. Also, it is said that the earth that was first translucent like a rainbow became progressively coarser, the external world systems developed, and the subtle particles of taste or the subtle particles of the element of space pervaded everywhere. Puṇḍarīka's great *Stainless Light* commentary on Kālacakra, for example, states:

> From among the subtle particles, first particles of wind coalesced with one another, and through their convergence, that which moves with lightness and mobility was called "wind." Likewise subtle particles of fire [422] coalesced, and the state merging lightning with wind was called "fire." So too subtle particles of water converged and coalesced, and this rain endowed with wind and fire was called "water." Similarly, the subtle particles of earth fully coalesced and the appearance of Indra's bow (rainbow) in space was called "the sustainer" (*dhāraṇī*). The subtle particles of taste however pervade everywhere.[389]

Taking this world system of a chiliocosm that is our abode as the center, then in each of the ten directions—east, south, west, and north, the four intermediate directions, and above and below—there exist up to one thousand world systems of a chiliocosm each with four continents. Taking these one thousand world systems as the basis, then up to one thousand times one thousand world systems constitute the one million world systems of a dichiliocosm. Further, taking this as the basis, then up to one thousand times one thousand times one thousand such world systems constitute the one billion world systems of a trichiliocosm. Then taking this as the basis, up to a great countless number of world systems constitute the world systems of a great trichiliocosm. "Great countless" in general is not taken as merely an immeasurable number but as the final measure to sixty decimal places counting from digits, tens, hundreds, thousands, ten thousands . . . and so on.[390] Further, the great *Stainless Light* commentary states: [423]

> "One thousand" refers to one thousand world systems above and below from the central world system of a chiliocosm. And just as they exist above and below, so too they exist in the east, west, north, and south, and the subdirections: void of truth, powerful, wind, and fire. So too for the dichiliocosm and trichiliocosm. That called "great trichiliocosm" refers to great countless world systems.[391]

## The Differences between the Abhidharma and Kālacakra Presentations

The shape of Meru is square in Abhidharma but circular in the Kālacakra system. The two sources differ also in the color of the four sides of Mount Meru. In the Abhidharma system Mount Meru is silver in the east, lapis lazuli in the south, red crystal in the west, and gold in the north, and since those colors pervade the local space, even the color of space adjacent to each of the four sides appears with those colors. In the Kālacakra system Mount Meru is black in the east, red in the south, white in the north, and yellow in the west. In the Abhidharma system Mount Meru rises 80,000 yojana above the water level and extends 80,000 yojana below the water. In the Kālacakra tantra it rises 100,000 yojana above the water. And 100,000 yojana explained in Kālacakra equal 200,000 yojana in Abhidharma.[392]

Similarly, in the Abhidharma system it is said that there are seven golden mountains encircling Mount Meru [424] and seven oceans that extend outward, and between them there are no continents. In the Kālacakra tantra Meru is surrounded by six continents, six oceans, and six mountains. Outside that are the seventh continent great Jambudvīpa, and the seventh ocean known as the external ocean or the salty ocean, and the seventh mountain described as the surrounding mountain, and so on. Thus there are differences in explanation in these two sources.[393]

## Multiple World Systems in Buddhist Sources

Further, in Buddhist texts it is said that not only does this world system that is our abode exist, but there are also limitless world systems in the ten directions. *Flower Ornament Sūtra* states:

> However far the limit of space extends,
> so too is the limit of all sentient beings.[394]

In the *Vajrapāṇi Initiation Tantra* too, there is the following:

> Children of good family, there definitely exists what is called a "Flower Bank Array of World Systems" where many myriads of world systems have been revealed. And in each of these myriads of world systems that have been revealed, you should know there exists many myriads of interconnected world systems, each with their own specific configurations.
>
> Children of good family, the measure of such interconnected configurations [425] is the billion world systems of a trichiliocosm that fully encompass one billion world systems of four continents, one billion Mount Merus, one billion great oceans, and one billion suns.
>
> That which fully encompasses one billion billion world systems of a trichiliocosm is that which is fully encompassed within the interconnected world systems of a Flower Bank Array.
>
> That which fully encompasses one billion interconnected world systems of a Flower Bank Array is one region of interconnected world systems of a Flower Bank Array.
>
> That which fully encompasses one billion regions of interconnected world systems of a Flower Bank Array is one intermediate region of interconnected world systems of a Flower Bank Array.
>
> That which fully encompasses one billion intermediate regions of interconnected world systems of a Flower Bank Array is one constellation of interconnected world systems of a Flower Bank Array.[395]

Up to one billion world systems from the four continents up to and including the Brahma heavens is [426] one billion world systems of a trichiliocosm. One billion such world systems is one *cluster* of myriads of a Flower Bank Array ocean of worlds. One billion such world systems is one *region* of world systems of myriads of a Flower Bank Array ocean of worlds. One billion such world systems is one *intermediate region* of world systems of myriads of a Flower Bank Array ocean of worlds. One billion such world systems is one *constellation* of world systems of a Flower Bank Array ocean of worlds.

Again, the scriptures speak of the world systems having different appearances and shapes, some possessing light and others devoid of light and so on. For example, the *Flower Ornament Sūtra* states:

> O children of the victors, the shapes of these oceans of world systems are varied. Of these oceans of myriads of world systems, some are spherical, some are square, some are spread across various sectors of space, some are spiral in shape, some have the shape of mountains of light, some [427] have the shape of a mass of various trees, some have the shape of different flowers, some have the shape of different houses, some have the shape of different sentient beings, some have the shape of various signs. Such world systems are fully revealed by their shape and fully manifest their subtlety through the particles of these oceans of world systems.[396]

And:

> In some fields there is no light,
> and they are dark and permeated with fear.
> When touched there is pain like a razor,
> and when seen there is unbearable dread.
> In some fields there is divine light,
> and in some the palaces are radiant,
> some possess the light of the sun and moon,
> the array of fields is inconceivable.
> Some fields possess their own light,
> and in some the trees possess stainless radiance.
> In some no suffering is seen
> and sentient beings arise from virtue.
> In some the mountains radiate light,
> and in some fields jewels emit radiance.
> Some have lamps that bestow light
> arising from the karma of sentient beings. [428]

Some fields possess the light of buddhas
and they are completely filled with bodhisattvas.
Some fields possess the light of lotuses
that emit a radiance of intense beauty.
Some fields possess the light of flowers,
also some are endowed with the light of fragrant water,
or that of perfume or burning incense,
that is purified by the power of prayer.[397]

In general world systems have no beginning or end, but specific world systems are accepted as having set, temporal stages of a beginning, middle, and end. Further, the world system in which we live has four successive temporal stages called the *eon of voidness*, the *eon of formation*, the *eon of endurance*, and finally the *eon of destruction*. Each stage lasts an extremely long period of time—twenty intermediate eons. This world system came into being from the five elements: the element of space and the basic elements of earth, water, fire, and wind. The element of space brings into existence the other elements, and space also facilitates their functions. As for the cause of destruction, the texts explain that, excluding the earth element and the space element, the other three—fire, water, and wind—act as agents of destruction. [429]

# 23

# How Worlds End

THERE ARE DIFFERENT explanations in the Buddhist texts on how these world systems come to cease. For example, *Sūtra on the Meeting of the Father and Son* states:

> The time when this world system is destroyed does exists. When this world system is about to be destroyed two suns will appear in the world, and owing to the appearance of two suns, springs and small streams will dry up. Then three suns will appear in the world and owing to the appearance of three suns, large springs and great rivers will dry up. Then four suns will appear in the world, and owing to the appearance of four suns the four great rivers and the great lake Anavatapta ("that which never heats up") into which the rivers flow will dry up and disappear owing to evaporation. Then five suns will appear in the world, and owing to the appearance of five suns the water of the great oceans will decline and disappear to a depth of one yojana.[398]

In Asaṅga's *Yogācāra Ground* there is the following: [430]

> When not even a single sentient being remains, their material possessions too will disappear, and when the possessions cease to exist there will be no longer any rainfall. Since rain would stop falling on this great earth, all the grass, the medicinal plants, and the trees will dry up. Drought will arise and the sun disc will thoroughly scorch the entire earth.
>
> Since there are six things that will be incinerated owing to the power of the destructive karma of sentient beings, six more

suns will appear. These suns will blaze, each with four times the force of the normal sun. Then by increasing sevenfold, they will each blaze with seven times their normal force.

What are those six things that will be incinerated? The small pools and great pools will evaporate with the rise of the second sun. Small streams and great streams will evaporate with the third sun. Lake Anavatapta will evaporate with the rising of the fourth sun. The great ocean will evaporate with the rising of the fifth and the sixth suns. Because Mount Meru and the great basis are extremely stable, they will only be incinerated by the rising of the sixth or seventh suns. Then wind fans the flames of this fire, which spreads up to and including the Brahma heavens and incinerates these realms. [431]

Physical things will undergo three types of change: Things derived from water such as grass and so on are desiccated by the first sun itself. Things composed of water evaporate owing to the other five suns. Physical entities that are stable and hard are desiccated by two suns.

As for how the enduring world will come to burn and will be incinerated, I shall elaborate on the explanations found in the sūtra itself. It will be incinerated without smoke or without even a trace like ash left behind. In this way, as the external environment is destroyed, the world is then said to have ceased. This lasts for twenty intermediate eons, so the destruction of the world lasts twenty intermediate eons.

What is destruction by water? Initially there are seven cycles of destruction by fire. Then the water element emerges from the second meditative absorption, and this water element dissolves external world systems just as salt dissolves in water, then the water element itself disappears along with those external world systems. Such destruction also takes twenty intermediate eons.

What is destruction by wind? Initially there are seven cycles of destruction by water, then there is a single destruction by fire. Then the wind element emerges from the third meditative absorption, and external world systems disappear like a body withering because of the wind. Then the wind element disap-

> pears along with those external world systems, just as for some in whom the wind [432] element is disturbed the flesh shrinks to the bone. This is the destruction of world systems, and such disintegration also takes twenty intermediate eons.[399]

According to the texts of the lower Abhidharma school, when this world system in which we reside is destroyed, the destruction takes place through fire and the sun progressively increasing in temperature. Thus in general when the external world system is destroyed it is necessarily destroyed by one of the three elements: fire, water, or wind. Furthermore, it is said that the world system is destroyed by fire seven times in succession, and then it will be destroyed by water seven times. After destruction by water seven times, the world system will be destroyed once again by fire seven times. Then it will be destroyed by wind. For example, the *Great Treatise on Differentiation* states:

> Some say when the world system is destroyed the sun will split into seven suns, and through the force of this the world will be destroyed. Some say that at the time of the final eon this single sun itself will increase in temperature sevenfold and incinerate the world system.[400]

In conformity with this view, the *Treasury of Knowledge* states:

> It is destroyed by fire seven times, then by water.
> Thus after destruction by water seven times, [433]
> it is destroyed by fire seven times.
> Then it is destroyed by wind at the end of that.[401]

Thus *Treasury of Knowledge* explains that after the world system is destroyed by fire fifty-six times, it will be destroyed by water seven times, making a total of sixty-three rounds of destruction. Then it will be destroyed by wind once. Asaṅga's *Facts of the Grounds*, on the other hand, states that after destruction by fire and water fifty-seven times, it will be destroyed by wind.

As for the explanation in the Kālacakra texts on how the world systems come to be destroyed, this is as follows. External world systems are first generated by empty particles of space, so when they are destroyed, coarser form progressively transforms into subtler form until everything exists naturally as subtle particles of empty space. The stages of dispersal are said to correspond with the order of earth, water, fire, and wind. First, external world systems containing earth particles—that coexist through mutually cohering—separate and penetrate water, then water particles separate and penetrate fire, then fire particles separate and penetrate wind, and then wind particles separate, and like particles prior to the formation of world systems, they function in space as discrete atoms beyond the range of the sense faculties. Further, the great *Stainless Light* commentary states:

> At the time of the destruction of this world system, "from earth" means [434] "water from earth" and so on, where those subtle particles of the elements that were fully conjoined disengage from the composite of subtle particles of earth and enter the composite of subtle particles of water owing to subtle particles of earth separating.
>
> So too, those disengaging from the water element enter the composite of fire, and those disengaging from fire enter the composite of wind, and those disengaging from wind fully function in space. Thus the world system fully contracts, and some deities called *fire of time* who exist beneath the ground incinerate the world and turn it to ash.[402]

In brief, there are statements in the Buddhist texts indicating there are more world systems than grains of sand in the River Ganges, and these too constantly undergo processes of formation and destruction, such that when some worlds are being formed, at the same time, other worlds are in the process of emptying. Also when the worlds undergo disintegration they are destroyed by the three elements of fire, water, and wind. Earth, water, fire, and wind themselves serially dissolve until finally they exist as empty space particles. [435] Āryadeva's *Four Hundred Stanzas* states:

> Just as one sees the end of a seed,
> and it has no starting point.[403]

Similarly, Bhāvaviveka's *Lamp of Wisdom* says:

> As it shall be explained below in relation to the lines "With regard to all things as well, there exists no starting point,"[404] vases and so on are states that arise in a linear stream of causation from one to the next from beginningless time. They do not exist as possessing a starting point.[405]

As such, all things—as represented by the seed of a sprout or a vase and so on—arise in a stream of causation progressing from one cause to the next without any starting point, and one cannot identify any initial cause. So too world systems arise within a stream of formation and destruction solely owing to the power of the wheel of dependent origination as a causal process without any starting point. Moreover, now and in future they will continue to arise in just such a stream of formation and destruction without limit. On this basis, Buddhist scholars assert as a general principle that one cannot posit that the formation and destruction of world systems began at a specific time nor will they end at some appointed moment. [436]

# 24

# Motion of the Celestial Bodies

## The "Space Particle" and the Movements of Celestial Bodies in Kālacakra Tantra

In the texts of Kālacakra it is stated that after the destruction of earlier external world systems empty space remains. The fundamental particles, which have not converged to form composite particles but remain isolated and dispersed, unobserved by the senses of ordinary beings, are referred to as "empty particles." Five such empty particles are identified—the subtle particles of earth, water, fire, wind, and space. The texts speak also of "six empty particles," referring to the most subtle aspects of six elements, these being the empty particles of earth, water, fire, wind, and wisdom. In the phrase "empty particle of pristine wisdom," the word "particle" refers to the ultimate nature of reality (*dharmatā*), thus it is not an actual subtle particle. The reason for designating it thus is to make known that every instance of formation and destruction of world systems is viable since every particle of the five elements by nature lacks inherent existence, and formation and destruction are not viable if such particles possess inherent existence. Puṇḍarīka's great *Stainless Light* commentary on Kālacakra states:

> Therefore, when it is said that "formation arises from empty particles according to the time of arising," empty particles in worldly convention are not objects of experience [437] of the eye sense faculty and so on but exist in their essential nature as subtle particles.

And:

> Voidness is the sixth quality that pervades the entire expanse of reality.[406]

In the system of Kālacakra, "space particles" are said to be the basis of other extremely subtle physical particles, and there is no difference between such subtle particles of space and space particles. Space that is termed the *subtle particles of space* is not the empty vacuity of nothingness, instead it is posited as the form of open space that is one of the five types of mental object form. Further it is free from obstructive contact and posited here as the whitish vacuity of empty space that appears to mental consciousness alone. The form of open space that can be differentiated by mind and the smallest discrete unit of form that appears to mind are designated *subtle space particles*. These space particles are also referred to by another name, "subtle taste particles." In brief, this earth [Jambudvīpa] was formed from five elements: the element of space as its basis and the four elements of earth, water, fire, and wind. Given that it is the space element that gives rise to the other elements and facilitates their specific functions, it is the basis of the formation and destruction of the four elements. [438]

Furthermore, this world has four phases: (1) formation, (2) endurance, (3) destruction, and (4) voidness until the emergence of a new world. It is said that in the void phase, space particles act as the basis for the emergence of all material phenomena that will exist in the newly formed world system, and during this phase of voidness it is these space particles that pervade every locus where the particles of earth, water, fire, or wind would come to exist. The great *Stainless Light* commentary states:

> The subtle taste particles pervade everywhere.[407]

As explained, after the destruction of the former world system, what are referred to as *empty particles* that reside as discrete subtle particles in empty space will assemble together and the wind maṇḍala will gradually come to emerge. Through the emergence and progressive expansion of the elements of earth, water, fire, and wind, the circular shaped maṇḍalas of wind, fire, water, and earth of the lower basis will come to be established. Then, stage by stage, the mountains and the continents will come to be formed. After that the zodiac band will come to be established, which consists of pathways supported by the wind in empty space above the globe of the earth. This band forms like an umbrella high in the center and lower at the extremes, and above that form the twenty-eight constellations that rotate anticlockwise. Then each of the moving bodies that comprise the ten planets arise: the group of seven planets, plus the planets Rāhu, Kālāgni, and Ketu.[408]

## The Twelve Astrological Houses

The twelve astrological houses are composed in the following manner. Since the two constellations of Śravaṇa and Abhijit comprise no more than one zodiac house, they are calculated as [439] one single zodiac house. Thus each sector of the lunar mansions comprising the twenty-seven constellations is divided into four parts, and each part is designated a "foot" (*pāda*). Nine such feet comprise two constellations, and the name of each zodiac house is applied to each separate part of the four feet of a single constellation. So the twelve astrological houses are: Aries, Taurus, Gemini, Cancer, Leo, Virgo, Libra, Scorpio, Sagittarius, Capricorn, Aquarius, and Pisces. Further, the house of Aries includes both the constellations of Āśvinī and Bharaṇī and the first foot of Kṛttikā. The house of Taurus includes three feet of Kṛttikā, Rohiṇī, and two feet of Mṛgaśīrṣa. The house of Gemini includes the latter two feet of Mṛgaśīrṣa, Ārdrā, and three feet of Punarvasū. The house of Cancer includes the last foot of Punarvasū and both Puṣya and Āśleṣā. The house of Leo includes both Maghā and Pūrvaphālgunī and the first foot of Uttaraphālgunī. The house of Virgo includes the latter three feet of Uttaraphālgunī, Hasta, and the first two feet of Citrā. The house of Libra includes the latter two feet of Citrā, Svāti, and the first three feet of Viśākhā. The house of Scorpio includes the latter feet of Viśākhā, Anurādhā, and Jyeṣṭha. The house of Sagittarius includes both Mūla and Pūrvāṣāḍhā, and the first foot of Uttarāṣāḍhā. The house of Capricorn includes the latter three feet of Uttarāṣāḍhā, Śravaṇa and Abhijit taken as one, and the first two feet of Dhaniṣṭhā. The house of Aquarius includes the latter two feet of Dhaniṣṭhā, Śatabhiṣā and the first three feet of Pūrvabhādrapadā. The house of Pisces includes [440] the latter two feet of Pūrvabhādrapadā, Uttarabhādrapadā, and Revatī.[409] Thus the great *Stainless Light* commentary states:

> Āśvina, Āṣāḍha, and the first foot of Kṛttikā is Aries. So too each of the nine feet should be known as the twelve houses, such as Aries and so on.[410]

From the perspective of their natural motion, the constellations of the zodiac and all the planets, except for Rāhu and Kālāgni, revolve counterclockwise. The zodiac band and the twelve great terrestrial divisions revolve clockwise, and the sun and all the planets and stars residing within the zodiac houses cycle daily in a clockwise direction through the twelve terrestrial divisions owing to the movement of the *gola* wind. Such motion is called *wind motion*.

The natural motion of the planets, except for Rāhu and Kālāgni, is to progress in a counterclockwise fashion through the lunar mansions. Moreover, due to the zodiac band and the planets moving in this way, the fact of the sun and moon being obscured by Rāhu during the full moon and new moon, as well as the appearance of Ketu, are unmistakably identified in the texts of Kālacakra. Furthermore, the fact of the individual progression of the planets to the south or north and the lack of any similar progression of the zodiac constellations to the south or north, is said to occur unfailingly due to the motion of the wind related to their specific path previously reached, at a precise point in the terrestrial divisions (*bhūmikhaṇḍa*). Also, the great *Stainless Light* commentary states: [441]

> Therefore the rotational force belongs to the houses, not to the sun that resides in the zodiac houses. As such the zodiac band revolves clockwise, while the planets move counterclockwise through the zodiac houses. Just as the zodiac band revolves, so too Rāhu appears in the west and revolves clockwise in its normal movement around Meru. The planets appear in the east and revolve counterclockwise around Meru in their normal movement through the zodiac band.[411]

Thus the sun is the main indicator of the divisions of time and so on. And when the sun resides in a specific house in any section of the twelve terrestrial divisions owing to its motion, it is said that it moves to the south or north and the length of the night and day is either the same or not. In that way the great *Stainless Light* commentary states:

> Here when the sun resides in any house such as Aries and so on, the division of the twelve zodiac houses into six seasons, twelve months, and twenty-four fortnights is due to the sun moving through the zodiac band.[412] [442]

## The Movements of Celestial Bodies in Āryabhaṭa's Work

The fifth-century scholar of astronomy at Nālandā known as Āryabhaṭa developed a presentation different from texts such as those of the Abhidharma and the Kālacakra systems.[413] He explained that at the center of a spherical constellation of stars exists space in which stars and planets revolve in circular paths, and at the center of that exists the earth, which is spherical in shape and characterized by four primary elements. For example, in *Āryabhaṭayam*, his text on astronomy, he writes:

> At the center of space is a spherical asterism [443]
> in which planets revolve in circular paths.
> At its center is the earth completely spherical,
> characterized by earth, water, fire, and wind.[414]

Again, for example, just as after boarding a ship and it moves forward the mountains that endure immovably appear to move in the opposite direction, so too by the power of the earth sphere itself circling from west to east, the stars that are stable appear as if they move from east to west from the brahma line in space. Āryabhaṭa also explains that in reality the stars that are stable do not move:

> Just as after boarding a ship that moves forward
> one sees the mountains moving backward,
> so too the stars that are stable appear to move
> to the west from the laṅka-line.[415]

Question: If the stars do not move, why do they appear to rise and set?

Response:

> The cause for their rising and setting
> is that the earth basis itself along with the planets
> circle the laṅka-brahma line [444]
> in dependence on the winds that are their constant
> mount.[416]

The stars rise and set because of the power of the earth-basis and the planets circling the brahma-line at the center, which acts as the navel, and progressing to both the southern and northern extremes in dependence on the wind.

Further, as for how the solar eclipse and lunar eclipse occur:

> They obscure, for the moon obscures the sun,
> also the great shadow of the earth basis obscures the moon.[417]

During the new moon the moon obscures the sun, and during the full moon the great shadow of the earth basis obscures the moon. Thus with respect to important questions of astronomy, such as the motion of celestial bodies, the difference between night and day and their length, solar and lunar eclipses, and so on, Āryabhaṭa addresses these explicitly from the standpoint of astronomy and in ways that resonate with the views of astronomy in modern science. [445]

# 25

## Measurement and Enumeration

WHEN DETERMINING the dimensions of external world systems, as well as the duration of the phases of formation, endurance, destruction, and voidness of external world systems and their inhabitants and so forth, in the Abhidharma texts as well as in the treatises of the Kālacakra system, extensive presentation on various measurements is provided. In brief, these measurements consist of two kinds: the measurement relevant to material entities and the measurement of time.

In accordance with two types of measurement, as discussed in an earlier section of this volume, Vasubandhu states in his *Treasury of Knowledge* as well as its *Autocommentary* that the smallest unit of matter is the subtle particle, and through their accumulation the measurements *krośa* and *yojana* are established. Similarly, the smallest unit of words is the letter, and through the accumulation of words, stanzas, chapters, and treatises are established. Likewise, the briefest unit of time is the shortest moment, and through the accumulation of moments, days, months, years, eons, and so on come to be established. [446]

### UNITS OF PHYSICAL MEASUREMENT

When explaining the measurements of material entities, according to *Treasury of Knowledge* the accumulation of seven *subtle particles* constitutes the measure of one *atom*, seven atoms are one *iron particle*, seven iron particles are one *water particle*, seven water particles are one *rabbit particle*, seven rabbit particles are one *sheep particle*, seven sheep particles are one *ox particle*, seven ox particles are *one light particle*, seven light particles are one *louse egg*, seven lice eggs are equivalent to *one louse* derived from that, seven lice are *one barley* seed, seven barley seeds are one *finger joint*, twelve finger joints are one *finger span*,[418] twenty-four finger joints are one *cubit*, four cubits are one *fathom*, five hundred fathoms are one krośa, eight krośa are one yojana. Vasubandhu's *Autocommentary* states:

> These subtle particles and so on should be recognized as a sevenfold progression: seven subtle particles are the measure of one atom, seven atoms are one iron particle, seven of those are one water particle, seven of those are one rabbit particle, seven of those are one sheep particle, seven of those are one ox particle, seven of those are one light particle, seven of those are one louse egg, seven louse eggs are the measure of that derived from it, and that derived from a louse egg is called a louse. Seven lice are one barley seed, seven barley seeds are one finger joint. [447] Three finger joints are a finger, but since this is well known it will not be elaborated. Obvious dimensions include twenty-four finger joints that are one cubit, where four cubits are one bow, which means one armspan. Five hundred armspans are one krośa, where a monastery is asserted to be one krośa from a town. Eight of those are one yojana. Eight krośa are called a yojana. Therefore the measure of a yojana has been explained.[419]

As for how units of measurement are also found in sources such as *Extensive Display Sūtra*, the Vinaya scripture *Vinayavibhaṅga*, and *Presentation on the World* and so on, this was discussed briefly in the earlier context of the presentation on atoms. *Condensed Kālacakra Tantra* states:

> Eight subtle particles are a minute particle, eight minute particles constitute the subtle tip of a single hair. Eight of those are a black mustard seed, eight of those a louse, and eight of those a barley seed. Barley seeds multiplied by a reptile constitute a finger. A finger breadth multiplied by a pair of suns make a cubit, and four cubits are a bow. Here two thousand bow lengths are a krośa, four krośas are a yojana, and that is the measure used for the divine realms, the earth and space.[420] [448]

Here we find an explanation that differs from that of the Abhidharma. In the above excerpt, the word "reptile" refers to the number eight and the term "sun" is a synonym for twelve, hence the phrase "a pair of suns" refers to twice twelve and should be understood as twenty-four.

## Units of Temporal Measurement

The measure of time will be discussed in terms of two measures: the year and the eon.

Now the duration composed of joining 120 shortest moments, as defined earlier, is one *second*, and sixty of those seconds equal one *minute*, thirty such minutes equal one *hour*, thirty such hours equal one *day*, thirty days make one *month*, and twelve months equal one *year*, and so on. The *Great Treatise on Differentiation* states:

> Why is it called an eon? How is an eon defined? It is called an eon because it differentiates temporal increments. Thus through differentiating time by seconds, minutes, and hours, days and nights arise. Through differentiating days and nights, fortnights, months, seasons, and years come to be formed. Through differentiating time as fortnights and so on, eons arise. [449] Eons are the ultimate differentiation of temporal increments, therefore it receives the generic name.[421]

In conformity with that view, Vasubandhu's *Treasury of Knowledge* states:

> One hundred and twenty such moments
> are a second. Sixty
> of those are one minute.
> The three periods of hours, days and nights, and months
> are each progressively thirty times the former.
> Twelve months with six
> omitted days make one year.[422]

In terms of a broad calculation, according to the Vinaya and Abhidharma systems a year consists of twelve months, which are made of six lesser months in which one day is deducted from every thirty-day month plus six full months that contain a full thirty days each.

Omitted days or subtracted days occur because lunar days cannot act as the exact equivalent for solar days. Thus since the lunar day is less than the solar day by half an hour, and when just fifty-nine solar days have passed, sixty lunar days have passed, one day is omitted every two months. Thus *Autocommentary* states:

> There are four months of winter, four months of spring, and four months of summer, and these twelve months, together with the omitted days, [450] make one year. In one year, there are six omissions of solar days. Why is that? "When one and a half months of winter, spring, and summer have passed, scholars discard the omitted day in the half months that remain."[423]

In Kālacakra's system of calculation, a *zodiac day* is the duration of the sun's transit through 1 degree of the 360 degrees of a complete rotation of the sun during its natural clockwise motion through the zodiac sphere, or its transit through 1 degree of the 30 degrees for each zodiac house. A *solar day* is from the time of seeing the lines of the hand in the early morning in a specific locality when the earth rotates owing to the counterclockwise motion of the sun's wind until one sees the lines of one's hand the next day. A *lunar day* is one waxing or waning part of either the fifteen waxing parts or the fifteen waning parts of the duration of the moon's transit through one rotation of the earth during the moon's natural counterclockwise motion.

Furthermore, to measure a year and so forth from another perspective, one *breath* is one exhalation and inhalation of a healthy person, six breaths are one *minute*, sixty minutes are one *hour*, sixty hours are one *day and night*, thirty days and nights are one *month*, and twelve months are posited as one *year*. Calculated in this way, one day and night is calculated to have 21,600 rounds of breath.

When twelve months are completed with respect to each of the three types of day—the zodiac, solar, and lunar—one has to posit [451] that one year is completed according to each of these perspectives. Therefore 360 lunar days are one lunar year, 360 solar days are one solar year, and 360 zodiac days are posited as one zodiac year. Among these three, a year composed of lunar days is the shortest, a year of composed of solar days is longer by six days than a lunar year, and a year composed of zodiac days is longer than a solar year by just five extra days. To bring these calculations together in a broad way, 371 lunar days, 365 solar days, and 360 zodiac days are equal in duration.[424]

Now to explain the length of an eon, earlier when discussing the formation and destruction of the world systems, we mentioned that there are the eons of destruction, voidness, formation, and endurance. The combination of these four is referred to, both in the sūtras and in the Abhidharma treatises, as a "great eon."

From among these, the *eon of destruction* is posited as beginning from when the new birth of sentient beings in Avīci hell ceases up to the termination of external world systems. The period of destruction of sentient beings who inhabit those worlds is nineteen intermediate eons, and the period of destruction of external worlds is one intermediate eon. Thus it lasts [452] twenty intermediate eons. The *eon of voidness* is said to be the period of abiding in a void state from the destruction of former worlds up until but not including the formation of future world systems. It also lasts for twenty intermediate eons. The *eon of formation* is posited from the initial development of the wind maṇḍala as the lower foundation up to the birth of the first sentient being in the hells. The formation of external worlds takes one intermediate eon, and the evolution of their inhabitants takes nineteen intermediate eons. Thus it lasts for twenty intermediate eons. The *eon of endurance* lasts one intermediate eon in its initial phase and one intermediate eon in its final phase, plus the eighteen intermediate epochs between them. By adding each of these intermediate eons the stage lasts for twenty intermediate eons. The initial phase is the period of one intermediate eon where the lifespan decreases from an incalculable period to a lifespan of ten years.

What is referred to as "eighteen intermediate epochs" consists of the following. One epoch is the period of an increase of lifespan from ten human years up to eighty thousand years, and the decrease from eighty thousand to ten human years. For each epoch there are two such periods of increase and decrease, and eighteen such epochs occur. The final phase occurs after the completion of the eighteen intermediate epochs where the lifespan increases from ten to eighty thousand years, and this is the duration of the twentieth intermediate eon. Since the two eons at the beginning and end of the eon of endurance progress more slowly and the intermediate epochs progress more quickly, [453] the two single eons at the beginning and end, and each of the intermediate epochs, have the same duration.

The durations of the four eons—of endurance, formation, destruction, and voidness after the destruction—are equal in being twenty intermediate eons each. When added together there are eighty intermediate eons, and this is the measure of a *great eon*. *Treasury of Knowledge* states:

> The eon of destruction is from the disappearance
> of the hells to the annihilation of the environment.
> Formation occurs first from the wind maṇḍala
> up to the existence of hell beings.
>
> An intermediate eon is from an incalculable
> lifespan to ten years.
> From there the remaining eons of increase
> and decrease are eighteen.
>
> There is increase for one intermediate eon.
> The lifespan in those eons increases to eighty thousand.
> Thus this formation of mundane states
> lasts twenty intermediate eons.
> Formation, destruction, and the aftermath
> of destruction are equal in duration.
> Eighty of those make a great eon.[425]

*Investigating Characteristics* states:

> In this way, since the two eons at either end progress slowly, and the eighteen eons in the middle progress quickly, all are equal in length.[426] [454]

"What then is the calculation in human years of the duration of such eons?" one might ask. Response: One intermediate eon is 3 quintillion, 397 quadrillion, 386 trillion, 240 billion human years. This number extends to nineteen digits: 3,397,386,240,000,000,000.[427]

One great eon is 271 quintillion, 790 quadrillion, 899 trillion, and 200 billion human years. This number extends to twenty-one digits: 271,790,899,200,000,000,000. There also appear to be different systems of calculation that assert that for one intermediate eon there are 9 quadrillion human years, or for one great eon there are 702 quintillion human years.

Daśabalaśrīmitra, on the other hand, speaks of one intermediate eon being the time it takes to empty a store of sesame or mustard grain filling a square iron enclosure one krośa in height and on each of its four sides by taking out one grain every hundred years. A total of eighty such eons are said to be the measure of one great eon. In *Ascertaining Conditioned and Unconditioned Phenomena* he states:

> As for eons there are four: the intermediate eon, [455] the eon of destruction, the eon of formation, and the great eon. The measure of an intermediate eon is the time it takes to exhaust all the sesame or oil grain completely filling a cubical natural iron enclosure one krośa in height and one krośa on each side when one grain is removed every hundred years. Eighty intermediate eons are a great eon. An eon of destruction is also twenty intermediate eons.[428]

The scriptural sources connected with the Kālacakra system also use terms such as "small eon," "intermediate eon," "supreme eon," and "great eon." As for their durations, the measure of one day of a *small eon* is the time it takes to utterly exhaust a mound of single hairs filling a square pit measuring one yojana on each of its four sides and in depth by extracting one single hair every hundred years. Three hundred and sixty such days are calculated as one year, and one hundred such years are the duration of one small eon. The duration of *one intermediate eon* is one hundred years, where each day of that intermediate eon is equal to one hundred years of a small eon, where every day is the time it takes to completely extract all the single hairs from the mound of single hairs filling a pit. [456] The duration of one *supreme eon* is one hundred years, where each day of that supreme eon is equal to one hundred years of an intermediate eon, where every day of a small eon is the time it takes to completely extract all the single hairs from the mound filling a pit. For example, the great *Stainless Light* commentary states:

> Regarding the measure of a small eon, the time it takes to empty a pit of one yojana in depth and breadth that is completely filled

> with fine single hairs, where every hundred years one single hair is removed, is the duration of one day of a small eon. Thirty such days and nights are one month. Twelve months are one year. One hundred years are one small eon. So too, a collection of those is an intermediate eon. A collection of those is a supreme eon.[429]

It is said that if a supreme eon is multiplied in a similar manner, it becomes the duration of a great eon. [457]

## Units of Numerical Measurement

These explanations of the measure of years, eons, and so forth must be comprehended through calculation based on a numerical system. So one might ask, "What then are the names of the units in such a numerical system?" On this point we find various systems of calculation in the sūtras and treatises. In his *Treasury of Knowledge Autocommentary*, Vasubandhu presents a sixty-digit decimal system that he sources in the sūtras:[430]

> *Miscellaneous Sūtra* states: "There is enumeration to sixty digits (or decimal positions)." What are these sixty? For one (the first digit), one takes the first not the second decimal position; for ten, one takes the second position; for one hundred, one takes the third position; for one thousand, one takes the fourth position; for ten thousand (*prabheda*), one takes the fifth position; for one hundred thousand (*lakṣa*),[431] one takes the sixth position; for one million (*atilakṣa*), one takes the seventh position, and so on. Thus each subsequent number is ten times the former: *koṭi, madhya, ayuta, mahāyuta, nayuta, mahānayuta, prayuta, mahāprayuta, kaṃkara, mahākaṃkara,* [458] *bimbara, mahābimbara, akṣobhya, mahākṣobhya, vivāha, mahāvivāha, utsaṅga, mahotsaṅga, vāhana, mahāvāhana, tiṭibha, mahātiṭibha, hetu, mahāhetu, karabha, mahākarabha, indra, mahendra, samāpta, mahāsamāpta, gati, mahāgati, nimbarajas, mahānimbarajas, mudrā, mahāmudrā, bala, mahābala,* [459] *saṃjñā, mahāsaṃjñā, vibhūta, mahāvibhūta, balākṣa, mahābalākṣa, asaṃkhya.* The names of the eight numbers between *mahābalakṣa* and

> *asaṃkhya* have not been found.[432] Thus the duration of eons lasting up to the sixtieth decimal position is referred to as the "countless eon."[433]

Enumeration begins from one and increases by a factor of ten for each shift in decimal place. The sixtieth number in this series is called "countless" (*asaṃkhya*). But it should not be understood as utterly incalculable, for in the series of calculations from ten times one is ten, ten times ten is one hundred, and ten times one hundred is one thousand, and so on, one eventually reaches ten times the fifty-ninth decimal, and this number is given the name "countless."

Among these decimal positions, the first nine positions from one to one hundred million are called the *single set enumeration*. From a billion up to, but not including, *countless* is "the enumeration of the great companion" and is called the *recurring enumeration*. But since no specific names appear for the eight enumerations from the fifty-second to the sixtieth decimal, it is said that the list must be supplemented with appropriate names. Therefore the tradition among scholars has been to insert the names of the four immeasurables, such as "loving-kindness" and so on, to expand the ten digits from *mahābalākṣa*. Therefore ten *mahābalākṣas* are loving-kindness, ten times loving-kindness is great loving-kindness, ten times great loving-kindness is compassion, [460] ten times compassion is great compassion, ten times great compassion is joy, ten times joy is great joy, ten times great joy is equanimity, ten times equanimity is great equanimity, ten times great equanimity is countless (*asaṃkhya*). There is, alternatively, also the tradition of using the names of the great elements of earth, water, fire, and wind, each followed by a great earth element and so on. There exists also a system of expanding from the sixtieth decimal, enumerated above, multiplying further progressively by factors of ten: immeasurable, incalculable, inexpressible, limitless, inestimable, beyond estimation, and inconceivable.

Other systems of decimal calculations are found in the Buddhist sources. For example, the *Extensive Display Sūtra* states:

> One hundred times ten million is a billion.
> One hundred times one billion is one hundred billion . . .

Up to:

> beyond that there is also a number following after the subtle particles.[434]

Thus it increases manifoldly from ten million up to "following after the subtle particles," which involves a system of calculation up to thirty-two decimals. *Flower Ornament Sūtra* states:

> One hundred times one thousand times one hundred is ten million . . .

Up to:

> thoroughly teaches manifold [increase] up to the inexpressible of the inexpressible.[435]

Thus many different systems of calculation are mentioned in the sūtras, including a way of calculating up to 120 decimals through a progressive expansion from one hundred to the inexpressible of the inexpressible. [461]

To summarize, what the preceding presentation makes clear are the following points. This world system is: *infinite*, in terms of its continuum; *formed* and *destroyed*, due to conditions; *dependent*, with respect to its essential nature; *threefold*, with respect to its types (namely the desire, form, and formless realms); and *limitless*, in terms of enumeration. Such Buddhist explanations elucidate what are the origins of this world system, what is its essential nature, and how it came into being. This said, within the explanation of the formation and the destruction of world systems, apart from recognizing space particles as the subtle source and describing its functions, there is little detailed inquiry into these issues. This is because in Buddhism the primary focus of inquiry centers on the question of how to overcome suffering and attain happiness, and the inquiry into objects of knowledge facilitates that aim. Therefore compared with the explanations found in Buddhist sources on the nature and origin of sentient beings, including especially the presentations on consciousness, the Buddhist presentations on the external physical world remain secondary and emerge as a byproduct.

At any rate, in Buddhist texts it is asserted that first world systems come to form, then they endure, then they are destroyed, then the void phase follows. After the phase of voidness new world systems come to be formed again from subtle particles of space or empty particles. Thus from voidness there is formation, then endurance, then destruction, and again voidness. This continuous cycle or wheel perpetuates without any absolute end.

In the texts of Kālacakra, subtle particles of vacant space in the void phase are called *empty particles*. From them arise space particles that exist as the cause of all physical phenomena, [462] but they are not asserted to be independent manifest physical particles since they mostly constitute mere potentialities. In dependence on the influence of such space particles the four primary elements facilitate the evolution from subtler to coarser physical states, and so too they facilitate the serial devolution from coarser to subtler states. Again, subtle particles of space act as the basic cause of the cyclic process of formation and destruction of world systems, and this is due to the very nature of the space particle.

However, according to the Kālacakra system, as long as such subtle empty particles do not act to cause the formation of new world systems, they maintain their essential empty nature prior to acting to establish specific worlds. Such empty particles are the basic constituents of specific world systems, since every component part that establishes the physical phenomena of that particular world system exists in the nature of empty particles. In addition, when the potency of the karma of sentient beings to be born in a specific world system matures, wind particles begin to assemble and progressively establish that world system owing to the impact of such karmic force in that specific location.

A world system comes to be destroyed in the following sequence: first the earth dissolves, followed by water, fire, wind, and finally the sphere of space. When they come into being, space gives rise to wind, and from wind, fire, water, and earth, and so on. Thus the four primary elements emerge from the sphere of space and, once again, dissolve back into space, so space [463] constitutes the basis for the arising and destruction of the four primary elements. Even so, when one traces the ultimate cause of the emergence of the shape of this world system, including the planets and stars within it, as well as the bodies of humans, animals, and so on, one arrives at the space particle as the basic source. So the question remains: How do these subtle space particles acquire the potential to give rise to such diverse configurations and so on, and what are the stages involved in this process? We feel this question requires further investigation.

Given that according to the Buddhist traditions both consciousness and karma are indispensible components for the formation of world systems, if one understands the notion of "karma" in an erroneous manner, there is the possibility of mistakenly believing that the world system and oneself are somehow bound by karma in a predetermined way. In general the term "karma" connotes action, so by karma one should understand it to mean an intentional act committed by a sentient being. Such an act is generated through body, speech, or mind, and irrespective of its force any such act leaves a residual imprint on the mind. And from motivation arises physical and verbal actions, and these actions in turn strongly affect one's mind and the course it takes. This mutually reinforcing process of cause and effect is true not just at the level of individuals but also for communities, and human society as a whole.

# Part 6

## FETAL DEVELOPMENT AND THE CHANNELS, WINDS, AND DROPS

## Embryonic Development Theory

Part 6 of our volume covers a range of topics that touch upon the challenging question of the mind-body connection. This should be distinguished from the so-called mind-body problem, which is a philosophical conundrum stemming from the Cartesian dualism of mind and body. For the Buddhist world, the link between mind and body is provided by what are referred to as *winds* (*vāyu*), subtle energy flowing within conduits known as *channels* (*nāḍī*), as well as *drops* (*bindu/tilaka*), vital essences that are understood to reside in specific points within the key channels. The most developed form of this subtle physiology, discussed in more detail below, is found in the Vajrayāna texts. On this view, the winds or subtle energies are conceived to be the support for all our cognitive activity, and in fact mind and subtle winds are understood to be inseparable. This inseparability of mind and body in the form of subtle energy is a unique Vajrayāna view.

In part 5, we saw that an important question for the Buddhists is "How do sentient creatures emerge from insentient material substances?" The texts identify four types of birth, (1) egg birth, (2) womb birth, (3) birth from heat and moisture, and (4) spontaneous birth, and give extensive examples for each. Of these, attention is focused on the second one, birth from a womb. One of the earliest Buddhist sources on the process by which a womb birth takes place is *Nanda's Sūtra on Entering the Womb*, a work dateable at the latest to around the first century and translated into Chinese in the third century. In this sūtra, the following conditions were deemed necessary for conception to take place: (1) the parents are having sex, (2) the woman is ovulating, (3) an intermediate-state being is nearby, (4) both the sperm and ovum are free of defects, and (5) the karmic connection between the intermediate being and the prospective parents has reached a maturation point (see page 349). Of these, conditions 3 and 5 contain an element that will be problematic from the current scientific view. The intermediate state is the interim after a sentient being's death and before his or her rebirth in a new form. This concept is broadly shared in Indo-Tibetan

Buddhist traditions, but it is rejected by the Theravāda Buddhist tradition. *Nanda's Sūtra* in fact provides quite a detailed description of the mental state of the intermediate being, who due to his or her past karma is said to experience all sorts of hallucinations, such as "I am entering inside a house," "I am entering a dense thicket," "I am burrowing into a hole," and so on.

From a history of ideas perspective, what is remarkable about the early Buddhist conceptualizations of the birth process is its detailed embryonic development theory, which has striking parallels with contemporary scientific understanding. In Western thought, the view until recently was that of the homunculus (Latin for a "little man") or preformation—namely, that a fully formed person is present in the sperm. On this view, the fetal development process was simply the increase in size of this homunculus until the child is ready to come out of the womb. In contrast, Buddhist sources such as *Nanda's Sūtra* present a week-by-week development and transformation process, beginning first with just the fusion of sperm and ovum—the fertilized egg—turning into something like a blob of rice broth, in the second week turning into something like a thick curd, something like worm entrails in the third week, and so on. In the seventh, protrusions for the four limbs appear, fingers in the eighth week, the orifices for the sense organs in the ninth week, and so on. The text then goes on to describe the formation of the various channels or nerves of the body, and a point of divergence from the modern account emerges when the text speaks of the appearance of the various winds or energies within these channels. In the thirty-eighth week, the unborn child is now ready to emerge from the birth canal and prepares to exit from the womb. As cited in our volume, we find a similar detailed account of the stages of the embryonic development in other Buddhist sources such as the *Kālacakra Tantra* and in early medical texts. In a medical text by a Buddhist author cited in our volume, the five limbs—head, two arms, and two legs—come to be formed in the third month with the arising of the sensations of pain and pleasure (see page 376).

Although all the specifics of the early Buddhist description of the stages of embryonic development may not pan out in the light of contemporary science, the basic presentation of the complex process of prenatal development is remarkably accurate. This view of embryonic development even made it possible for someone like Vasubandhu to conceive the possibility of surrogate motherhood in the fourth century. In a discussion of matricide in the *Treasury of Knowledge*, a question is raised: "When the embryo of a woman falls and another woman puts the embryo in her womb, which of these two women is to be recognized as the mother whose killing constitutes a heinous crime?" Vasubandhu responds, "The mother is the woman from whose embryo one is born" (4:103d). Although I have only circumstantial evidence, I am convinced that much of the Buddhist understanding of the embryonic development comes from early medical knowledge, which must have involved postmortem examinations with dissection of human bodies as well as perspectives obtained from experiential knowledge of midwives. For many traditional Buddhists, of course, the detailed descriptions found in *Nanda's Sūtra on Entering the Womb* come from the Buddha's enlightened wisdom that sees everything as they are.

## THE BRAIN IN AN EARLY BUDDHIST MEDICAL TEXT

Historians of Indian Buddhist thought have noted that, despite the tremendous sophistication of the science and philosophy of mind in the classical sources, one finds hardly any recognition of the centrality of the brain in human experience. The language of mental life in the Indian Buddhist sources is predominantly from the perspective of the mind and not the brain. Consequently, even the Abhidharma sources, so detailed in their treatment of many aspects of human physiology, are rather silent when it comes to the brain. A notable exception is classical medical texts, some of them composed by Buddhist authors. Our volume draws attention to the description of the brain found in two distinct classes of medical texts. One is from the Four Medical Tantras, a foundational corpus for Tibetan medical science, whose Indian origin remains a matter of debate. The second text is an intriguing apocryphal work entitled *Somarāja*, attributed to Nāgārjuna and translated into Tibetan, reputedly by the noted Tibetan translator Vairocana, sometime in the eighth or ninth century.

In these medical texts, we find a remarkable discussion of the brain, beginning with how different parts of the body emerge: bones, brain, and spinal cord from father's semen, and flesh, blood, and organs from the mother's ovum (see page 406). *Somarāja* speaks of how the brain is linked to the nerves of the five sense organs as well as other organs such as the liver, kidneys, and heart. The text describes different parts of the brain, asserting, for instance, that the cerebellum is the shape of a half horsehoof and is the basis for our experience of bliss (such as in sex). In brief, this text describes the brain as an "ocean of nerves" from which suspend a whole series of nerves, like threads shooting downward. Three main parts of the brain are identified: (1) the spinal cord, (2) the cerebral membrane, and (3) something called the "brahma-conch cakra," referring to a network of nerves wired in a specific manner (see pages 412–15). This third part roughly equals what in modern science one calls the cerebrum, consisting of the cortical and subcortical structures of the brain. This same text also asserts that there exists a nodule the size of a human thumb at the point where cerebellum and the skull come into contact.

The remarkable aspect of this Buddhist medical knowledge of the brain is that its understanding of the brain and its functions reflects some understanding of the brain's anatomy. Here, too, my guess is that these authors were either themselves engaged in dissection or drew from other's insights based on dissection of human bodies. From a history of ideas point of view, a critical study of the *Somarāja*, based on careful comparison against today's knowledge of brain anatomy and specific functions, has the potential to reveal that, when it comes to the brain, perhaps classical Indian medical knowledge was more advanced than the West prior to the dawn of the modern scientific era.

## The Subtle Body

For those interested in the human body, perhaps one of the most intriguing concepts in Buddhist thought is its unusual Vajrayāna physiology—of channels, winds, and drops—as well as the postulation of degrees of subtlety of the human body. On this view, the human body is said to possess 72,000 channels, all stemming from three main channels where they intersect at the heart—a central channel and two side channels. In these channels flow the energy-like winds, and at important junctures within these channels reside "drops," or vital essences. These junctures are where the two side channels intersect with the central channel to form channel centers (*cakras*), such as at the crown, throat, heart, navel, and sex organ. There are said to be ten types of winds or energies that flow within these channels—five primary and five secondary winds. According to Vajrayāna physiology, it is these winds that perform the functions of various bodily organs, for example, the life-sustaining wind (*prāṇa*) enables the other winds to move to the specific organs to perform their functions, the downward voiding wind (*apāna*) helps bowel movements as well as the flow of fluids within the body, and so on. The texts even describe with specificity the number of movements of these winds within the body in any given day.

One might question the inclusion of this particular Vajrayāna physiology in our volume, which is supposed to be on the physical sciences from the Buddhist sources. Setting aside the question of evidence for the time being, it is clear that the Buddhist sources themselves understood these claims about physiology as statements of fact. And each of the three main features of this physiology—channels, winds, and drops—are explained with respect to their functions and effects. At least in principle, one could argue that the claims are scientific, which is not the same as asserting that they are scientifically proven. At this point, these claims are best treated as constituting an interesting hypothesis coming from the Buddhist sources.

One point I think needs to be taken more seriously by contemporary scientists is the observed effects experienced at highly specific points of the body by meditators who use the body map portrayed in this unique physiology. Through *tumo* meditation, a heat-generating practice involving visualization, for example, Buddhist adepts have demonstrated the ability to project heat over the surface of their body even when exposed to extreme cold. These kinds of meditation are premised upon employing and manipulating energies within specific channel centers within the body—the navel center in the case of *tumo* practice. Finally, there is the observed phenomenon of *thukdam*, when a person is clinically dead yet his or her body remains without decomposition for many days after the death. The Vajrayāna tradition sees this as yet another indication of the presence of subtle states of body and mind, where an individual can still remain "alive" even though all functions of the gross body have ceased. At this point, I know no scientific theory that explains phenomena such as these.

Our volume contains a brief section on the difficult topic of the relationship between body and mind. Since this is an issue that is closely connected with how one defines mind and consciousness, as opposed to what is material, it is addressed more fully in volume 2 in the series.

## Microorganisms within the Human Body

The final topic in part 6 of our volume is the curious thesis found in some of the early Buddhist texts on the role of microorganisms within the human body. Although the idea of microorganisms within the human body was found also in Western thought, until recently their presence was almost always understood in terms of parasites. The idea that these microorganisms—microbes in contemporary scientific language—might in fact have a crucial role in regulating our health, in a constructive way as well, emerged extremely late in the West. In contrast, some of the early Buddhist texts, such as the *Twenty-Five Thousand Verse Perfection of Wisdom*, speak of eighty thousand types of microorganisms, while the *Sūtra on the Foundation of Mindfulness* describes specific health-related functions of these organisms located in different parts of the body. The text even speaks of how the activities of the microorganisms affect the mental states of a person, giving rise to sadness and downcast states of mind (see page 427). The picture that emerges is a concept of body health that is defined, to a large extent, by the level of equilibrium with these internal microorganisms. Some of the detailed explanations are such that one can't help marvel at the strikingly contemporary-sounding tone. Again, from an epistemological point of view, one wonders how early Buddhist authors came up with these ideas!

## Further Reading in English

For a contemporary essay on embryonic development as described in *Nanda's Sūtra on Entering the Womb* and its comparison with Indian medical sources, see Robert Kritzer, "Life in the Womb: Conception and Gestation in Buddhist Scripture and Classical Indian Medical Literature" in *Imagining the Fetus: The Unborn in Myth, Religion, and Culture*, edited by Vanessa R. Sasson and Jane Marie Law (New York: Oxford University Press, 2008), 73–89.

For a succinct explanation of the Vajrayāna physiology of channels, winds, and drops according to the Kālacakra tantra, see Khedrup Norsang Gyatso, *Ornament of Stainless Light*, translated by Gavin Kilty (Boston: Wisdom Publications, 2004), 173–94.

For a recent scientific paper on findings from research on Tibetan *tumo* meditation, see Maria Kozhevnikov et al., "Neurocognitive and Somatic Components of Temperature Increases during g-Tummo Meditation: Legend and Reality," *PLOS/One*, March 29, 2013; http://journals.plos.org/plosone/article?id=10.1371/journal.pone.0058244.

# 26

# The Birth Process

IN BUDDHIST TEXTS there are explicit discussions about how a new life emerges and the stages of fetal development in the womb. Like the other great classical Indian schools of thought, the standpoint of Buddhist philosophy is to maintain with logical reasoning that this present life does not occur adventitiously but arises from a previous life. Thus existence is presented in terms of four states: (1) "birth existence," which represents the moment of conception, (2) "intermediate existence," which is the state between two lives, (3) "prior existence," which begins from the second moment after conception and lasts until the moment of death, and (4) "death existence," which is the moment of death itself. All four states are necessarily understood on the basis of past and future births. [466] Here we shall present how the human body comes to be formed in the stages of fetal development within the womb as described in the Buddhist sūtras and tantras as well as in treatises on medical science.

First, the texts speak of four types of birth: (1) birth from an egg, (2) birth from the womb, (3) birth from heat and moisture, and (4) spontaneous supernatural birth. For example, ducks, cranes, and so on are egg-born; elephants, horses, oxen, and so on are womb-born; microorganisms, butterflies, and so forth are born from heat and moisture; and celestial beings, hell beings, intermediate beings, and so on are examples of spontaneous birth. These points are extensively discussed in Maudgalyāyana's *Presentation on Causation*, from which we cite the following:

> As for the types of birth there are four: birth from an egg, birth from a womb, birth from heat and moisture, and spontaneous supernatural birth. What is birth from an egg? It is any sentient being who is born from an egg, those who reside within an

egg, those who are covered by albumen within the shell, those who are born and fully arise through breaking the eggshell, those who develop, manifestly develop, fully develop, [467] and emerge. Who are they? They are ducks, cranes, peacocks, parrots, grouse, cuckoos, partridges, and so on. There are some serpentine nāgas,[436] some garuḍas,[437] some humans, and other sentient beings who are born from eggs, who reside in eggs, who are covered by albumen within the shell, who are born and fully arise through breaking the eggshell, who develop, manifestly develop, fully develop, and emerge. This is called *birth from an egg*.

What is birth from the womb? It is any sentient being who is born from the womb, those who reside in the womb, who are covered by amniotic fluid in the womb, who are born and fully arise through piercing the amniotic sac in the womb, who develop, manifestly develop, fully develop, and emerge. Who are they? They are elephants, horses, oxen, donkeys, water buffalo, deer, pigs, and so on. There are some nāgas, some garuḍas, some ghosts, some humans, and other sentient beings who are born from the womb, who exist in the womb, who are covered by amniotic fluid in the womb, who are born and fully arise through piercing the amniotic sac in the womb, who develop, manifestly develop, fully develop, and emerge. This is called *birth from the womb*. [468]

What is birth from heat and moisture? It refers to any sentient being who is born from collective heat, interaction of particles, combined harm, and compression of mounds of waste, or sewers, or the presence of excrement, or rotten meat, or warm rotten food, or dense grass, or dense forests, or straw huts, or leaf huts, or putrid water, or flimsy dwellings, or beehives, whereby a being comes to be born and fully born, develops, manifestly develops, fully develops, and emerges. Who are they? They are microorganisms, beetles, butterflies, mosquitoes, insects who feed on sesame, chaff, herbs, and so on. There are some nāgas, some garuḍas, some humans, and other sentient beings who are born from shared warmth, shared particles, shared harm, and shared compression, who are born and fully arise, who develop, manifestly develop, fully develop, and emerge. This is the called *birth from heat and moisture*.

> What is spontaneous supernatural birth? It refers to any sentient being who is instantaneously born in possession of and complete in all major and minor limbs and who does not lack any sense faculty, those who are born and fully arise, who develop, manifestly develop, [469] fully develop, and emerge. Who are they? They are all the gods, and all hell beings, and all intermediate beings. There are some nāgas, some garuḍas, some ghosts, some humans, and other sentient beings who are instantaneously born in possession of all major and minor limbs and not lacking any sense faculty, who are born and fully arise, who develop, manifestly develop, fully develop, and emerge. This is called *spontaneous supernatural birth*.[438]

## How Consciousness Enters the Womb

Within these four types of birth, how does consciousness enter the womb when a new life occurs? There are six causes that lead the intermediate being to enter the mother's womb: (a) the father and mother generate sexual desire for each other and engage in sexual intercourse, (b) the mother is fertile, (c) there is the presence of the intermediate being nearby, (d) the mother's womb as well as the father's sperm and the mother's ovum are healthy, and (e) there is a convergence of the past karma of the parents and the intermediate being. When all these causes assemble and are complete, the texts state that the intermediate being will enter the mother's womb. *Nanda's Sūtra on Entering the Womb* states:

> If the mother and father have generated minds of sexual desire, if this coincides with the fertile stage of the menstrual cycle, [470] if the aggregates of the intermediate being are present, if they possess few of the above faults, and if there exists the condition of karma, then the intermediate being will enter the womb.[439]

Further, it is said that it is not necessary that both the father and mother engage in sexual intercourse in order for the fetus to form in the womb. *Detailed Explanations of Discipline* states:

> When Gupta (Tib. *sbad ma*) thought of her former spouse with a mind of intense desire recalling their separation, she inserted his semen into the entrance of her womb. Then, owing to the inconceivable nature of the karma of sentient beings, a sentient being entered her womb.[440]

This is an account of Gupta, who when overcome by the mind of desire for her former spouse, who was now the monk Udayin, inserted cloth smeared with his semen into her vagina. A fetus then developed in her womb and a child was born.

A statement in *Treasury of Knowledge Autocommentary* refers to a child being conceived when a viable fetus from the womb of one woman is transplanted into the womb of another woman. The text says:

> When the kalala embryo from one woman has been inserted into the womb of another woman, which one is the mother the killing of whom would constitute a heinous crime? "The woman from whom the ovum came," for that sentient being from whom the ovum emerged is the mother of that child.[441] [471]

Thus if a child is born from inserting the kalala embryo from the womb of one woman into the womb of another woman, it is the killing of the first woman that accumulates the karma of the heinous act of killing one's mother.

## How the Intermediate Being Enters the Womb

As for how the intermediate being enters the womb, when the intermediate being sees beings of similar type on the way to the site of conception and it desires to look at them and play with them and so on, then it proceeds to its site of conception. It sees the sperm and ovum of the mother and father distortedly, and though the father and mother have not yet engaged in sex, it sees them engaging in sex as in an apparition and develops attachment. It is said that if it is to be born male, it generates attachment to the mother and aversion toward the father. If it is to be born female, it generates attachment to the father and aversion toward the mother. Further, it sees the male or female to whom it feels attached and does not see the other. And with the thought to enter into union as it approaches, instead of seeing the body of that male or female, only the sexual organ appears, which causes anger to arise in the intermediate being. Because of the arising of anger the intermediate being dies and instantly its consciousness enters the combined red and white substances of the ovum and semen of the mother and father at the center of the mother's womb. *Nanda's Sūtra on Entering the Womb*, for example, states:

> Further, when that intermediate being enters the mother's womb its mind becomes distorted. If it is to be born male, it generates attachment to the mother [472] and aversion toward the father. If it is to be born female it generates attachment to the father and aversion toward the mother. Owing to the formations of past karma it generates distorted thoughts due to false discernment.[442]

*Yogācāra Ground* states:

> That being who exists as an intermediate being sees other beings of the same status and desires to engage in play and so on; it then generates the desire to proceed to the site of conception. Also at that time it distortedly sees the sperm and ovum derived from the father and mother. But this is not [the main] distortion, for it sees both the future father and mother engaging in sex when the father and mother are not engaged in sex, and the intermediate being sees this with a distorted mind, like seeing an illusion.[443]

## The Posture of the Intermediate Being upon Entering the Womb

With respect to the physical posture of the intermediate being after it enters the womb, if the being that entered the womb is to be male, it resides on the right side of its mother's womb facing the mother's spine in a squatting position. If it is to become female it resides on the left side of its mother's womb, facing the way its mother faces in a squatting position.

If it is to become a neuter it will sit in accordance with whether desire for male or female is stronger. [473] This is stated in *Nanda's Sūtra on Entering the Womb*:

> Nanda, if the body of such a being is to become male, it squats to the right of its mother's womb, two hands covering its face, facing its mother's spine. If it is to become female it lacks one rib, it squats to the left of its mother's womb, two hands covering its face, facing the way its mother faces.[444]

*The Treasury of Knowledge Autocommentary* also states:

> However, if a neuter, it resides in accordance with the desire it possesses.[445]

It is said that there are three doors of entry for the consciousness of that intermediate being to initially enter the mother's womb. The first involves entry through the crown. Because that intermediate being sees the two organs of the father and mother joined in sex, it discards its physical body in order to act on its desire. Then its subtle innate mind mounts the subtle innate wind as if riding a horse and enters the womb of its mother via the Vairocana[446] door at the crown of the father. This is mentioned in Nāgabodhi's *Presentation of the Guhyasamāja Sādhana*:

> Since it sees the two organs joined in sex, it discards the body of the intermediate being to facilitate its desire. Then, as if riding a horse, the mind vajra[447] that is the lord of consciousness rides the wind as its mount and swiftly reaches the door of Vairocana and enters like a wisdom being, in a moment, [474] an instant, or a brief period.[448]

The second entry of consciousness is through the mouth. As in the previous method, the intermediate being's subtle innate mind along with the wind that is its mount enters the womb of the mother via the mouth of the father. *Saṃvarodaya Tantra* states:

> The sexual union of the father and mother and so on
> is seen by the intermediate being with craving.

> Consciousness rides its mount, the wind,
> as if riding a horse,
> and it arrives extremely swiftly
> in just a moment or in an instant.
> Owing to the power of intense joy
> it enters via the avenue of the mouth.[449]

*Vajra Garland Tantra* states:

> The wind that enters the mouth of the father
> becomes a seed at that time.
> From that seed a sentient being arises.[450]

When the intermediate being enters the mother's womb after its death by way of its subtle innate mind along with the wind that is its mount, first it enters the mouth and in stages reaches the father's secret place. From there it exits and then enters the mother's womb. As stated in the passage cited earlier from *Presentation of the Guhyasamāja Sādhana*, one can infer that the crown being referred to as the initial door of entry is the father's crown. [475]

The third entry of consciousness is via the mother's womb. *Investigating Characteristics* states:

> For it seizes a functioning womb, and it [consciousness] enters via the door of the birth site.[451]

Since the intermediate being is not obstructed by mountain rocks or walls it does not necessarily need to have an opening that facilitates its entry. As such, examples are mentioned of living beings, such as spiders and frogs and so on, that are observed to exist inside rocks that have no openings or cracks at all when cleaved apart.[452]

Medical texts and so forth have a different way of explaining how a being becomes male, female, or neuter when it enters the womb. In accordance with Matṛceṭa's *Core Summary of the Eight Branches*, when the father and mother engage in sex and the last drop of white or red substance of the semen and blood of the father and mother is emitted, if the father's semen is greater in quantity the intermediate being will become a male, if the mother's blood is greater in quantity it will become a female, and if equal in measure it will become a neuter. Alternatively, when the wind separates the semen and blood of the father and mother into white and red substances, the intermediate being will become male if the father's sperm is greater, it will become female if the mother's blood is greater, and it will become male twins or female twins if the semen and blood are equal and so on. The text states:

> Therefore, through there being more semen
> a male will probably be born.
> Through there being more menstrual blood a female.
> Through both being equal it will become a neuter.
> Through wind fully separating the semen and blood
> twins will be born from it.[453] [476]

In this context "probably" indicates uncertainty since it is possible that the outcome will change owing to the influence of different conditions. Further, in some texts of secret mantra it is said that when the final drop of the father's semen is deposited in the mother's womb, the fetus will develop as a male if the wind moves from the right of the composite of both white and red substances of semen and blood of the father and mother. Alternatively, it will be female if the wind moves from the left, and if the wind moves from the center the fetus will develop as a neuter. *Saṃvarodaya Tantra* states:

> Those who are learned correctly indicate
> the temporal moments of the stages of the flow of the seed.
> Any movement of wind from the right
> generally indicates it will become a male.
> Any movement of wind from the left
> indicates it will definitely become a female.
> Any seed movement between them both
> always indicates it will become a neuter.[454]

Moreover, there is no certainty since it is possible that the outcome will change owing to the influence of different conditions. [477]

# 27

## Fetal Development in the Sūtras

THERE ARE THREE explanations of how birth occurs after entry into the womb and of the stages of the formation of the fetal body: the explanation in the sūtras, in the Kālacakra tantras, and in the medical treatises.

With regard to the sūtras, in general when the fertilized composite of the father's sperm and mother's ovum progressively develops in the womb, there are five stages of development: (1) arbuda embryo, (2) kalala embryo, (3) peśin embryo, (4) ghana embryo, and (5) praśākhā embryo.

In the first week after the intermediate being has entered the mother's womb, it is called an *arbuda embryo*, and the appearance of the arbuda embryo is similar to rice broth or yogurt scum. In those seven days it transforms through incubating the white and red substances of the semen and blood. The arbuda embryo is said to manifest from the basis of the solidity of the earth element, the moistness of the water element, the heat of the fire element, and the mobility of the wind element that are a composite of the embryo's body sense faculty. The way the body of the embryo forms in the womb and develops each week is clearly stated in *Nanda's Sūtra on Entering the Womb*. Given that these ancient presentations derived from Buddhist sources bear similarity with the views of contemporary scientists based on empirical observation, we shall discuss these presentations in some detail here. [478] The sūtra states:

> In the first week it is called an *arbuda embryo* since it sleeps in its first abode, its mother's womb, which is unclean, and the physical faculty and consciousness together experience unbearable suffering as if cooked and roasted in an extremely hot copper pot. Its appearance is like rice broth or yogurt scum and this primitive embryo clearly manifests the solidity of the earth

> element, the moistness of the water element, the heat of the fire element, and the mobility of the wind element owing to its internal and complete incubation for seven days.[455]

In the second week a wind called "pervasive contact" arises from within the mother's womb, and because of that wind impacting the womb the earlier arbuda embryo becomes like stretched curd and is called a *kalala embryo.* The appearance of the kalala embryo is like thick curd or congealed butter. During those seven days the kalala embryo incubates within the fertilized ovum (zygote) and the four elements, such as the earth element that comprises its body faculty, [479] develop beyond their earlier stages. *Nanda's Sūtra on Entering the Womb* declares:

> Nanda, in the second week the embryo resides in the unclean abode of its mother's womb. Thus when the physical faculty and consciousness together experience unbearable suffering as if cooked and roasted in an extremely hot copper pot, a wind called "pervasive contact" arises from within the mother's womb owing to the power of prior karma, and that which it contacts in the womb is called the *kalala embryo.* Its appearance is like thick curd or congealed butter, and since it internally incubates and fully incubates for seven days, the four elements more clearly manifest.[456]

In the third week a wind called "repository door" arises from within the mother's womb, and because of that wind impacting the womb the earlier kalala embryo becomes oblong and is called a *peśin embryo.* The appearance of the peśin embryo is like an iron spoon or worm entrails. During those seven days the peśin embryo incubates within the fertilized ovum (zygote) and the four elements, such as the earth element that comprises its body faculty, develop beyond their earlier stages. *Nanda's Sūtra on Entering the Womb* states:

> Nanda, in the third week the embryo develops as before. A wind called "repository door" arises within the mother's womb owing to the power of prior karma, [480] and that which it contacts in the womb is called a *peśin embryo.* Its appearance is like an

> iron spoon or worm entrails. Since it internally incubates and fully incubates for seven days, the four elements more clearly manifest.[457]

In the fourth week a wind called "differentiator" arises within the mother's womb, and because of that wind impacting the womb intense pain arises and the earlier peśin embryo congeals and is called a *ghana embryo*. The appearance of the ghana embryo is like a clod of earth or a pestle. The sūtra states:

> Nanda, in the fourth week the embryo develops as before. A wind called "differentiator" arises within the mother's womb owing to the power of prior karma. Intense pain arises in the embryo abiding in the womb and that which expels this pain is called the *ghana embryo*. The shape of the ghana embryo is like a clod of earth or pestle.[458]

In the fifth week a wind called "perfect gatherer" arises within the mother's womb, and because of that wind impacting the womb five protrusions arise: the protuberances of the two shoulders, the two thighs, [481] and the head, just like the growth of branches and leaves of trees and forests in the summer rain. The sūtra states:

> Nanda, also in the fifth week the embryo develops as before. A wind called "perfect gatherer" arises within the mother's womb, and when that wind impacts the womb the five limbs arise: the two shoulders, the two thighs, and the head, just like the growth and increase of branches and leaves of trees and forests due to the summer rain.[459]

In the sixth week a wind called "great expanse" arises within the mother's womb, and because of that wind impacting the womb two forearms arise from the two shoulders and two calves arise from the two thighs, like the growth of grass, trees, and branches in the summer rain. The sūtra states:

> Nanda, in the sixth week a wind called "great expanse" arises within the mother's womb, and when that wind impacts the

> womb the four limbs arise: the two forearms and the two calves. Thus the [482] four limbs appear just like the growth of grass, trees, and branches in the summer rain.[460]

In the seventh week a wind called "circulator" arises within the mother's womb, and because of that wind impacting the womb the upper part of the two hands arise from the two forearms and the upper part of the two feet arise from the two calves, just like bubbles forming in water or those occurring in fish gills. The sūtra states:

> Nanda, in the seventh week a wind called "circulator" arises within the mother's womb, and when that wind impacts the womb four limbs arise: the upper part of the two hands and the upper part of the two feet. Thus that embryo comes to possess those four limbs just like bubbles forming in water or fish gills.[461]

In the eighth week a wind called "that which stops and transforms" arises within the mother's womb, and when that wind impacts the womb the twenty digits arise: the ten fingers and the ten toes, just like the first emergence of the roots of trees in the summer rain. The sūtra states:

> Nanda, in the eighth week a wind called "that which stops and transforms" [483] arises within the mother's womb, and when that wind impacts the womb the twenty digits arise. Thus the ten fingers and the ten toes first appear, just like the first emergence of the roots of trees in the summer rain.[462]

In the ninth week a wind called "specific separator" arises within the mother's womb, and when that wind impacts the womb nine features arise: the orifices of the two eyes, the two ears, the two nostrils, the mouth, the orifice for excrement, and the orifice for urine. The sūtra states:

> Nanda, in the ninth week a wind called "specific separator" arises within the mother's womb, and when that wind impacts the womb nine features appear: the two eyes, the two ears, the two nostrils, the mouth, the orifice for excrement, and the orifice for urine.[463]

In the tenth week, a wind called "solidifier" arises within the mother's womb, and because of that wind the body of the embryo residing in the womb becomes firm and stable. Also a wind called "thorough arising" arises, and because of that wind the body of the embryo residing in the womb grows by expanding, just like a bag becomes larger by expanding. The sūtra states:

> Nanda, in the tenth week a wind called "solidifier" [484] arises within the mother's womb, and because of that wind the body of the embryo abiding in the womb becomes firm and stable. During the seventh week a wind called "thorough arising" also arises in the womb, and because of that wind the body of the embryo abiding in the womb grows by expanding, just like a bag becomes larger by expanding from its center.[464]

In the eleventh week, a wind called "appearance of orifices" arises in the mother's womb, and because of that wind impacting the womb internal openings are produced within the body and the nine orifices, such as the two eyes and so on, are created, just as bellows blow air above and below and the air creates orifices. The sūtra states:

> Nanda, in the eleventh week a wind called "appearance of orifices" arises within the mother's womb, and when that wind impacts the womb it creates internal openings for the embryo abiding in the womb and nine orifices fully appear, just as when a blacksmith or his apprentice operate the bellows air blows above and below and the air creates orifices.[465] [485]

In the twelfth week a wind called "crooked gate" arises within the mother's womb, and because of that wind expanding, the bowels, intestines, and colon form and the intestines and colon come to reside together within the belly like the roots of a lotus. Again, a wind called "platted hair" arises, and because of that wind 130 joints and 101 physical points are fully generated in the body of the fetus. The sūtra states:

> Nanda, in the twelfth week a wind called "crooked gate" arises within the mother's womb, and because of that wind expanding, the intestines, colon, and bowels form to the right and left.

> Thus, for example, they reside together in dependence on the body like the roots of a lotus. During that seven days a wind called "platted hair" also arises, and because of that wind 130 joints are fully generated [in the fetus] in the womb. Also, by the power of that wind 101 physical points arise without omission.[466]

In the thirteenth week, because of the power of the two winds called the "crooked gate" and "platted hair" that arose earlier in the mother's womb, the embryo abiding in the mother's womb comes to know hunger and thirst. [486] Therefore any nutrients derived from what is consumed by the mother benefit the body of the embryo abiding in the womb via the umbilical cord at the navel. The sūtra states:

> Nanda, in the thirteenth week that fetus comes to know hunger and thirst owing to the power of those earlier winds in the mother's womb. Any nutrients derived from what is consumed by the mother benefit the body of that fetus via the navel.[467]

In the fourteenth week a wind called "tip of the thread" arises within the mother's womb, and because of that wind 1,000 sinews are generated in the body of that fetus abiding in the womb. Further, 250 sinews exist in the front of that body, 250 sinews exist in the back of that body, 250 sinews exist in the right of that body, and 250 sinews exist in the left of that body. The sūtra states:

> Nanda, in the fourteenth week a wind called "tip of the thread" arises within the mother's womb, and because of that wind 1,000 sinews are generated in the body of that fetus abiding in the womb: 250 sinews exist in the front of that body, 250 sinews exist in the back of that body, 250 sinews exist in the right of that body, and 250 sinews exist in the left of that body.[468] [487]

In the fifteenth week a wind called "lotus" arises within the mother's womb, and because of that wind twenty types of intrachannel movement are generated in the body of that fetus abiding in the womb. Further, five types of movement exist in the channels in the front of that body, five types of movement exist in the channels in the back of that body, five types of movement exist in the channels in the right of that body, and five types of movement exist in the channels in the left of that body. The names of intrachannel movements are saga (*viśākha*), strong, stable, and powerful.

Again, because of that initial wind generating eighty thousand types of movement within the channels of that fetus who abides in the womb, tastes are inhaled or drawn in. Further, twenty thousand types of channel movement exist in the front of that body, twenty thousand types of channel movement exist in the back of that body, twenty thousand types of channel movement exist in the right of that body, and twenty thousand types of channel movement exist in the left of that body. Furthermore, eighty thousand types of such movement in the channels occur in one or two up to seven openings, and each of these openings also possesses from one or two up to seven pores, and each of these pores are connected with other pores, just as there are many openings in the root of a lotus. The sūtra states:

> Nanda, in the fifteenth week a wind called "lotus" arises within the mother's womb, and because of that wind twenty types of intrachannel movement are generated in the body of that fetus abiding in the womb: five types of movement in the channels in the front of that body, five types of movement in the channels in the back of that body, five types of movement in the channels in the right of that body, and five types of movement in the channels in the left of that body. Also they possess various names and [488] various colors. Their names are saga, strong, stable, and powerful.
>
> There are twenty thousand types of channel movement in the front of that body, twenty thousand types of channel movement in the back of that body, twenty thousand types of channel movement in the right of that body, and twenty thousand types of channel movement in the left of that body. Nanda, these eighty thousand types of channel movement have many openings. Therefore they have from one or two up to seven openings, also each of these openings possess from one or two up to seven pores, and each of these pores are connected with other pores, just as there are many openings in the root of a lotus.[469]

In the sixteenth week a wind called "moving nectar" arises within the mother's womb, and because of that wind the two eye sense bases, the two ear sense bases, the two nose sense bases, the mouth opening, the throat orifice, the heart cavity, the passages for food, drink, consumables, and those to be tasted are generated, and the external and internal movement of the breath of that fetus abiding in the womb is cleared of obstruction, just as a potter or his apprentice places a lump of clay on a spinning wheel and makes various vessels of different types according to his wish. The sūtra states: [489]

> Nanda, in the sixteenth week a wind called "moving nectar" arises within the mother's womb, and because of the application of that wind the two eye sense bases, the two ear sense bases, the two nose sense bases, the mouth opening, the throat orifice, the heart cavity, and the passages for food, drink, consumables, and those to be tasted are established, and the external and internal movement of the breath of that fetus abiding in the womb is cleared of obstruction. Thus, for example, just as a potter or his apprentice takes a lump of clay and places it on the spinning wheel, and in accordance with his will he casts and molds variously shaped vessels, so too the bases of the eye and so on are established by the power of the action wind according to its will, and the internal and external movement of the breath is rendered unobstructed.[470]

In the seventeenth week a wind called "drawn face" arises within the mother's womb, and because of that wind the eyes, ears, nose, the mouth opening, the throat, the heart cavity, and the food pathways become transparent, and the external and internal passages for movement of the breath of that fetus abiding in the womb are cleaned, just as a mirror covered with dust is cleaned with grain oil and made clear. The sūtra states: [490]

> Nanda, in the seventeenth week a wind called "drawn-face" arises within the mother's womb, and because of that wind the eyes, the ears, the nose, the mouth opening, the throat, the heart cavity, and the pathways in which food is inserted are made transparent, and the external and internal passages for

> movement of the breath of that fetus abiding in the womb are cleaned. Thus, for example, just as a skillful male or female child takes a mirror covered with dust and wipes it with grain oil, ash, and sand and it becomes clear, so too the sense bases are established by the power of the action wind, and they are made unobstructed.[471]

In the eighteenth week a wind called "stainless" arises within the mother's womb, and because of that wind the six sense bases of that fetus abiding in the womb are purified, just as the sun and moon may be covered by great clouds but radiate light and shine purely when a strong wind arises. The sūtra states:

> Nanda, in the eighteenth week a wind called "stainless" arises within the mother's womb, and because of that wind the six sense bases of that fetus abiding in the womb are purified. Thus, for example, just as the sun and moon [491] may be covered by great clouds but radiate light and shine purely when a strong wind arises and the clouds are dispelled and scattered in the four directions, so too Nanda the six faculties of that fetus abiding in the womb are fully purified by the power of that action wind, and it should be understood in this way.[472]

In the nineteenth week that fetus abiding in the womb establishes four more faculties, such as the eyes and so on, in addition to the three faculties that were obtained when it first entered the womb: the body faculty, the life-force faculty, and the mental faculty. The sūtra states:

> Nanda, in the nineteenth week that fetus abiding in the womb establishes the four faculties of the eyes, ears, nose, and tongue. When it first entered its mother's womb it initially obtained just three faculties: the body faculty, the life-force faculty, and the mental faculty.[473]

In the twentieth week a wind called "extremely stable" arises within the mother's womb, and because of that wind twenty bones of the toes related to the left foot, twenty bones of the toes related to the right foot, four bones of [492] the heel of the foot, two bones of the lower leg, two bones of the knee, two bones of the thigh, three bones of the base of the spine, three bones of the ankle, eighteen bones of the vertebra, twenty-four bones of the ribs, twenty bones of the finger joints related to the left hand, twenty bones of the finger joints related to the right hand, four bones of the forearm, two bones of the upper arm, seven bones of the chest, seven bones of the shoulder, four bones of the neck, two bones of the jaw, thirty-two bones that are teeth, and the four bones of the head of that fetus abiding in the womb are generated. By analogous example, a statue maker or his apprentice first carves the shape of gods or humans in solid wood and then wraps it with thread, after which he spreads clay on it and forms the image. Further, at that time two hundred large bones are generated, excluding the fine bones. The sūtra states:

> Nanda, in the twentieth week a wind called "extremely stable" arises within the mother's womb . . . except for the fine bones, two hundred large bones are generated.[474] [493]

In the twenty-first week, a wind called "perfect producer" arises within the mother's womb, and because of that wind the flesh of that fetus abiding in the womb is generated. Just as a plasterer or his apprentice prepares the plaster well and then applies it to the walls, that karmic wind similarly generates flesh. The sūtra states:

> Nanda, in the twenty-first week a wind called "perfect producer" arises within the mother's womb, and because of that wind the flesh of the fetus abiding in the womb is generated. By example, just as a plasterer or his apprentice properly applies mud to walls, so too flesh is generated by that karmic energy.[475]

In the twenty-second week, a wind called "total victor" arises within the mother's womb, and because of that wind the blood of the fetus abiding in the mother's womb is generated. The sūtra states:

> Nanda, in the twenty-second week a wind called "total victor" arises within the mother's womb, and because of that wind the blood of the fetus abiding in the mother's womb is generated.[476]

In the twenty-third week a wind called "perfect inherer" arises within the mother's womb, and because of that wind the skin of the fetus abiding in the [494] mother's womb is generated. The sūtra states:

> Nanda, in the twenty-third week a wind called "perfect inherer" arises within the mother's womb, and because of that wind the skin of the fetus abiding in the mother's womb is generated.[477]

In the twenty-fourth week, a wind called "fully soaring" arises within the mother's womb, and because of that wind the skin of the fetus abiding in the mother's womb comes to possess elasticity and color. The sūtra states:

> Nanda, in the twenty-fourth week a wind called "fully soaring" arises within the mother's womb, and because of that wind the skin of the fetus abiding in the mother's womb comes to possess elasticity and color.[478]

In the twenty-fifth week a wind called "maintaining the city" arises within the mother's womb, and because of that wind the flesh and blood of the fetus abiding in the mother's womb become extremely transparent. The sūtra states:

> Nanda, in the twenty-fifth week a wind called "maintaining the city" arises within the mother's womb, and because of that wind the flesh and blood of the fetus abiding in the mother's womb become extremely transparent.[479]

In the twenty-sixth week a wind called "fully manifesting birth" arises within the mother's womb, and because of that wind the hair, body hair, and nails [495] of the fetus abiding in the mother's womb are generated, and all these are mutually related to their root causes. The sūtra states:

> Nanda, in the twenty-sixth week a wind called "fully manifesting birth" arises within the mother's womb, and because of that wind the hair, body hair, and nails of the fetus abiding in the

> mother's womb are generated, and all these are mutually related to their root causes.[480]

In the twenty-seventh week a wind called "physician's great fee" arises within the mother's womb, and because of that wind the hair, body hair, and nails of the fetus abiding in the mother's womb are fully established.

> Nanda, in the twenty-seventh week a wind called "physician's great fee" arises within the mother's womb, and because of that wind the hair, body hair, and nails of the fetus abiding in the mother's womb are fully established.[481]

In the twenty-eighth week distorted minds whose mode of apprehension does not conform with reality arise in the fetus abiding in the womb, generating such recognitions as: "This is a home," "This is riding," or "This is a garden," and so on.

> Nanda, in the twenty-eighth week, eight types of distorted recognition arise in that fetus abiding in the womb. [496] What are those eight? They are the recognitions: "This is a home," "This is riding," "This is a garden," "This is a multistory house," "This is a pleasure grove," "This is a throne," "This is a stream," and "This is a pool." Such false discernments are generated and their objects do not exist.[482]

In the twenty-ninth week a wind called "flower garland" arises within the mother's womb, and because of that wind the color of the skin of the body of that fetus abiding in the mother's womb becomes completely clear, fully white, and thoroughly pure and transparent.

> Nanda, in the twenty-ninth week a wind called "flower garland" arises within the mother's womb, and because of that wind the color of the skin of the body of that fetus abiding in the mother's womb becomes completely clear, fully white, and thoroughly pure and transparent. Owing to the power of karma, some come to have dark skin, some come to have bluish

> skin, some come to have skin of varied color, some have rough skin that lacks luster, some have white skin or black skin, for many types arise.[483] [497]

In the thirtieth week a wind called "iron door" arises within the mother's womb, and because of that wind the hair, body hair, and nails of that fetus abiding in the mother's womb grow and increase, and various types of white and black skin arise owing to the power of past karma. The sūtra states:

> Nanda, in the thirtieth week a wind called "iron door" arises within the mother's womb, and because of that wind the hair, body hair, and nails of that fetus abiding in the mother's womb grow and increase, and there appear as before various types of white and black skin that arise owing to the power of prior karma.[484]

From the thirty-first to the thirty-fourth week that fetus abiding in the mother's womb progressively grows larger and larger. The sūtra states:

> Nanda, in the thirty-first week, that fetus abiding in the mother's womb develops larger and larger in size. So too in the thirty-second, thirty-third, and thirty-fourth weeks it grows larger.[485]

In the thirty-fifth week that fetus abiding in the mother's womb has now fully developed all its major and minor limbs. The sūtra states:

> Nanda, [498] in the thirty-fifth week that fetus abiding in the mother's womb has fully developed all its major and minor limbs while in the womb.[486]

In the thirty-sixth week that fetus abiding in the mother's womb develops the desire not to remain in the womb. The sūtra states:

> Nanda, in the thirty-sixth week that fetus abiding in the mother's womb desires not to remain in the womb.[487]

In the thirty-seventh week that fetus abiding in the mother's womb develops the undistorted thought that is in conformity with its actual state of affairs, such as the recognition of the womb being unclean. Here, the sūtra states:

> Nanda, in the thirty-seventh week that fetus abiding in the mother's womb develops three types of undistorted recognition: the recognition of the womb as unclean, the recognition of its smell as unpleasant, and the recognition of its darkness. These arise together.[488]

In the thirty-eighth week a wind called "retracting the limbs" arises within the mother's womb, and because of that wind the posture of the body of the fetus that resides in the womb changes, [499] the head now points downward, the two arms retract, and the head is established at the opening of the birth canal. Also at that time a wind called "facing downward" arises, and because of that wind the fetus abiding in the womb turns its head downward with its legs raised upward and prepares to exit from the birth canal. The sūtra states:

> Nanda, in the thirty-eighth week a wind called "retracting the limbs" arises within the mother's womb, and because of that wind the body of that fetus that resides in the womb changes, the head points downward, the two arms retract, and the head is established at the opening of the birth canal. Further, at that time a wind called "facing downward" arises, and by the power of karma that fetus faces its head downward with both legs upward at the opening of the birth canal.[489]

Thus the *Nanda's Sūtra on Abiding in the Womb* presents in extensive detail how the fetal body evolves in the womb from a subtle to a coarse state, how the various parts of the body come to develop, and how the sensory organs emerge and so on. [500]

# 28

## Fetal Development in the Kālacakra Tantra

THE KĀLACAKRA TANTRAS present the stages of the formation of the body in the womb in the following way. The extremely subtle mental consciousness of the intermediate being enters the center of the composite of the white and red substance of semen and blood of the parents who have engaged in sex, and the coarse body of the sentient being comes to form within the mother's womb from the coalescence of the three substances: semen, blood (ovum), and consciousness. Further, it is said that this composite of seed or fluid that is the mixture of semen, blood (ovum), and consciousness enacts the following five functions in the mother's womb: its earth element supports, its water element binds, its fire element matures, its wind element increases, and its space element or the empty space within the womb accommodates the development of the fetus. For example, the *Condensed Kālacakra Tantra* states:

> The seed residing in the lotus is supported by the retainer (i.e., earth), bound by water; it is matured by fire, consumed by taste, increased by the wind; and it is space that accommodates development.[490]

First the father and mother engage in sex, and the consciousness of the intermediate being enters the center of the composite of white and red substances of semen and blood in the womb. Then to this is added the explanation of the four phases of fetal development prior to birth such as the fish phase and so on.[491] [501] Puṇḍarīka's great *Stainless Light* commentary states:

> After the days of the fish phase pass, then the tortoise phase arises, where the fetus develops five protuberances: the head, the two arms, and the two legs.

And:

> In the fifth month, during the boar phase, the fetus obtains flesh, blood, and 360[492] bones as well as joints.

And:

> At the end of the boar phase, when the fetus has completed ninth months, ten months, eleven months, or twelve months in the womb, intense suffering arises at the time of birth. Birth is now certain, the man-lion fetus is crushed as it passes through the birth canal and then emerges from the womb.[493]

Further, it is said that in the first month that mixture of semen, blood (ovum), and consciousness within the womb is merely a mass of semen and blood (ovum) without channels. Then in the second month, at the site of the development of the future heart, ten extremely subtle channels like single hairs develop and act as the causal basis of the arising of the ten types of wind, such as *life-sustaining wind* and so on. At the site of the development of the navel, twelve extremely subtle causal channels arch upward and act as the causal basis of sixty-four channels, the eight times eight channels of the navel cakra. At the site where both the legs and arms and the head develop, subtle channels spread there, acting as their causal basis. It is said that mere symbolic indicators of these will emerge in the fetus's second month in the womb. [502] Thus *Condensed Kālacakra Tantra* states:

> For one month the blood and seed in the womb transform into the taste of nectar and fully develop. Later at the heart, the seed develops into the ten types of extremely subtle channels. At the navel, eight times eight channels extend to the legs and upward to the arms and face.[494]

The term "the fish phase" is applied to the first and second months of gestation because the physical shape of the embryo in the womb in the second month is oblong and naturally red in color from semen and blood, just as the rohita fish has an oblong shape and is red in color. At the completion of the second month and the first dawning of the third month, the protuberances of the arms, legs, and neck begin to mutually extend in the fetal body in the womb. At the completion of the third month and at the dawn of the fourth month, the main protuberances of the feet, arms, and neck increase in bulk. From the beginning of the fourth month until completion, the five protrusions that act as causes of the legs, arms, and neck serially develop. At the sites of the six great joints of the legs and arms subtle channels simultaneously arise, and at the site of the neck extremely subtle complete-enjoyment channels develop. The *Condensed Kālacakra Tantra* states:

> Those who have completed the second month [503] possess the preliminary signs of legs, arms, and face. When the third month is completed the protuberances of the legs, arms, and neck arise together. Then in the fourth month the subtle channels develop at the site of the legs, arms, face, and neck.[495]

The third to the fourth month of gestation is considered "the tortoise phase" because just as a tortoise has five protrusions—the two legs, two arms, plus the head—the body of the fetus has five protrusions: the four legs and arms plus the head from the second month. In the fifth month 360 bones as well as small joints of the same number begin to develop in dependence on the flesh. In the sixth month the flesh and blood separate and the fetus experiences bliss and suffering. At the completion of the seventh month, the eyebrows, hair, body hair, cavities, and bases of the eyes and so on complete their development, having previously begun their development from the fifth month. In the first month of entering the womb, known as the "the womb month," the channels have yet to arise, that is to say, none of the channels have arisen at all.[496] But with the completion of that month and from the beginning of the second month, two hundred channels, referred to by the phrase "empty empty eyes,"[497] develop each day [504] for the next twelve months, altogether leading to the development of seventy-two thousand channels. In the eighth month the natural essence of 360 bones develop, and from among these bones the legs develop, and in order to apprehend tastes or generate the tongue sense power the sense bases such as the eyes that previously formed now develop, the tongue faculty develops the ability to apprehend tastes, and excrement and urine come to be formed. The *Condensed Kālacakra Tantra* states:

> In the fifth month flesh, blood, 360 bones,[498] and joints develop.
> In the sixth month flesh and blood arise, also the feelings of pain and pleasure.
> Eyebrows, hair, body hair, and orifices arise, and with regard to the supreme sage the remaining channels.
> By the end of that month two hundred[499] channels arise each day.
> Bone essence and bones become legs, taste is apprehended, and feces and urine arise in the eighth month.[500]

From the fifth month of gestation up to the time of birth is called "the boar phase." Just as the boar lives by eating unclean matter such as feces, so too the fetus in the womb survives solely on the unclean products of the mother's food and drink, or because the hunched figure of the fetal body resembles the shape of a boar. [505] For some fetuses the orifices do not form until the end of the ninth month, for some not until the tenth, for some not until the eleventh, and for others not till the twelfth month when it exits the womb. It is said that there is great, unbearable suffering, for the birth process involves being squeezed owing to the skeletal structure of the hips, as if being drawn through a narrow tunnel. *Condensed Kālacakra Tantra* states:

> It is called an orifice, and at the time of birth, owing to the birth site, intense suffering arises from being squeezed.[501]

Development in the womb before the ninth month is completed is called the "man-lion" phase. Just as the emanation of Viṣṇu (i.e., Narasiṃha), whose lower body resembles a human and upper body a lion, disemboweled the king of the demigods with his claws, so too the fetus at the completion of nine months and the time of birth opens the birth canal of the mother with its two hands.

As the Kālacakra texts have it, during the period in the womb the seed-like state of the zygote (composite semen and blood) develops in stages.

There is a serial development of the channels and winds, the establishment of the secondary limbs of the body, and the protrusions of the sense faculties and so on divided into four stages: from the first month to the end of the second, [506] from the third month to the end of the fourth, from the fifth month until just prior to birth, and the final stage of the birth itself. These correlate, in their respective order, to the fish, tortoise, boar, and man-lion phases. The texts also give clear presentations on how the fetus transforms each month inside the womb.

# 29

# Fetal Development in Buddhist Medical Texts

EXPLANATIONS OF FETAL DEVELOPMENT in the womb are also found in medical texts by Indian Buddhist masters. For example, Matṛceṭa's *Core Summary of the Eight Branches* states:

> The seed itself and the great elements
> are subtle and develop in stages in the womb
> through following the mind and the
> refined essence of food consumed by the mother.[502]

Thus it is in dependence on the following that the fetus inside the womb develops in stages, such as the phases of the kalala embryo and so on: (1) the white and red seed or fluid produced by the union of the father and mother, (2) the great elements such as space and so on, (3) consciousness that is subtle and lies beyond the ordinary sense faculties but arises from its preceding similar-type continuum, [507] and (4) the refined essence of the food consumed by the mother.

1. On the stage of the kalala embryo the same medical text states:

> In the first month it does not clearly manifest,
> but after seven days it becomes a kalala embryo.[503]

In the first week of the first month of gestation the fetus abiding in the womb does not clearly manifest as a kalala embryo and so on, but after the first seven days have passed it resides as a kalala embryo for up to one month through mixing that which is viscous with nonviscous.

2. On the stage of the ghana embryo and so on, the same text states:

> Then in the second month the kalala embryo becomes
> a ghana embryo, or an arbuda embryo, or a peśin embryo.
> Through that it serially becomes male, female, or neuter.[504]

Then in the second month that earlier kalala embryo becomes a ghana embryo, or an arbuda embryo, or a peśin embryo. Further, that earlier kalala embryo becomes male through becoming a ghana embryo, becomes female through becoming an arbuda embryo, or becomes neuter through becoming a peśin embryo. [508]

3. On the stage of the protrusions of the limbs and so on, the text states:

> In the third month the head, two hips, and
> two upper arms of that body develop.
> The five limbs become clear protrusions.
> Also the subtle limbs arise.
> At the same time the head and so on arise,
> knowledge of pleasure and suffering arise.[505]

In the third month of gestation the head, two hips, and two upper arms of the body of that fetus abiding in the womb develop. The five limbs become clear protrusions and the secondary limbs arise serially. At the same time as the head and so on form, knowledge of pleasure and knowledge of suffering arise. The text states:

> Because there is a relationship between the navel of that fetus
> abiding in the womb and the heart channel of the mother,
> that fetus develops owing to that,
> similar to irrigating a field.[506]

Since there is a connection between both the navel of the fetus and the channels at the mother's heart, the body of the fetus develops through the nutrients of the food and drink consumed by the mother and conveyed via the umbilical cord. It is like enhancing crops in the field through water conveyed via an irrigation channel. [509]

4. On the stage of clear protrusion of all limbs and so on, the text states:

> In the fourth month the limbs clearly protrude.[507]

All the limbs of the fetus abiding in the womb clearly protrude where before they did not clearly protrude.

5. On the stage where the mind becomes much clearer, the text states:

> In the fifth month the mind becomes clear.[508]

In the fifth month of gestation the mind of that fetus abiding in the womb becomes extremely clear where previously it was not extremely clear.

6. On the stage where the secondary limbs grow and evolve and so on, the text states:

> In the sixth month ligaments, channels, body hair,
> nails, skin, complexion, and strength arise.[509]

In the sixth month of gestation, ligaments or tendons, channels, body hair, nails, skin, complexion, and strength arise in that body of the fetus abiding in the womb. [510]

7. On the stage where all limbs are complete and developed and so on, the text states:

> In the seventh month all the limbs
> of the fetus are complete.[510]

In the seventh month of gestation the shape of the major limbs and secondary limbs of the body of that fetus abiding in the womb are fully complete and developed.

8. On the stage of mutual transference of the mother's vital essence containing all she has physically assimilated, the text states:

> In the eighth month the refined vital essence
> progressively transfers between the mother and fetus at
> different times.
> Owing to that, one declines and the other is satiated.
> Because of the vital essence abiding in the fetus and not
> the mother,
> doubt arises as to whether the mother
> will lack nourishment at the time of birth.[511]

In the eighth month of gestation the refined vital essence of the physical strength of the body transfers at different times between the mother and the fetus in her womb. Furthermore, when the vital essence transfers to the fetus, it is fully satisfied and its body becomes vibrant while the mother's body experiences decline. Alternatively, when the vital essence transfers to the mother, the mother's body is fully sated and her body becomes vibrant while the body of the fetus experiences decline. Moreover, when the [511] vital essence transfers to the mother, and if the fetus is born at that time, then the fetus will not be nourished. But when the vital essence transfers to the fetus, and if the fetus is born at that time, doubt remains as to whether the mother's life-force will receive nourishment.

9. On the stage of giving birth to the fetus and so on, the text states:

> Beginning from the first day of the ninth month
> the mother should enter the house to give birth
> that is in an auspicious location, with quality chattels,
> on a day marked as virtuous.[512]

This preparation for birth occurs from the first dawn of the ninth month of gestation. On a day when the planets and stars align favorably, the pregnant woman should enter the house where the birth will take place, possessing an auspicious location, and endowed with excellent conditions conducive to giving birth, and so on.

In brief, one can discern the existence of extremely extensive explanations in classical Indian Buddhist texts concerning the stages of fetal development in the womb. The broad categories of texts in which these explanations are found include: (1) sūtras such as *Nanda's Sūtra on Abiding in the Womb* and the Abhidharma treatises that take these sūtras as their basis, (2) the texts of the highest yoga tantra in general and the Kālacakra texts in particular, and (3) medical texts such as the Matṛceṭa's *Core Summary of the Eight Branches.*

There are slight differences in the way the various stages of fetal development in the womb are differentiated in these sources. [512] For example, *Nanda's Sūtra on Entering the Womb* principally delineates changes week by week. In the Kālacakra texts, on the other hand, although in general the explanation is in terms of monthly changes, the structure is chiefly four main stages: (1) the first two months, (2) the third and the fourth month, (3) from the fifth month up to birth, and (4) the birth itself. These four periods are correlated to the fish, tortoise, boar, and man-lion. In the medical texts each month is identified as a specific stage within the womb and explicit explanations are offered on what kind of changes occur during each stage. These kinds of minor differences can be discerned from the three different sources. This said, there is no difference at all on the fundamental point that at first merely the vital essence that is a mixture of the semen and blood of the parents exists within the womb, and this is what progressively develops and gives rise to the formation of the body limbs and the protuberances of the sense organs and so on. [513]

# 30

## The Subtle Body of the Channels, Winds, and Drops

THE INDIAN BUDDHIST TEXTS of the highest yoga tantra as well as some medical texts speak of a body that exists during subtle states of existence that is not the coarse material body of flesh and bones that we possess. The texts explain the activity of the subtle body in relation to our coarse body on the basis of (1) the channels that reside within the body, (2) the winds that move within these channels, and (3) the drops that adorn the vital points of the body. Below we shall offer a brief explanation of these topics as an introduction.

To explain first how the subtle channels and winds are presented in the highest mantra texts, in general, the Buddhist texts discuss three types of body: (1) the *coarse* body, which refers to the body that is composed of flesh and bones, (2) the *subtle* body, referring to the channels, winds, and drops, as well as the dream body and the body of the intermediate state, and (3) the *extremely subtle* body, which refers to the extremely subtle wind that is the medium of the mind on the four stages of emptying, especially the wind that is the mount of the fourth empty, which is the clear light mind. Similarly, as will be explained below, with respect to the mind, the five sense consciousnesses are said to be *coarse*, the minds concomitant with the root and branch afflictions as well as the eighty indicative conceptions are *subtle*, and states of mind that constitute the [514] four empties are said to be *extremely subtle*.

As such the composite of coarse body and mind is referred to as the "coarse common reality of the body and mind," and the indivisible unity of extremely subtle mind-body is referred to as the "subtle common reality of the body and mind."[513] For example, Nāropa's *Clear Compilation of the Five Stages* states:

> Entities have a twofold mode of being:
> that of the mind and that of the body.
> In terms of their states,
> there is the coarse, the subtle, and the extremely subtle.
> Their common mode of being is their indivisibility."[514]

Now to explain briefly the channels, winds, and drops—which constitute the subtle body—as taught in the highest yoga tantra texts. "Channels" refer to parts of the body within which the winds and the elements reside and flow and which also serve as the basis for consciousness. Etymologically, the word *channel*[515] connotes something fundamental, since the channels act as a basis of movement and connection by which the winds and vital blood of one's being move and connect through all external and internal open pathways. They act initially as channels to establish and enhance the body, in the intermediate period to maintain and sustain the body for an extended period, and in the end they transform into the channels of death and so forth. *Nanda's Sūtra on Entering the Womb* states: [515]

> In the fourteenth week as the fetus resides in the mother's womb, the action wind that manifests is called "head of the thread." By the power of that wind nine hundred channel pathways are generated that together cover the front, the back, the right, and the left of the body. In the fifteenth week as the fetus resides in the mother's womb, the action wind that manifests is called "lotus." By the power of that wind twenty types of movement [of wind] in the channels are generated. The taste of food and drink enters this channel and benefits the body. What are those twenty? In the front, back, right, and left of the body there are five channels each, for each of these channels there are forty subtle branch channels, and for each of those channels there are one hundred further branches. Twenty thousand [channels] in the front of the body are called "saga," twenty thousand in the back of the body are called "power," twenty thousand in the right of the body are called "stable," and twenty thousand in the left of the body are called "forceful." Thus eighty thousand extremely subtle channel branches arise in that body. Also these

> channels possess various types of color. Thus there is blue, yellow, red, white, the color of butter, curd, and grain oil. For each of these eighty thousand channels there is a root channel, and each root has various openings, for they possess from one or two to seven openings. [516] Each is related in common to body-hair pores, for example, like the many openings in a lotus root.[516]

In brief, in that sūtra twenty thousand channels are mentioned for each of the four parts of the body—left, right, back, and front—and in total eighty thousand channels exist in the body.

According to the system of the highest yoga tantra, there are seventy-two thousand channels in the body. Of these, the first three channels to form at the heart are: (1) the right channel or *rasanā*, (2) the left channel or *lalanā*, and (3) the central channel or *avadhūtī*. *Saṃvarodaya Tantra* states:

> Channels such as left channel and so on
> are the main channels among all the channels.[517]

So too *Vajra Garland Tantra* states:

> They are given the names
> lalanā, rasanā, and avadhūtī.[518]

The first five channels to form at the heart are, in addition to these three, the *traivṛtta* channel to the east of the heart and the *kāminī* channel to the south of the heart. These five channels develop together. After these, three more develop, which are the *gehā* channel to the west, the *cāṇḍālī* channel to the north, and the *māradārikā* channel that runs parallel to the central channel. These last three develop together at the same time, and [517] by adding them to the five there are eight. *Vajra Garland Tantra* states:

> Owing to the process of proliferation,
> serially they each divide into eight, eight times.[519]

The "eight channel spokes at the heart" refer to, in addition to the four (*traivṛtta* in the east, *kāminī* in the south, *gehā* in the west, and *cāṇḍālī* in the north), the southeastern channel spoke branching from *traivṛtta*, the southwestern channel spoke branching from *kāminī*, the northwestern channel spoke branching from *gehā*, and the northeastern channel spoke branching from *cāṇḍālī*. Thus the "eight channel spokes at the heart" are not the same as the "eight channels that first form at the heart."

When the eight channel spokes at the heart, such as *traivṛtta* in the east and so on, are differentiated in terms of body, speech, and mind, there emerge twenty-four channels. These are called "the channels of the twenty-four places." *Saṃpuṭa Tantra* states:

> Within the heart of the body
> reside perfectly the five channels.
> Differentiated in terms of body, speech, and mind
> they are called the "twenty-four."[520] [518]

When the eight channels that first form at the heart are combined with the channels of the twenty-four places, there are "the thirty-two channels in which bodhicitta flows." When each of these channels of the twenty-four places again branch into three, there are seventy-two channels. By dividing each of these into one thousand there are seventy-two thousand channels. If all such channels are summarized, they are subsumed in three channels: the left channel, right channel, and central channel. The *Hevajra Tantra* states:

> The Bhagavan stated, "The channels are thirty-two in number, for from these descend the thirty-two bodhicittas to the site of great bliss."[521]

Similarly the explanatory Guhyasamāja tantra *Vajra Garland* states:

> If enumerated consecutively there are seventy-two.[522]

And:

> Again, owing to the process of proliferation
> they are also each divided into one thousand,
> and their count always
> remains as seventy-two thousand.[523]

Thus the seventy-two principal channels each divide into one thousand branch channels, grouped into three sets of twenty-four thousand. These are also designated as twenty-four thousand in which semen descends, twenty-four thousand in which blood descends, and twenty-four thousand in which wind descends. *Vajra Garland Tantra* states: [519]

> The twenty thousand channels,
> are channels that continuously amaze,
> know that bodhicitta descends in them,
> all levels of bliss are greatly enhanced.
> Channels that continuously amaze
> twenty-four thousand in number,
> know that blood descends in them;
> they have the character of sun's movement.
> Channels that continuously amaze
> twenty-four thousand in number,
> know that wind moves in them.[524]

Among these, the channels in which white bodhicitta mainly descends are referred to as "the body channels," the channels in which red substance mainly descends are referred to as "the speech channels," and the channels in which wind mainly descends are referred to as "the mind channels." All other channels are described as subsumed in these three. With respect to the main channels the *Saṃvarodaya Tantra* states:

> The channels associated with
> the body are seventy-two thousand.
> The channels and the secondary channels,
> these are dependent on those sites.
> Twenty plus one hundred channels
> are described as the main channels.[525]

The *Ḍākā Ocean Tantra* states:

> In the lotuses of the four cakras,
> there are one hundred and twenty channels.[526] [520]

This refers to the channel spokes at the crown, throat, heart, and navel. Here the crown cakra, Great Bliss wheel, has thirty-two channel spokes. The throat cakra, Perfect Enjoyment wheel, has sixteen channel spokes. The heart cakra, the Reality wheel, has eight channel spokes. And the navel cakra, the Emanation wheel, has sixty-four channel spokes. Thus there are one hundred and twenty in number. Since these channels act as the basis of the winds and consciousness they are referred to as the "main channels." If summarized, the channels of the thirty-two places can be called the "main channels" since they act as the basis of the elements in terms of generating teeth, nails, and so on. The channels of the twenty-four abodes can be called the "main channels." The three channels—the left channel, right channel, and central channel—can be called the "main channels" since the great innate bliss of clear light arises in dependence on the winds of the right channel and left channel, entering, abiding, and dissolving into the central channel. Even within these three, the central channel constitutes the main channel. The central channel at the level of the heart is the main channel since it is the basis of the three actions of entry, abidance, and transference of consciousness on the ordinary level and the site where the perfect clear light dawns.

## Channel Centers

The crown great-bliss cakra exists between the cranial membrane and the skull with four channel spokes that branch out from the point where the right channel and the left channel wrap around the central channel, forming a knot. Eight channel spokes branch out from those, sixteen from them, and thirty-two from them. Their color is diverse and their dimension is that of a large straw. Its center is triangular and its shape is like an upright opened umbrella. [521] It is called the *great-bliss cakra* since kunda-like bodhicitta that is the basis for generating bliss resides at the crown cakra.

The throat perfect-enjoyment cakra has four channel spokes that branch out from the point where the right channel and the left channel form a knot circling the central channel at the throat at the level of the larynx. Eight channel spokes branch out from those and sixteen from them. Their color is red, the center of the cakra is circular, and its shape is like an inverted open umbrella. It is called the *perfect-enjoyment cakra* since the six tastes are enjoyed in the throat cakra.

The heart reality cakra has four channel spokes that branch out from the point where both the right channel and the left channel encircle the central channel three times at the central point between the nipples. Eight channel spokes branch out from them. Their color is white, the center of the cakra is circular, and its shape is like an upright open umbrella. It is called the *reality cakra* since the indestructible drop of extremely subtle wind and mind that is the source of reality resides in the heart cakra.

The navel emanation cakra has four channel spokes that branch out from the point where the right and left channels encircle the central channel at the navel aperture. Eight channel spokes branch out from these, sixteen from them, thirty-two from them, and sixty-four from them. They are red in color, the center of that cakra is triangular, and its shape is like an inverted open umbrella. It is called the *emanation cakra* [522] since the main basis for emanating or generating great bliss is the fire of cāṇḍālī that resides in the navel cakra.

The secret bliss-sustaining cakra has four channel spokes that branch out from the point where both the right and left channels constrict the central channel at the point of demarcation of the white and black body hairs of the penis. Eight channel spokes branch out from those. In each of the four directions there are five (making twenty), four in the intermediate directions, and eight of the basis that branches, making thirty-two. Their color is red, the center of that cakra is triangular, and its shape is like an open umbrella. It is called the *bliss-sustaining cakra* since the innate bliss of the forward and reverse order is mainly sustained in the secret cakra.

There are also three minor cakras, which when added to the five great cakras brings the number of cakras to eight. They are the wind cakra at the center of the eyebrows with six channel spokes, the fire cakra between the throat and heart with three channel spokes, and the midjewel cakra with eight channel spokes. The roots of all the channel spokes are described as piercing the central channel, and also all channels abutting the right or left channel are said to pierce through to the central channel by directly piercing the right or left channel.

As for how the winds and bodhicitta descend in these channels, the fire-wind moves and blood descends in the right channel, the wind-wind moves and semen descends in the left channel, and both the earth-wind and water-wind move and descend in both the right and left channels. On the ordinary level, it is only at the point of death that, due to the power of karma, the knots at the heart loosen and the winds move in the central channel. [523] Apart from this, without meditating on the path, there is no occasion when the knots at the heart are released.

How the central channel is wrapped in a continuous circle of knots by the right and left channels, making them so difficult to loosen, is as follows. The central channel at the heart is constricted by the right channel circling it three times to the right and the left channel circling it three times to the left, forming a triple knot. At the other sites—secret place, navel, throat, and crown—there are only single knots. The *Vajra Garland Tantra* states:

> It is bound by three knots at the center
> that are difficult to loosen.[527]

Though both the right and left channels are similar in having two openings—one above and one below—the explanation that the left channel faces downward and the right channel faces upward is from the perspective of whether semen and blood are emitted or retained. From among the twenty-four channels, those channels where a greater amount of white bodhicitta descends are called *body channels* or *courage* (*sattva*) *channels*.[528] Those channels where a greater amount of red substance descends are called *speech channels* or *particle* (*rajas*) *channels*. Those channels where a greater amount of wind descends are called *mind channels* or *darkness* (*tamas*) *channels*.

In the channel spokes in the four cardinal directions the winds of the four elements flow, while in the channel spokes of the four intermediate directions the winds of form, sound, aroma, taste, and tactility descend. Hence they are called *offering channels* or *channels of the descent of the five sense objects*. The texts explain that from those eight channel spokes the five nectars—feces, urine, semen, blood, and phlegm—descend as well. [524]

The *māradārikā* channel, which is one of the eight channels that initially form at the heart, is called *cessation channel* or *time channel*. On the ordinary level, except at the time of death, wind and bodhicitta do not descend in that channel. This channel in fact hinders the descent of wind and bodhicitta in the other channels. However, at the time of death wind and bodhicitta will descend in this channel, so it is called *cessation channel* or *time channel*.

## The Order of Formation of the Channels

After consciousness first enters the center of semen and blood, which are the white and red substances of the father and mother, a mass resembling curd scum comes to form. From this, over time the following five channels develop simultaneously: the right channel, the left channel, and the central channel at the level of the heart, and both the traivṛtta channel and kāminī channel, which constitute two of the eight spokes of the heart cakra. After this, the three channels—gehā, cāṇḍālī, and māradārikā—develop. Then the four channels of the intermediate directions of the heart cakra develop. Next, gradually, the channels of the twenty-four places (from the division of each of the eight channel spokes at the heart branch into three) and the seven-two channels (from the branching off of each of the twenty-four channels into three), and the seventy-two thousand channels that branch off from these will come to develop.

In the Kālacakra tantra system, however, the three channels above the navel—the right, the left, and the central channel—are referred to by the following names: The right channel is the "sun channel," "orange channel," "*rasanā*," and "the sun's path." The left channel is the "moon channel," "*igasa*," "*lalanā*," and "the moon's path." And the central channel is "Rāhu channel" and the "supreme channel." The three channels below the navel—the right channel, the left channel, and the central channel—are referred to by the following names: the right channel is called the "urine channel," the left channel is the "feces channel," [525] and the central channel is the "*śāṅkinī* channel," "the *kalāgni* channel," "the semen-descent channel," and "the supreme channel."

The six great channel centers (*cakras*) exist at the sites where the right and left channels encircle the central channel: (1) the secret cakra with thirty-two spokes, (2) the navel cakra with sixty-four spokes, (3) the heart cakra with eight spokes, (4) the throat cakra with thirty-two spokes, (5) the forehead cakra with sixteen spokes, and (6) the crown-protrusion cakra with four spokes. Puṇḍarīka's *Stainless Light* states:

> It is known by the names of four by dividing the crown protrusion into four sections . . . up to . . . the thirty-two channel spokes at the secret place.[529]

## The Winds that Move in Those Channels

The presentation of the winds is most extensive in the texts of the highest yoga tantra of Guhyasamāja and its commentaries. There when the classifications of winds is presented the following ten winds are mentioned, which are the five root winds: (1) life-sustaining wind, (2) downward-voiding wind, (3) upward-flowing wind, (4) pervading wind, and (5) accompanying wind;[530] plus five branch winds: (1) moving wind, (2) roving wind, (3) perfectly flowing wind, (4) intensely flowing wind, and (5) definitely flowing wind.[531]

The explanatory tantra of Guhyasamāja, *Tantra Predicting Realization*, states:

> The life-sustaining, upward-moving, and downward-voiding,
> the accompanying, and the pervading wind, [526]
> and those known as nāga, turtle, lizard,
> as devadatta, and dhanaṃjaya winds.[532]

Nāgārjuna's *Five Stages* states:

> This is the mount of consciousness;
> embodied in five, possessing ten names.[533]

The "mount of consciousness" refers to wind. *Vajra Garland Tantra* states:

> Continuously, day and night,
> one hundred and eight are exceedingly clear.[534]

These lines refer to one hundred and eight winds composed of fifty-four object-associated winds and fifty-four subject-associated winds, all of which are subsumed in the ten root and branch winds.

When consciousness first enters the womb there is already by nature the presence of the extremely subtle life-sustaining wind. Then in the first month, from this extremely subtle life-sustaining wind, the coarse life-sustaining wind arises. From this coarse life-sustaining wind, in the second month the downward-voiding wind arises. In the third month, from this downward-voiding wind the accompanying wind arises. In the fourth month, from this accompanying wind the upward-moving wind arises. In the fifth month, from this upward-moving wind the pervading wind arises. In the sixth month, from this pervading wind the moving wind arises. In the seventh month, from this moving wind the roving wind arises. In the eighth month, from this roving wind the perfectly flowing wind arises. In the ninth month, [527] from this perfectly flowing wind the intensely flowing wind arises. In the tenth month, from this intensely flowing wind the definitely flowing wind arises. Although these winds arise during the womb state, they lack the function of entering and exiting. But from the moment birth occurs the winds exhibit the activity of exiting and entering through the nostrils and so on.

As for where these winds reside, the life-sustaining wind resides at the heart, the downward-voiding wind at the secret place, the upward-moving wind at the throat, the accompanying wind at the navel, and the pervading wind throughout the body. With regard to their specific functions, the life-sustaining wind guides the other winds to the doors of the sense faculties and acts to prolong life and so on. The downward-voiding wind directs urine, winds, feces, semen, and so on to the lower openings. The upward-moving wind directs the winds to the upper body and facilitates eating food and speaking. The accompanying wind separates the nutrition and waste of food and drink and guides nutrition to the body. And the pervading wind facilitates movement and motion of the body.

Thus *Lamp on the Compendium of Practice* states:

> The five winds such as the life-sustaining and so on are dependent on the five aggregates and facilitate the function of the aggregates.[535]

Given that the five secondary winds are mostly branches of the life-sustaining wind, the main site where they both arise and subside is the heart. Thus via the eight spokes at the heart, the moving wind flows into the eyes, the roving wind into the ears, the perfectly flowing wind into the nose, the intensely flowing wind into the tongue, and the definitely flowing wind moves pervasively in the body, [528] and in this way they help the senses to engage their respective objects. That is, the moving wind facilitates eye consciousness seeing form, the roving wind facilitates ear consciousness hearing sounds, the perfectly flowing wind facilitates nose consciousness smelling aromas, the intensely flowing wind facilitates tongue consciousness tasting flavors, and the definitely flowing wind facilitates body consciousness apprehending tactile objects.

*The Lamp on the Compendium of Practice* describes these functions in a detailed manner:

> The five winds arise and move and so on and engage in external action through abiding in the sense faculties.[536]

The ten winds move thus. Each day is composed of twelve periods or twenty-four half periods. Now during a half period, counting each cycle of exiting and entering as one movement, there are 900 such movements of each of the winds except for the pervading wind. Thus, excluding the pervading wind, there are in total 21,600 movements each day for each of the nine root and branch winds. Furthermore, when the life-sustaining wind moves, it moves in both the right and left channels evenly through both nostrils without interruption, and for each half period there are 900 slow movements. So too when the downward-voiding wind moves, it moves in both the right and left channels through both nostrils without interruption, and for each half period there are 900 strong movements. When the ascending wind moves, it moves in the right channel through the right nostril, and for each half period there are 900 uninterrupted movements. When the accompanying wind moves, it moves in the left channel through the left nostril, [529] and for each half period there are 900 uninterrupted movements. The pervading wind, apart from the time of death, does not move through the nostrils at the time of the basis (that is, while alive). Yet according to the Kālacakra texts, even the pervading wind is said to be similar to the other winds in possessing the capacity to move through the nostrils.

As for the branch winds, the moving wind moves to the eye sense door, the roving wind moves to the ear sense door, the perfectly flowing wind moves to the nose sense door, the intensely flowing wind moves to the tongue, and the definitely flowing wind moves through all the body pores, transmitted from within all the channels of the body.

## Drops

A "drop" refers to a composite consisting of four things—wind, consciousness, and the vital essences of semen and blood—measuring about the size of a mustard seed. And it is within the empty capsule at the center of the eight channel spokes of the heart cakra that the *indestructible drop* of white and red bodhicitta received from both the father and mother resides. At the very center of the indestructible drop is the extremely subtle innate mind and the wind that is its mount, and these remain stable as long as one lives. From the white drop at the heart an emergent part moves to the crown and mainly resides inside the center of the great-bliss cakra at the crown. Then gradually in incremental steps the drop moves to other sites, such as the forehead, the lower tip of the central channel, and so on. One emergent part from the red drop at the heart proceeds to the navel and mainly resides [530] at the center of the emanation cakra at the navel. Through gradual increment the drop proceeds to other sites of the body, such as the throat, secret place, and so on.

Drops are of two types: the white drop and the red drop. Or they are of four kinds: (1) the drop generating deep sleep, (2) the drop generating dreams, (3) the drop generating the waking state, and (4) the drop generating the fourth state. All of these drops are composites of three things—semen, blood, and wind. Both the white and red drops initially develop at the heart, then they separate and the white drop obtained from the father proceeds to the crown, its main location, while the red drop obtained from the mother proceeds to the navel, its main location.

(1) The drop generating deep sleep resides at the heart and at the center of the jewel. When the upper winds gather at the heart and the lower winds gather at the center of the jewel, deep sleep arises. (2) The drop generating the dream state resides at the throat and secret place.

When the upper winds gather at the throat and the lower winds gather at the center of the secret place, dreams appear. (3) The drop generating the waking state resides at the crown and at the navel. When the upper winds gather at the crown and the lower winds gather at the navel, one wakes from sleep. (4) The drop generating the fourth state resides at the crown [531] and at the secret place. Bliss is generated when the male and female engage in sex.[537]

# 31

## Channels and Winds in Buddhist Medical Texts

WE SHALL PRESENT below a brief summary from the extensive descriptions of the channels and winds in the texts on the science of healing. As to when the body first develops in the womb, at what specific weeks the channel cakras develop according to the medical texts, *Explanatory Tantra* states:

> During the second month in the fifth week
> the navel is the first part of the body to develop in the womb.
> In the sixth, the life-force channel develops in dependence on
> the navel.
> In the seventh, the protruding form of the eye organs.
> In the eighth, the shape of the head develops in dependence
> on that.[538] [532]

Thus in the fifth week during the second month, the first part of the body of the sentient being abiding in the womb to develop is the four channel spokes of the emanation cakra at the navel. In the sixth week the *life-force channel* develops at the center of the channel spokes at the navel. In the texts of secret mantra that is designated the *central channel*. The life-force channel moves upward, and the phenomena cakra at the heart with four channel spokes develops at its upper opening. Though it is not explicitly stated here, in the seventh week the life-force channel moves upward from the heart and the channel spokes of the enjoyment cakra at the throat develop, and again the life-force channel moves upward and the great-bliss cakra at the crown develops along with the protruding form of the eye faculties. At the conclusion of the eighth week the mere shape of the head develops in dependence on that cakra along with the eye sense faculties. At the same time as the development of the channels at the throat, the life-force channel descends from the navel and the channel spokes of the bliss-sustaining cakra at the secret place develop.

In explaining the presentations of the channels in the texts on the science of healing, *Explanatory Tantra* states:

> In teaching the basic state of channels that are interconnected,
> there are four channels: developmental, existence, related, and
> life.[539]

Thus among all the channels, the presentation of channels is made on the basis of first explaining the four channels—developmental channels, [533] existence channels, related channels, and life-sustaining channels. All of these channels are subsumed within the three classes of wind channels, blood channels, and water channels. Similarly, although the terminology of central channel, right channel, and left channel and so on does not explicitly appear in the root texts of medical science, such as the four medical tantras, still when understood in the terminology of Buddhist tantra all wind channels are extensions of the central channel, all blood channels are explained as extensions of the right channel, and all water channels are explained as extensions of the left channel.

## Developmental Channels

From among those four channels, the *developmental channel* is the navel channel that first develops when the body initially develops in the womb. It branches into three different tips, with one channel in which the water element mainly moves proceeding to the crown. How the brain develops will be explained later in the discussion of the principles of the brain as derived from the medical texts. Further, from those three channel branches, one channel in which blood or the fire element moves pierces the channel conveying vital essence to the fissures of the liver in the middle of the body. Two branches of that channel proceed to the thirteenth vertebra, and the black life-force channel and its branches develop in dependence on them. With blood acting as the cooperative condition, aversion arises, and with blood acting as the substantial cause, bile arises.[540] Therefore aversion manifests in dependence on the black blood channel and blood. Since aversion generates bile, disturbance is seen to arise from the middle of the body when aversion spontaneously arises. Thus *Explanatory Tantra* states:

> Owing to one channel piercing the middle, the life-force channel develops. [534]
> Aversion manifests in dependence on the life-force channel and blood,
> and from that bile develops and resides in the middle.[541]

One tip of that channel that has branched into three channels pierces downward, and from that the secret abode develops in which the blissful element moves. Further, owing to the semen and wind acting as the substantial cause and attachment acting as the cooperative condition, attachment is seen to reside in the secret place of both the father and mother, and wind arises from that attachment. Thus *Explanatory Tantra* states:

> Owing to one channel piercing downward the secret place develops.
> Attachment manifests in the secret place of the father and mother,
> and from that wind is generated and resides in the lower place.[542]

Thus when explaining the developmental channel, *Explanatory Tantra* states: "Since one channel proceeds upward, the brain develops," this explains the left channel. "Since one channel pierces the middle, the life-force channel develops," this explains the right channel. "Since one channel pierces downward, the secret place develops," this explains the central channel below the navel. [535]

## EXISTENCE CHANNELS

In this body there are four principal channels of existence or abidance. They are:

1. The class of channel called "the spiral" that resides in the brain and generates the five faculties.

2. The class of channel called "the pure mind" that resides in the heart and illuminates the faculty of memory or that of the mind and generates the sixfold collection of consciousness.
3. The class of channel called "the support" that resides in the navel and supports enhancement of the seed that develops and increases this aggregate of the body.
4. The class of channel residing in the secret place called "the authentic" that supports the element of bliss and generates and enhances the lineage of sons without degeneration.

Each of these classes of channel is said to be surrounded by five hundred minor channels. Since these great channels branch out to all parts of the body—above, below, and horizontally or vertically and so on—the physical aggregate is sustained and nourished for long periods through this network of channels. Thus *Explanatory Tantra* states:

> The great existence channel has four types:
> The channel that makes objects appear to the sense faculties;
> this existence channel in the brain is surrounded by five hundred minor channels.
> The channel that makes clear the faculty of memory;
> this existence channel at the heart is surrounded by five hundred minor channels.
> The channel that develops the aggregate of the body; [536]
> this existence channel at the navel is surrounded by five hundred minor channels.
> The channel that increases the lineage of sons;
> this existence channel at the sexual organ is surrounded by five hundred minor channels.
> They maintain the entire body, above, below, and vertically.[543]

## Related Channels

The *Explanatory Tantra* states:

> As for the related channels, they are two: white and black.[544]

The channels related to the left and right channels have two great channel types: the "white life-force channel" and the "black life-force channel," so named because of the power of the movement of white and red constituents in them.

The black life-force channel is related to the right channel like a tree trunk that is the basis for generating all branch blood channels. From it the 24 great channels that increase flesh and blood branch upward in the manner of branches. The 24 comprise the 8 great concealed channels related internally to the organs[545] and the 16 manifest channels related externally to the limbs. There are 77 channels branching from those that can be bled by a doctor and 112 fierce channel points that cannot be bled, thus in total making 189 minor channels. From those there are 120 related to the external skin, 120 related to the organs internally, and 120 related to [537] the marrow between, thus in total they branch into 360 minor channels. From those there are 700 minor channels that branch in three ways: externally, internally, and between. And there are channels branching out from those that are subtler channels related to the network of channels within the body. In summation, the body or physical aggregate is established as an extensive aggregation of channels. Thus the *Explanatory Tantra* states:

> The life-force channel is the trunk of the channels;
> from this in a manner of branches growing upward
> are the 24 great channels that increase flesh and blood:
> they are the 8 great concealed channels internally related
> to the organs,
> and 16 manifest channels related externally to the limbs.
> From these emerge 77 channels that can be bled
> and 112 fierce channel points,
> in total 189 minor intertwined channels.
> From those emerge 120 each—external, internal, and
> the in-between.
> Thus 360 minor channels come to branch out.
> From these 700 minor channels branch out.
> From these emerge subtler channels linking the body into
> a network.[546] [538]

The white life-force channel related to the left channel is the following. The brain that is generated through the upper opening of the left channel is the basis from which all white channels are generated like a great ocean. The spinal cord that pierces downward from it is the white life-force channel, and how the minor channels branch out from it will be explained below since this is closely related to the presentation of the brain.

## Life-Sustaining Channels

Here the name *channel* is applied to the life or life-force that proceeds or moves to various parts of the human body. The life-force has three such types of channel: One conveys the life-force alternately to the various elements possessing the aspect or shape of letters and pervades the entire head and body at the time of the new moon and full moon. One moves the life-force in association with breath from the nostrils and exits externally no more than sixteen finger widths and then reenters the body. One resembling vital energy that is commonly known as "that which spreads" passes from between the ligaments of the hand and moves externally. Since all these channels are components conveying the life-force, they are given the name "life." *Explanatory Tantra* states:

> Humans have three types of life channel:
> one that pervades the entire head and body,
> one that moves in association with the breath, [539]
> and one that resembles vital energy that spreads.[547]

Alternatively, there is another way of identifying the life channel. Here three main types of channel are identified that endure long and sustain the life-force. The first is related principally to the liver and is the main trunk channel branching from the liver that conveys blood, while minor channels branching from that channel pervade the entire body. The second branches from the heart and is the main channel for the coalescence of both wind and blood. That channel and its minor branches assist the breath and enhance it. The third is the water channel or the main white trunk channel branching from the great ocean of channels in the brain. Minor channels that are branches of it constitute something like "vital essence" in popular parlance[548] that spread throughout the body. These are the three life channels. Moreover, a latter section of the four tantras states: "Examine the ordinary channels in terms of the three root channels."[549] On this view, the channels residing within the body of an ordinary being unaffected by disease can be examined by touch on the basis of the three classes of male channels, female channels, and bodhicitta channels. "Male channels" refer to those that are broader and pulsate in a coarser manner. "Female channels" refer to those that are narrower and pulsate in rapid manner. "Bodhicitta channels" [540] refer to those that are long and pulsate in a supple and gentle manner. These are but brief representative accounts drawn from the extensive enumerations of channels that are mentioned in texts on the science of healing.

## The Essential Nature of the Channels

The *Explanatory Tantra* states:

> Because they are openings for the movement of winds and blood,
> and connect all external and internal components of the body and generate and maintain the body, and since they are the root of the life-force, they are called "channels."[550]

They are called "channels" because they are the subtle and coarse openings through which flow all the winds, blood, vital essence, and waste, they connect all external and internal components of the body, and they generate the body and sustain it for a long period of time. They are thus like the root of the body as well as the root that ensures the maintenance of the life-force.

As a brief example of the winds that move in those channel conduits, *Moonlight Commentary on the Essence of the Eight Branches* states:

> It is said: "In that regard wind is coarse and light, cold and hard, subtle and moving." "In that regard" means "definitely," and from among those faults, wind is coarse, light, cold, and subtle. It enters and moves in [541] the subtle channel conduits, but it does not adhere to just one type of movement.[551]

*Somarāja: A Medical Treatise* explains the essential nature of the wind that flows in these channel conduits:

> The definition of wind is lightness.[552]

Wind may be classified as two types: winds that are beneficial and winds that are harmful. Beneficial winds may be classified as five types: (1) life-sustaining wind, (2) all-pervading wind, (3) fire-like digestive wind, (4) downward-voiding wind, and (5) upward-moving wind.

As for their abodes and function, the life-sustaining wind resides at the navel and maintains internal breathing. The all-pervading wind pervades the entire body, facilitates the body rising and sitting, and makes clear the five faculties. The fire-like digestive wind aids in the digestion of food and separates the nutrition and waste in food. The downward-voiding wind facilitates elimination and retention of feces and urine and maintains the lungs and intestines. And the ascending wind assists the five elements and moves through the left and right nostrils.

*Somarāja: A Medical Treatise* states:

> If the winds are differentiated there are two types:
> beneficial winds and harmful winds. [542]
> Owing to beneficial winds one's strength grows; owing
> to harmful winds it declines.
> In terms of the faculties there are five:
> life-sustaining wind, all-pervading wind,
> fire-like digestive wind, ascending wind,
> and downward-voiding wind, making five.
> Life-sustaining wind resides at the navel
> and maintains breathing internally.
> All-pervading wind pervades the entire body,
> facilitates the body rising and sitting, and illuminates the
> five faculties.
> Fire-like digestive wind facilitates digestion of food
> and separates nutrition from waste.
> Upward-moving wind moves in the left and right nostrils
> by accompanying the five elements.
> Downward-voiding wind eliminates feces and urine;

since it resides in the lungs and intestines,
it enhances the body and its strength.[553]

As for the harmful winds, through the combination of a spreading wind and hot wind illness is generated, and through the combination of a spreading wind and cold wind illness is increased. *Somarāja: A Medical Treatise* states that there are three kinds of harmful winds:

The harmful winds are of three types:
the spreading winds, the hot winds, and
the cold winds and their combinations.

Also:

All diseases are increased by the winds.

*Four Medical Tantras* also presents the classifications of winds, their abodes, and functions. For example, the *Explanatory Tantra* states:

If winds are classified they are five: life-sustaining, upward-moving, [543]
pervading, fire-accompanying, and downward-voiding wind.[554]

The *Explanatory Tantra* also states:

In particular, the life-sustaining wind resides at the crown;
it moves in the windpipe and chest, it facilitates swallowing food and drink;
inhalation and exhalation, spitting, sneezing, belching;
it makes awareness and the faculties clear, and it supports the mind.
The ascending wind resides in the chest;
it moves in the nose, tongue, and larynx, it produces speech;
it clarifies one's strength, beauty, color, exertion, and memory.
The pervading wind resides in the heart;
it pervades the entire body, it facilitates ordinary coarse physical

action such as rising, sitting, going, extending, contracting, opening, closing.
The fire-accompanying wind resides in the stomach,
it moves throughout the intestines, and facilitates digestion of food,
separating nutrition from waste, and maturing what is harmful.
The downward-voiding wind resides in the anus;
it moves in the abdomen, bladder, secret place, and thighs, [544]
and regulates semen, blood, feces, urine, and processes inside the womb.[555]

# 32

## The Brain in Buddhist Medical Texts

In general although there are extensive discussions about the stages of fetal development and formation of the body in the womb in early Indian Buddhist texts translated into Tibetan, such as *Nanda's Sūtra on Entering the Womb* as well as the texts of the highest yoga tantra, there seems to be no explicit presentation on the brain. Similarly, in the Abhidharma and epistemological (*pramāṇa*) works where presentations on cognition and the means by which the mind engages objects are found, except for the discussion on the role the sensory faculties play as the base of sensory consciousnesses, there seems to be no explicit discussion of the role of the brain. In the texts of highest yoga tantra too, in the context of their extensive presentation of subjective consciousness, explanation of the channels and winds occurs in the context of being the basis of consciousness. There, except for the explanation that the channels at the crown—described as the great-bliss cakra—are the basis for the arising of bliss, there appears to be no clear discussion about the brain. [545]

This said, in the early medical texts on the science of healing there exists some detailed presentation on the brain. So we shall provide here a brief discussion of the brain based on the explanations found in the *Explanatory Tantra* and *Instruction Tantra* (from the *Four Medical Tantras*) and *Somarāja: A Medical Treatise*, known also as *Somarāja*. This latter work is attributed to Nāgārjuna and was translated into Tibetan sometime around the eighth century by the Chinese monk Hashang Mahāyāna and the great translator Vairocana.

### Brain Development in the Womb

In general the medical treatises explain that although the substances that compose the body that are obtained from the father and mother remain like water and milk mixed together, they can be separated. Bones, brain, and spinal cord develop from substances obtained from the father, and flesh, blood, and the five organs (heart, lungs, liver, spleen, kidney), as well as the six receptacles (stomach, intestines, entrails, gallbladder, bladder, seminal vesicle), develop from substances obtained from the mother. Therefore the main cause for the development of the brain is said to be male semen. *Explanatory Tantra* states:

> Owing to the father's semen, the bones, brain, and spinal cord develop.
> Owing to the mother's blood, flesh, blood, and organs develop.[556] [546]

*Somarāja* states:

> Inside the mother's womb,
> first the action wind flows for seven days,
> from this develops the arbuda embryo.
> Through this the foundational wind
> arises in the womb on the first day.
> On the second day the penetrating wind arises in the womb
> and the arbuda embryo develops.
> On the third day the agitating wind arises in the womb
> and the first-stage ghana embryo develops.
> On the fourth day the fully contracting wind arises
> and the second-stage ghana embryo develops.
> On the fifth day owing to the wind called "consequential"
> arising, the first-stage peśin embryo develops.
> On the sixth day the wind called "condenser" arises
> and the second-stage peśin embryo develops.
> On the seventh day the wind called "unobstructed" arises
> and the embryo comes to possess orifices.[557]

Thus each day within the first week the body of the embryo transforms to an arbuda embryo and so on due to the power of different winds arising. *Somarāja* states:

In the second week the wind called "saṃsāra"
arises and resides at the navel.
With the presence of channels supporting the winds,
what is called "the sun disc" comes to develop.[558]

In the second week the "saṃsāra" wind arises and the sun-disc channel develops at the navel. Thus the text explains how each week up until the forty-third [547] there emerges a different wind, and it is these winds that develop the body. These explanations from the medical sources are slightly different from the stages of body development inside the womb in the sūtra and tantra treatises discussed earlier.

As for the development of the brain, *Explanatory Tantra* states:

With respect to developmental channels, three branch off
from the navel and
one proceeds upward, from which develops the brain.
Delusion resides supported by the brain, and
since phlegm is produced from the brain, it resides in the
upper body.[559]

When the body develops in the womb, the first of all the channels is the developmental channel that branches into three different tips from the center of the navel: (1) the channel for the movement of blood or the fire element, (2) the channel for the movement of the moon or the water element, and (3) the channel for the movement of the wind element.

Among these the white channel for the movement of the moon element proceeds from the left side of the body in stages to the heart, throat, and then to the center of the crown cakra, and the brain develops in dependence on that channel proceeding upward and branching into other minor channels. With the brain acting as its cooperative condition, delusion arises, and with the brain acting as its substantial cause, phlegm arises. For this reason when one experiences confusion, mental heaviness, and so on, one feels as if those states are arising from the brain itself. [548]

## Classifying Brain Regions

In general the medical texts speak of the brain in terms of the masculine brain and the feminine brain. This distinction is drawn on the basis of solidity or liquescency of the parts; all solid or harder parts of the brain are masculine and all liquescent or pliable parts are feminine.

The texts also speak of different types of brain based on the number of cranial fissures, the shape of the head, the personality, and so on. There are brains without any cranial fissure up to those with sixteen cranial fissures. A head that is spherical and high mostly has one or two fissures, one that is high and square mostly has four fissures, a crown that is flattened has six fissures, one that excessively protrudes and is uniformly elevated like a beehive probably has eight fissures. All these belong to the category of "male brains." In contrast, persons who are healthy and have protruding eyes, are moderate in speech, speak firmly, and have sharp eyes probably have healthy brains. A crown that is flat and oblong probably has ten fissures, one that is uniformly flat and oblong probably has twelve, fourteen, or sixteen fissures. All these belong to the category of "female brains." Thus *Somarāja* states:

> The brain has two kinds, the masculine brain and feminine brain.[560]
> All masculine brains are solid,
> all feminine brains are pliable.
> There are those without fissures, spherical heads,
> those with two, four, five, and
> six, eight, and ten, [549]
> twelve, fourteen, sixteen fissures.
> A crown that is spherical and high
> probably has one or two fissures,
> one that is highpeaked and square has four,
> a crown that is flattened probably has six.
> One that excessively protrudes and is
> uniformly elevated like a beehive,
> such brains are masculine brains.
> Those who are healthy with protruding eyes,
> are moderate in speech, speak firmly, and have sharp eyes

> probably have healthy brains.
> A crown that is flat and oblong has ten.
> One that is uniformly flat and oblong
> has twelve, fourteen, or sixteen.
> Such brains are included in the category "feminine brains."[561]

There is also a system of differentiating the brain in terms of seven types. For example, *Instruction Tantra* states:

> In teaching the nature of the ocean-like brain,
> there are these seven—the flesh brain, butter brain, beehive brain,
> curd brain, yogurt brain, milk brain, water brain.[562]

A person with an excess of phlegm has a brain similar to hard and solid flesh. A person with an excess of bile has a brain similar to hard and solid butter. A person with an excess of compaction has a brain that becomes spongy like a beehive. A person possessing an excess of phlegm and bile has a brain similar to the consistency of butter curd that is a mixture of milk and butter milk. [550] A person with an excess of phlegm and wind has a brain similar in consistency to yogurt. A person with an excess of wind and bile has a brain similar in consistency to milk. And, finally, a person susceptible to wind disease has a brain pliable like water. These are characterized on the basis of the robustness and relative maturity of the brain.

As for the size of a brain in its entirety, *Explanatory Tantra* states that it would fit inside one's palms:

> An adequate size is two palms full.[563]

## Energy Pathways, or Nerves, in the Brain

In treatises on medical science, when explanations are given of the nerves within the brain, the body itself is labeled "the great nerve of existence" with four principal groups. Each of these groups contains twenty-four channels and five hundred minor channel nerves.[564] There are twenty-four channels each in relation to the sensory organs of the eyes, ears, nose, tongue, and bodily sensations, which facilitate, respectively, the experience of their associated objects—visible form, sound, smell, taste, and texture. Of these, the principal basis of the five branch channels associated with the five senses is the "great spiral ocean of nerves" at the brain. It is from here the minor nerves branch out in a series of circles, from which sub-branches form, and from which emerges a complex of five hundred minor nerves resembling a network. *Explanatory Tantra* states:

> The great existence channel has four types. [551]
> Channels facilitate the appearance of objects to the senses.
> In the brain the existence channel is surrounded by five hundred minor channels.[565]

*Somarāja* states:

> That is the cerebellum
> associated with the five sense faculties.[566]

This passage indicates that the channels of the five sense faculties are connected with the cerebellum. *Somarāja* states:

> Of the five parts, the head is the brain
> that is pervaded by the five sense faculties.
> The brain and the cerebellum are the basis
> of the bodily organ; the seed descends from them.
> It has eight channels: that of the liver, eyes,
> nerves of the ears and the kidney,
> the channels of the nose and lungs,
> and the channels of the tongue and heart.[567]

*Somarāja* states:

> The cerebellum is like a half horse-hoof.
> The eyes, ears, nose, and tongue,
> are related through nerves, tendons, and ligaments.
> The four pulsating nerves and four accompanying nerves
> are connected to the cerebellum
> through four tendons and the four ligaments.

> Some call it the breadth of the brain.
> At the center is the nerve of the seed nutrient,
> which enhances bliss within the body.
> From the heart both the brain and [552] life-force departs,
> bringing the cessation of the bliss of the body.[568]

There are two parts of the brain that remain below the cerebral membrane: (1) the part that resides toward the front of the skull, and (2) the cerebellum, which is located toward the back of the skull. Since the nerves of the five sense faculties are connected to the cerebellum, it constitutes the principal basis of the body faculty. Furthermore, medical texts explain how the eight nerves—such as those of the liver and so on—and the four pulsating nerves and the four accompanying nerves are connected with the cerebellum, how the shape of the cerebellum is like a half horse-hoof, and how the nerves of seed nutrition at the center of the brain help increase or diminish the bliss of the body. *Explanatory Tantra* states:

> From the great ocean of nerves that is the brain,
> extending downward in the manner of a root
> are nineteen water nerves that perform functions;
> further, there are thirteen concealed nerves, like suspended threads,
> that connect the organs to the interior,
> and six manifest nerves that connect the limbs to the exterior.
> From these branch off sixteen minor water nerves.[569]

The brain is similar to a great ocean from which arise all the white nerves. From it the spinal channel pierces downward like a root, and this is the white life-force nerve. It in turn is made of nineteen coarse water nerves that generate and nurture the water element. [553] Further, these nerves spread internally from the brain to the throat as nineteen nerves, thirteen concealed nerves called the *thirteen suspended thread-like veins*, which connect internally to the organs, and six secondary manifest patterned channels that connect to the exterior. Such texts state that these nerves primarily function to give rise to physical motion and, more specifically from those nineteen, sixteen minor water nerves are connected to the head, legs, arms, and so on.

## Brain Components

The spinal cord or spinal column is the white life-force channel that is the main controller and regulator of the external and internal parts of the body. It extends from the brain to the coccyx and is connected through the spinal cavity of the neck and vertebra like the roots of a tree. *Somarāja* states:

> The spinal cord develops from the base of the head.
> If up to the thirteenth vertebra of the spine
> is crushed or damaged, the seed dissipates.
> Then, like lifeless matter, one could die.[570]

The text explains where the spinal cord branches from, where it extends to, what it is connected to, its functions, and what defects lead to illness. [554]

As for the cranial membrane or cerebral membrane, it fully covers the brain and covers the physical faculties, the cerebellum, and the spinal cord with its external layer. *Somarāja* explains that the cerebral membrane is clear and transparent and does not obscure external and internal states. It is bright maroon in color since it possesses the movement of blood and so on, and in terms of pliability it is flexible and the mere thickness of a delicate deer hide. It protects against weapons and harmful microbes. *Somarāja* states:

> The basis of the physical sense faculties is the brain.
> The cerebral membrane is like a coil of white silk.[571]

The *Instruction Tantra* states:

> The cerebral membrane, clear maroon, just a deer hide's
> breadth,
> exists wrapped in a network of channels.[572]

The *brahma-conch cakra* is a network of channel nerves configured in the following manner. It includes four veins connected to the brain from the life-force channel in the manner of the nodes of four trees extending upward, and the water and the faculty veins that branch downward from the brain and connect to the organs and so on. At the center of the brain, the wind that is the mount of consciousness supports all drops, channels, winds, and minds without exception, and the center of the brain acting as the conduit for the nerves of the brain facilitates the flow of action winds and the movement of drops and facilitates the flow of drops and winds to other areas. [555] The medical texts also say that there exists a nodule the size of a thumb at the point where the cerebellum and skull meet. *Somarāja* states:

> That which meets at the crown
> is the brahma-conch cakra.

And:

> From that, that which contacts the crown
> is the brain that is the basis of the physical faculties.
> The cerebral membrane is like a coiled white silk,
> connected to ligaments, muscles, and the sagittal suture.
> The external and internal
> veins, like a tree with four roots,
> converge at the crown
> and sustain the seeds of the body through the winds.
> This is the brahma-conch wheel;
> it is the body's vital essence and the seed nerve.[573]

*Somarāja* also states:

> The cerebellum at the upper limit of the skull
> exists as a nodule about the size of a thumb.[574]

Furthermore, regarding the relationship between the nerves within the brain and the five organs[575] and six receptacles within the body, *Somarāja* states:

> The nerves of the organs and receptacles
> are related to the crown-aperture channel.[576] [556]

That the sense organs and the brain interact with each other with respect to both health and injury is stated in the *Somarāja*:

> From the nerves at pivotal points of the brain,
> there extends a finger's breadth to the back of the neck,[577]
> from the brain it coils behind the eyes;
> when severed, trachoma and spinal paralysis.[578]

*Somarāja* also speaks of how the parts of the brain are connected to the nerves of the body, and how the root and branch winds, such as the life-supporting wind, perform their specific functions:

> At the door of the brahma-conch wheel at the crown
> the demon nerve and the black-essence nerve
> are connected and the winds flow;
> these are the upward moving winds,
> for they help make awareness lucid.[579]

The *Explanatory Tantra* states:

> In particular, the life-sustaining wind resides at the crown;
> moving in the windpipe and chest it regulates the swallowing of food and drink,
> inhalation and exhalation, spitting, sneezing, belching;
> and it makes awareness and the senses lucid and supports the mind.[580] [557]

The brain is the controller, and what is controlled are the components of the body, such as the sense faculties, the internal organs, and the vessels.[581] Therefore the brain is described in the early medical texts as the "body organ." The brain controls in this manner: organs such as the heart perform their specific function through their dependence on the brain, and the internal organs, the vessels, and the sense organs engage in their specific functions when instigated by the nerves emerging from the brain. Likewise, the brain is dependent on the cerebellum, the organs, the internal organs, the vessels, the water nerves, and the great elements, just as they, in turn, are dependent on the brain. They all perform their specific functions interdependently. The subtle physical components of the brain alone could not perform any function whatsoever without mutual interdependence. These points in brief constitute the perspective of the medical texts *Explanatory Tantra* and *Instruction Tantra*, and the medical treatise *Somarāja*. [558]

# 33

## The Relation of Body and Mind

### Mind-Body Relation According to the Sūtras

In general there is a connection between the stages of formation of the external world and the beings that inhabit that environment. The physical bodies of the inhabitant beings that we can perceive at present are related to stages of development that exist at subtler levels. Ultimately, it is said, these subtler states emerge from the extremely subtle level of indivisible wind and mind. Thus according to the texts of Guhyasamāja tantra, this world system is an expression of the basic reality composed of wind and consciousness that we experience as a manifest reality. Given that ultimately wind and consciousness are inseparable, there exists a profound connection between the elements that compose our bodies and the natural elements of the external world. This subtle relationship can be perceived by someone who has realization born of tantric meditation focused on the channels and winds or by certain types of individuals who possess innate intuitive capabilities.

We shall present, just as a sample, the nature of this relationship between the body and the mind as explained in the texts of the sūtra system and tantric traditions.

The relationship is described in the sūtra texts as follows. When consciousness enters the womb, at that time the new body of this present life begins. Simultaneously, the aggregates of feeling, discernment, formation, and consciousness of this present life—the four aggregates of name—are established. [559] These aggregates remain for as long as one lives, like a three-pole tent where each pole is dependent on and mutually supportive of the others. Nāgārjuna states in his *Extensive Explanation of the Rice Shoot Sūtra*:

> These and the four great elements assemble together as a three-pole tent and they are called "name and form."[582]

Asaṅga too says in *Facts of the Grounds*:

> Thus one can say "Owing to consciousness as a condition, name and form exists," or "Owing to name and form as a condition, consciousness exists." So at the present time and for as long as one lives they function in the manner of a three-pole tent.[583]

Whenever the body and mind of a person sever their connection as that which supports and that which is supported, then the aggregate of form is discarded, heat dissipates, and life-force ceases, hence the person dies. *Sūtra on the First Dependent Origination and Its Divisions* states:

> What is death? From any class of sentient being, any sentient being who has passed on, who is passing on, who has perished, who is internally undone, whose heat has dissipated, whose life-force faculty has ceased, whose aggregates are discarded, who dies, or whose time has come, this is called death.[584] [560]

In general the mind or consciousness can be differentiated in two categories: sensory consciousness and mental consciousness. Sensory consciousnesses, given their close dependence on their physical sense organs, are contingent on the presence or absence of physical sense organs as well as on how clear or unclear the senses are. Therefore it is the physical sense organs that are recognized as the unique dominant conditions of sensory consciousness. As for mental consciousness, even though mental consciousness is indirectly dependent on the physical sense organs, it is not the case that every mental consciousness necessarily depends on physical sense organs or aspects of the brain. For example, there are observed cases where at the time of death breathing stops, blood flow within the brain ceases, and all the functions of the brain come to an end, yet the individual remains for many days without the cessation of subtle processes of consciousness.[585] Therefore the body acts merely as a condition of mental conception that is its result, it does not act as the productive cause that transforms mental conception. In fact it is mental conception that acts as the specific productive cause that transforms the body. Dharmakīrti's *Exposition of Valid Cognition* states:

> When any of the sense faculties are damaged,
> mental consciousness is not damaged;
> but we see that when it is affected,
> those sense faculties too are affected.[586] [561]

One can also observe, for example, when generating the mind of loving-kindness wishing to help others, the blood in the body flows better, one's face becomes more radiant, and one has a more appealing appearance. That such observable changes occur at the level of one's body is stated in *Sūtra on the Foundation of Mindfulness*:

> Those who desire to benefit all sentient beings, their blood becomes extremely fluid. Because their blood becomes extremely fluid, their faces become radiant. Because their faces become more radiant, they become more appealing in their appearance. Because of this all sentient beings will come to like them. These benefits exist as observable facts.[587]

Furthermore, given that the nature of the physical body and of consciousness are so different—one characterized by obstructive resistance, the other by nonresistance and clarity and knowing—there is no possibility that one can become a substantial cause of the other. This said, however, body and mind do act as each other's cooperative conditions, as stated in Dharmakīrti's *Exposition of Valid Cognition*:

> Through acting as cooperative causes
> the results generated reside together.[588]

This text explains that the present mind arises from the prior moment of mind acting as its substantial cause and the prior moment of the body acting as its cooperative condition. Similarly, the present body arises from the prior moment of the body acting as its substantial cause and the prior moment of mind acting as its cooperative condition. It is in this manner that the two—the body and the mind—coexist. [562] That they do not act as each other's substantial cause is stated in *Exposition of Valid Cognition*:

> That which is not itself consciousness is not a substantial cause
> of consciousness; from this too it is established.[589]

Devendrabuddhi's *Commentary on the Exposition of Valid Cognition* explains these lines:

> Consciousness itself is the substantial cause of consciousness, but that which is not consciousness is not, and that which is by nature not consciousness, such as the body and so on, is not the substantial cause.[590]

Moreover, although the body can act as the cooperative condition of the mind through benefiting the mind, this in itself does not make the body a special cause of consciousness, such that the force that stops the body also acts to stop the mind. For example, although fire and so on may help produce specific features, such as the reddish color of an earthen jar, this does not make the fire a special cause of the jar itself, such that when the fire is extinguished the earthen jar would also cease to exist. *Exposition of Valid Cognition* states:

> Granted that in certain cases the body does benefit
> the states within the mind's continuum;
> but this is like fire and so on in relation to a jar and so forth;
> this alone does not ensure invariability.[591] [563]

In brief, therefore, the body does not act as the substantial cause or the specific dominant condition of the mind. *Exposition of Valid Cognition* states:

> In some cases increase in attachment and so on
> can occur from increase in pleasure and pain;
> this, in turn, results from the proximity of inner conditions,
> such as the elements resting in equilibrium and so on.
> This also explains fainting
> that occurs from typhoid;

but it is in fact due to changes in sensory cognitions
brought about by specific internal objects.
For example, some are stupefied and so on
simply through hearing the phrase
"There is a tiger!" or on observing blood and so on.[592]

The text states that although on occasion attachment and so on may increase because of the waxing or waning of the body's strength, it is in fact from feelings of pleasure and pain that they directly arise. It is not the case that the body itself acts as the direct dominant condition of attachment and so on. Feelings of pleasure and pain also arise from internal perceived objects of contact, such as through balanced or unbalanced elements, and the body itself is not the dominant condition of feelings of pleasure or pain. Likewise, fainting that occurs because of a typhoid epidemic is due to the presence of inner contact, which then brings about a change in sensory cognition. For example, some individuals with a weak constitution are stupefied through hearing "There is a tiger!" or through seeing the blood of someone who is killed [564] and so on. Such reactions are results enhanced by internal feelings and are not directly dependent on the body. Kamalaśīla's *Commentary on the Compendium on Reality* states:

> Thus by the power of seeing fear-inducing objects such as a tiger or blood and so on, at that time minds characterized by cowardice and dullness become transformed. It is because of that alone that the mental faculty transforms, but this alone does not indicate that the mental faculty is contingent on such things.[593]

For example, cowards faint because of intense fear upon seeing a tiger or the blood of someone killed. The fearful state arises when mental consciousness is transformed by focusing on these objects, but it does not arise because those objects act as the substantial cause.

Furthermore, qualities of mental consciousness like compassion and wisdom do not depend for their increase or decrease on the enhancement or diminishment of the body. Rather they are functions of the force of familiarity with mental states themselves within the same substantial continuum. Such enhancement or diminishment, which is due to the power of cultivating familiarity within a continuum, does not exist for states that are contingent on physical substantial causes, such as the light of a butter lamp and so on. Thus *Exposition of Valid Cognition* states:

> With no relation to waxing and waning of the body,
> but through the functions of the mind itself, [565]
> increment in wisdom and so on
> as well as their decline come to occur.
> This is absent in those contingent on matter,
> such as the light of the butter lamp, for example.[594]

If the mind and body exhibited a relationship typical of a substantial cause or dominant condition, then the enhancement or diminishment of something like attachment would necessarily follow the enhancement or diminishment of the body. That it is not the case is stated by Kamalaśīla in his *Commentary on the Compendium on Reality*:

> When changes in one thing always directly give rise to changes in another, that thing is properly regarded as its substantial cause. But changes in characteristics such as attachment do not arise because of any increase or decrease in physical strength, because attachment and so on do not exist in those possessing critical insight even when their physical strength increases. So too, it is observed that a person with an extremely weak body who is mentally dissolute, and also those of the animal kingdom with extremely weak bodies, may have very strong attachment and so on.[595] [566]

It is possible that some individuals who lack analytical wisdom increase in attachment through the enhancement of the body and generate hatred through the decline of the body. However, because of the power of the physical elements being in balance or not, the physical feeling of pleasure or suffering arises through taking specific internal contact as the object, and because of that perception as a condition, attachment or hatred arise. But attachment and hatred are not directly generated from the body and they do not arise by the body acting as the specific indispensable cause. The "specific indispensable cause" is explained in *Exposition of Valid Cognition*:

> Because the effect always follows it,
> and by it existing the effect is benefited,
> it is a cause.[596]

A cause benefits its effect by merely existing prior to its effect, and the effect must invariably follow the cause, such that without the cause arising the effect cannot occur. For example, the butter lamp is the specific indispensable cause of the light of a butter lamp.

There is the story of how some individuals remained alive during a famine for as long as they believed that a bag of sand they had was barley flour, but when one of them realized that it was actually a bag of sand, his hopes faded and he died. Similarly, there is the story of eggs laid on the shore of an ocean that do not decay for as long as the mother turtle does not forget them, but when she does they begin to decay. From statements such as these we can understand the nature of the relationship between mind and body. Thus *Treasury of Knowledge Autocommentary* [567] states:

> Even mental intention appears to prevail here, for it is said that a father hard pressed by a bad year filled a bag with dust and said, "It is barley meal." His two sons believed this and through thinking it was barley they lived on for a long time. But then one son opened the bag, and when he saw it was dust he lost hope and died.[597]

The same text cites *Statements on the Types of Beings*, which is a subdivision of the Seven Treatises of Abhidharma:

> *Statements on the Types of Beings* says: "Adult turtles emerge from the great ocean and reach dry land, and then on that sandy shore they lay eggs, cover them with sand, and again return to the ocean. Because those mothers remember the eggs they laid, the eggs are not forgotten and they do not decay. But those that are forgotten do decay." This does not state that the eggs do not decay because they receive the food of another. It speaks of the eggs not decaying since she remembers and does not forget where the eggs were laid, and she remembers when to make contact with them.[598] [568]

## Mind-Body Relation in Highest Yoga Tantra

In the texts of highest yoga tantra, however, although there is the view that even the extremely coarse body, which is composed of the primary elements, is the basis of consciousness, it is the relationship of the subtle body, namely, the winds, and consciousness that is explained with greater emphasis. Further, it is said that consciousness is mounted on the wind and the mind engages its object by riding on the wind. So the mind's movement toward the object as well as its movement from the preceding to subsequent moments are understood to be the function of the wind. A lame person with sight and a blind person with able legs can, through mutual support, navigate the path and reach their destination. In the same manner, consciousness is able to move to its object because of the wind that is its mount, and since seeing the object is achieved by consciousness, the two can never be separated. Thus the movement of the body because of consciousness occurs only when consciousness is associated with wind, for if consciousness were not linked with wind it would lack the capacity to affect the body in that way. That this is so is stated in the Kāśmīri thinker Lakṣmi's *Clarifying the Meaning of the Five Stages:*

> The statement "this is the mount of consciousness" means that the wind itself is the mount of the six engaging consciousnesses and acts as their support. For consciousness mounts on the wind and, through six senses such as the eyes, fully ascertains the six objects such as form and so on. Therefore consciousness relies on the wind activating it, otherwise it would be like a blind man who cannot see anything. Therefore, [569] like the symbiotic relationship of a cripple and a blind man, if the two are interconnected it can fulfill its purpose. Just as a horse and its rider enter a path and accomplish their specific roles, so too when wind and consciousness are connected they possess the capacity to perceive objects. Thus these two, consciousness and wind, possess a relationship that is intrinsic in character.[599]

In particular, the extremely subtle fundamental innate body and mind are referred to as the "innate body and mind." For example, *Vajra Garland Tantra* states:

> Just as when fire burns fuel is consumed,
> so too without a cause the winds dissolve.
> Again, when the life-sustaining wind
> as well as the multiple action winds
> arise together with consciousness,
> so once again one resides in the three worlds.
> From that arises karma, and from that birth,
> from that attachment and so on, and their imprints,
> and from these again death and birth.
> It is thus like a turning wheel.[600]

Just as when fire burns the entity of wood becomes nonexistent, so too at the time of death coarse winds up to the life-sustaining wind dissolve in their natural order. At the time of birth the action winds arise from the clear light of death, and [570] both that and consciousness take birth in the abode of the three worlds together. Also from that action winds, conceptions such as attachment and so on, arise, and due to that, both good and bad karma are accumulated and death and birth arise, as in the turning of a wheel.

Thus when the clear light of death manifests, the coarse body and mind dissolve into the extremely subtle wind and mind, which are indivisible. In highest yoga tantra texts, such subtle winds and minds are described as being intrinsically related as a single entity, thus they can never be separated and neither is there any end to their continuum. Furthermore, because the wind and mind have the same point of engagement, the texts explain that when the mind is held single-pointedly on a drop at a specific cakra within the body, one gains controls over the wind at that specific site. Similarly, the texts state that as long as the wind, which is like a horse, is not controlled, then the mind, which is like a rider, will also not be controlled. Furthermore, the texts state that if a body is aligned the channels will be aligned, if the channels are aligned then the winds will be aligned, and if the winds are aligned then the mind will be aligned and so on. Thus tantric treatises explain with special focus and emphasis the unique relationship between the subtle body and the mind. In brief, although a differentiation can be made between body and mind at a coarse level, at the subtle level, given that wind and consciousness exist as a single entity, no differentiation can be made between the two in terms of their reality, except in conceptual terms with respect to their distinct identities. This point is established with great emphasis in tantric sources. [571]

## Mind-Body Relation through Activity of the Microorganisms within the Body

There are also statements in the sūtras of how the body and mind effect change in each other through the activity of the microorganisms that reside within the body. To provide a brief sample, the *Twenty-Five Thousand Verse Perfection of Wisdom Sūtra* states:

> Subhūti, again within the body of those who are human, there are eighty thousand types of microorganisms that feed on the body.[601]

Thus many different types of microorganisms exist in the human body. Similarly, *Sūtra on the Foundation of Mindfulness* says:

> Ten types of microorganisms reside in one's head.[602]
>
> Moreover, ten types of microorganisms make the throat swell.[603]
>
> Within the channels there are ten movements of semen, the movement of sweet intoxication, complete intoxication.
>
> The ten types of microorganisms that arise from the blood are those that reside together with the flesh, they eat body hair.
>
> Microorganisms arising from the blood have no eyes and are characterized as causing skin ailments.
>
> The ten types of microorganisms abiding in the flesh give rise to sores.
>
> The ten types of microorganisms abiding in the gallbladder cause tsuratha.[604]
>
> Microorganisms abiding in all the limbs and secondary limbs lick the bones. . . .

> The ten types [572] of microorganisms that move in the feces are transparent.
>
> The ten types of microorganisms that move in what is soft reside in the body hair.[605]

These passages mention in detail the types of microorganisms that reside throughout the body, their names, their body shape, essential nature, and function, and how they cannot be seen by ordinary eye consciousness and so on but must be perceived by mental cognition born of study, contemplation, and meditation. To give a brief example of how the human mind undergoes change because of the enhancement or diminishment of these types of microorganisms, the same text says:

> As for microorganisms that cause enhancement, they enhance my body in two specific ways. Thus if the microorganisms within my body fail to flourish because of the food I consumed, this would utterly deprive me of joy. [606]

Thus because of the force of the body's microorganisms not thriving, the mind of that person too will be deprived of joy. The same text states:

> If "microorganisms that are involved in the control and movement of the joints" are starved because of nutritional faults, my hands will eventually contract. We also observe that owing to such faults the limbs of the body become diseased, heart ailments arise, and cities come to be emptied. Mucus is stirred up, [573] continuous mental sadness is created, and one takes no joy in objects of contact and form. Why is that? Because this starving of the microorganisms can cause strong sensations.[607]

Starving the microorganisms causes the limbs of the body to contract and the mind to experience a vivid, empty state, sadness, and a lack of joy in form and so on. The same text says:

> The microorganism called "experiencer" thrives in my bones. How does it cause the maintenance of natural equilibrium and

> disequilibrium within me? Equilibrium is wisdom derived from study or the divine eye. Due to faults in my nutrition, the microorganism called the "experiencer" will be starved and cause pustules in me. Pustules move throughout the limbs and secondary limbs of the entire body, and it is this movement through my body that makes me stiff. It will also make my heart become void and my body devoid of sensation. It will block my bladder and bowels. It will also lead to sleeplessness.[608]

So if certain types of microorganisms within the body are starved, the body will lose [574] sensation and there will be blockage and sleeplessness and so on. The same text states:

> The microorganism called "skin eater" resides in the bones. If it becomes starved it will cause the lips to swell, the mouth to swell, the eyes to swell, puss to flow, the muscles to contract, intense thirst, the throat to dry, the ears to swell, stiffness in the neck, "grinding of the skull,"[609] early onset of gray hair, pressure on the neck, untimely sleep, craving for unsuitable foods, dislike of a place, preference for places or gardens that are empty, the mind to become distracted, indulgence in excessive chatter over trivial matters, or one's body to appear as a dust storm because of continuously scratching the major and minor limbs. These are flaws associated with dysfunctional microorganisms that eat skin.[610]

Owing to the activity of some microorganisms within the body, the hair turns gray and one comes to like improper food and so on.

This completes the first volume, *The Physical World*, the first of two volumes on the sciences in the four-volume series *Science and Philosophy in the Indian Buddhist Classics*. [575]

# Appendix

## The Eighteen Topics of Chapa Chökyi Sengé

1. Color—white and red and so on (*kha dog dkar mar*). This topic introduces the student to objects experienced by the five sense consciousnesses. Though the title indicates only colors, the topic includes all five sensory objects: visible form, sound, smell, taste, and tactility. In turn these sensory objects form the referent basis for detailed analysis by the philosophical categories outlined in the remaining seventeen topics. See chapter 6 of this volume, the "The Five Sense Objects."

2. Substantial phenomena and abstract phenomena (*rdzas chos ldog chos*). Substantial phenomena (*rdzas chos, anvayavyāpti*) refer to the attributes of impermanent things that are in essence types of substances that possess functionality. This category includes material phenomena such as wood or fire, mental phenomena such as the emotions, and nonassociated formative factors such as time, the person, impermanence, and so on. Abstract (or excluding) phenomena (*ldog chos, vyatirekavyāpti*) refer to those that appear to conception as substantial phenomena but in reality are not types of substances (*dravya*), nor are they impermanent. For example, the conceptual isolate of a pot (such as the object-universal or generic image of a pot) appears to conception as a substantial (or impermanent) phenomena, but in fact is an abstract image (*ldog cha*) that appears to conception only because of the exclusion of its specific negandum (or object of negation). See the subheading "Substantial Phenomena and Abstract Phenomena" in chapter 13.

3. Contradictory and noncontradictory (*'gal ba dang mi 'gal ba*). Contradictory phenomena (*'gal ba, viruddha*) refer to two or more phenomena

that are distinct and share no common locus. They are contradictory by being incompatible (*lhan cig mi gnas 'gal*) or by mutual exclusion (*phan tshun spang 'gal*), or they are directly contradictory (*dngos 'gal*) or indirectly contradictory (*brgyud 'gal*). Noncontradictory phenomena (*mi 'gal ba, aviruddha*) refer to two or more phenomena that are distinct but share a common locus, such as the pair a pot and impermanence. See the subheading "Reasoning with the Laws of Contradiction and Relation" in chapter 2 and the subheading "Explaining Contradiction" in chapter 13.

4. Universals and particulars (*spyi dang bye brag*). Universal (*spyi, sāmānya*) refers to a phenomenon that encompasses its specific instantiations, such as knowable object or thing. For example, *thing* is a universal since it encompasses or subsumes its instances, such as a pot, for *thing* is a set that includes all pots. In general universals have three types: type-universal (*rigs spyi, jātisāmānya*), composite-universal (*tshogs spyi, sāmagrisāmānya*), and object-universal (*don spyi, arthasāmānya*). A particular (*bye brag*) is a phenomenon that has a type that operates as its pervader. For example, a pot is a particular of *thing* since *thing* operates as the pervader of pot, for pot is a subset of *thing*. See the subheading "Universals and Particulars" in chapter 13.

5. Relation and absence of relation (*'brel ba dang ma 'brel ba*). Relation speaks of mutual connection and interdependence. In terms of the meaning of relation, one thing [*x*] is related to another [*y*] when *x* and *y* are distinct at least nominally, and when *x* does not exist then also *y* necessarily does not exist. If differentiated, relation consists of two types: intrinsic relation and causal relation. The absence of relation between any two phenomena such as *x* and *y* indicates that either *x* and *y* are not differentiable even on a nominal level or that even when differentiable the absence of one does not entail the absence of the other. See the subheading "Explaining Relation" in chapter 13.

6. Distinction and nondistinction (*tha dad dang tha mi dad*). Distinction refers to multiple phenomena such as those that appear separately to conception, for example, a pot and a pillar. Such phenomena are distinct entities or types and exist as distinct subsets of phenomena. Nondistinction refers to phenomena that are not multiple, that cannot be distinguished nor separated. See the subheading "One and Many" in chapter 13.

7. Presence and absence (*rjes su 'gro ldog, anvayavyatireka*). *Anvayavyatireka* indicates both positive concomitance and the exclusion of its opposite. In Buddhist logic the three modes (*tshul gsum, trairūpya*) must be present to establish a correct reason (*rtags yang dag*): the attribute of the subject (*phyogs chos, pakṣadharmatā*), the forward pervasion (*rjes khyab, anvayavyāpti*), and the reverse pervasion (*ldog khyab, vyatirekavyāpti*). "Presence and absence" signifies the latter two modes. For example, in the syllogism "sound is impermanent because it is produced," the forward pervasion is established because "produced" must be *present* in its similar class (i.e., the class of impermanent things), as described in the logical entailment "if produced it must be impermanent," and "produced" must be *absent* from its dissimilar class (i.e., the class of permanent phenomena), as described in the logical entailment "if permanent it must not be produced." See the subheading "Explaining Relation" in chapter 13.

8. Cause and effect. In general a cause (*rgyu, hetu*) is that which produces an effect and, conversely, an effect is that which is produced by a cause. Moreover, causes, effects, and functional things are mutually inclusive. There are two types of cause: substantial cause (*nye len gyi rgyu, upādānahetu*) and cooperative condition (*lhan cig byed rkyen, sahakāripratyaya*). Also there are two types of effect: substantial effect (*nye len gyi 'bras bu*) and cooperative effect (*lhan cig byed 'bras*). See chapter 11.

9. First stage, intermediate stage, later stage (*snga btsan bar btsan phyi btsan*). Some say these three terms refer to the proponent, adjudicator, and opponent in a debate. According to the *Sé Collected Topics*, these three stages refer to the three elements of a syllogism: subject (*chos can*), predicate (*bsal ba*), and reason (*rtags*). Thus the first stage (*snga btsan*) refers to the subject (*chos can*), intermediate stage (*bar btsan*) refers to the predicate (*bsal ba*), and the later stage (*phyi btsan*) refers to the reason (*rtags*). See Ngawang Tashi 2001, 232.

10. Definitions and definiendum (*mtshan nyid dang mtshon bya*). The definition of a *definiens* is "that which possesses the three properties of a referent existent (*dravyasat*)" where the three properties are: (1) it is a definition, (2) it is established with respect to its basis, and (3) it does not serve as a definiens of any other object defined. The definition of a *definiendum* is that which possesses all three properties of an imputed existent

(*prajñāptisat*) where the three properties are: (1) it is an object defined, (2) it is established with respect to its basis, and (3) it does not serve as the definiendum of any other definition. See the relevant sections of this book. See the opening section of chapter 13.

11. Multiple reasons and multiple predicates (*rtags mang bsal mang*).

12. Forward negation and reverse negation (*dgag pa 'phar tshur gnyis*).

13. Direct contradiction and indirect contradiction (*dngos 'gal dang brgyud 'gal*). The definition of "direct contradiction" is "those which are mutually and directly incompatible," such as impermanent and not impermanent. The definition of "indirect contradiction" is "that which is indirectly incompatible," such as cold and hot tactility, or grasping at a self and the wisdom realizing emptiness. See the subheading "Explaining Contradiction" in chapter 13.

14. Two types of logical entailment (*khyab mnyam rnam pa gnyis*).

15. Being and not being (*yin gyur min gyur*). This topic examines a special case related to the nature of being something or not being something. For instance, if it is said "If *x* has something that is permanent that is it, it follows that *x* is permanent," then this is not necessarily so, for the instance "permanent alone" has something permanent that is it, such as space, for space is permanent alone, but "permanent alone" is not permanent because "permanent alone" does not exist, since "impermanent things" exist. This reasoning is then applied to "thing alone" or "impermanent alone."

16. Negation of being and not being (*yin log min log*). This topic indicates that "the negation of not being a vase" and "being a vase" are equivalent, that any multiple of "the negation of not being a vase" is always equivalent to "being a vase," and this applies to all similar instances. Also "the negation of being a vase" and "not being a vase" are equivalent. Any even multiple of "the negation of being a vase" is equivalent to "being a vase." Any odd multiple of "the negation of being a vase" is equivalent to "not being a vase." See the subheading "Explaining the Negation of Being and the Negation of Not Being" in chapter 13.

17. Understanding existence and understanding nonexistence (*yod rtogs med rtogs*). This topic establishes the premise that if something exists then valid cognition that comprehends its existence must exist, and if something does not exist then valid cognition that comprehends its existence necessarily does not exist. Further, these two categories form a dichotomy. See chapter 4.

18. Understanding permanent phenomena and understanding things (*rtag rtogs dngos rtogs*). This topic establishes the premise that if something is permanent then valid cognition that comprehends it to be permanent must exist, and if something is a thing (i.e., impermanent) then valid cognition that comprehends it to be a thing must exist. Further, these two categories form a dichotomy. See chapter 4.

# Notes

1. In its original Tibetan, the text of His Holiness's introduction covers both volumes on science and therefore appears only at the beginning of this first volume. In our translation, however, we have moved the section of the introduction relevant to the mind sciences to volume 2 so that each of the two volumes now carries an introduction from the Dalai Lama. All notes to the introduction are from its translator.
2. For an accessible and personal account of the Dalai Lama's encounter with science, and especially his reflections on the convergence between science and Buddhism with respect to their method, see Dalai Lama 2005, especially chapter 2.
3. The proceedings of many of these Mind and Life Dialogues are available in books. For details, visit https://www.mindandlife.org/books/.
4. The Dalai Lama outlines his views on what he understands by secular ethics in two important books, *Ethics for the New Millennium* and *Beyond Religion: Ethics for a Whole World*.
5. In Mahāyāna sources, buddhahood is characterized by two dimensions—the truth body, which is the enlightened being's ultimate nature, and the form body, which refers to the manifestation of that ultimate reality. When further elaborated, the truth body has two aspects—the wisdom aspect and its underlying natural expanse—while the form body is differentiated into an enjoyment body (a celestial or subtle form) and an emanation body (a form visible to ordinary humans).
6. These two are typically accompanied by a third criterion of existence, that "it must not be invalidated by any analysis pertaining to the ultimate reality of things." These criteria are implicit in the Indian Madhyamaka sources, but they are made explicit and discussed extensively in the writings of Tsongkhapa. For a brief contemporary explanation, see Jinpa 2002, 155–57.
7. This is a reference to *Bodhicaryāvatāra* 9.122, where in the context of negating the viewpoint that the world is created by a transcendent being, Śāntideva writes, "Beginningless happiness and suffering come from karma, / so what is it that he created? / Since there is no beginning, / how can there be an end?"
8. See the essay preceding part 1 for a brief explanation of these two systems of classification.

9. A unique feature of the Kālacakra, or "wheel of time," tantra is its highly developed system of astronomy and cosmology. See part 5, especially chapter 22.
10. The five characteristics of life being referred to here are most probably that (1) living things are composed of cells, (2) they use energy, (3) they respond to and adapt to their environment, (4) they grow, and (5) they reproduce.
11. See chapter 19 for the citations from this sūtra in Nāgārjuna's *Compendium of Sūtras*. This passage presents a series of examples where, depending on one's levels of spiritual realizations, a moment of time can be shorter or longer for different people. See also the essay that introduces part 4.
12. Nāgārjuna's analytic corpus logically presenting the view that nothing possesses intrinsic existence includes his (1) *Fundamental Treatise on the Middle Way* (*Mūlamadhyamakakārikā*), (2) *Sixty Stanzas of Reasoning* (*Yuktiṣaṣṭikā*), (3) *Seventy Stanzas on Emptiness* (*Śūnyatāsaptati*), (4) *Treatise on Pulverizing [the Views of Opponents]* (*Vaidalyaprakaraṇa*), and (5) *Averting Objections* (*Vigrahavyāvartanī*). Sometimes his *Precious Garland* (*Ratnāvalī*) is also added, making the corpus a collection of six works.
13. For a succinct explanation by the Dalai Lama on emptiness and the two selflessness, see chapter 9 of his *Essence of the Heart Sutra*.
14. Candrakīrti was an eighth-century Indian thinker whose interpretation of Nāgārjuna's Madhyamaka philosophy acquired an important status of authority in Tibet, with his *Entering the Middle Way* (*Madhyamakāvatāra*) becoming a key textbook in many Tibetan monastic centers of learning.
15. See the essay that introduces part 1 for a brief explanation of the four schools of classical Buddhist philosophy.
16. Dharmakīrti develops these two presentations of the twelve links of dependent origination, in sequential and reversal orders, in chapter 2 of his *Exposition of Valid Cognition* (*Pramāṇavārttika*).
17. "Trailblazers" translates the Tibetan phrase *shing rta srol 'byed*, literally "chariotway makers," and conveys the idea that Nāgārjuna and Asaṅga paved the great pathways for the flourishing of Mahāyāna Buddhism. In Indian Mahāyāna sources, both these masters came to be deeply revered and mythologized with widespread recognition of both having been in fact prophesized by the Buddha himself.
18. Most Tibetan historical sources on Indian Buddhism contain confusing accounts of this Buddhist council. Butön, for example, lists two different views on the identity of the third council. In one view, the third council took place in the 137th year of the Buddha's final nirvāṇa in Pāṭaliputra and was supervised by the elder Vastīputra. In the second view, the third council is thought to have taken place in the 300th year of the Buddha's final nirvāṇa. It is this second account that is cited here as presenting the viewpoint of the Sarvāstivāda school. Since archaeological evidence puts Kaniṣka in the middle of the second century CE, the Tibetan sources dating this council are chronologically problematic.

19. See *Engaging in the Bodhisattva's Deeds* 9.40–53, where, while making the case that the realization of emptiness is indispensable for the attainment of true freedom, Śāntideva presents a series of arguments establishing the authenticity of the Mahāyāna teachings.
20. These three texts are, respectively, Toh 267 (*Dpang skong phyag brgya pa*), 116 (*Karaṇḍvyūha-sūtra*), and 212 (*Pratītyasamutpāda-sūtra*) within the Dergé edition of the Kangyur.
21. *Sad mi mi bdun.* These were Ba Salnang, Vairocanarakṣita, Khön Lui Wangpo Sung, Ma Rinchen Chok, Ngenlam Gyalwa Chökyang, Rinchen Lekdrup, and Ba Yeshé Wangpo. They came to be known as the "seven who are awake" because the potential for joining the monastic order had been awakened in them, and they were thus chosen to be the first monastics ordained in Tibet.
22. These two are, respectively, Toh 4346 (*Bye brag rtog byed chen mo*) and 4347 (*Sgra sbyor bam po gnis pa*) in the Dergé edition of the Tengyur. Although a short version of *Vyutpatti* seems to have been compiled as well, today no version of this work seems to be extant.
23. *Bam po*, literally meaning a "bunch," became a technical Tibetan term used in measuring the size of texts. One *bampo* is equivalent to three hundred four-lined stanzas; for those texts in prose, however, a bampo is measured at around 8,400 syllables, based on the calculation of each stanza having four lines of seven syllables each. The bampo system was used so that no adulteration through addition or deletion could occur in the Indian texts that were translated into Tibetan.
24. This is a reference to Apabhraṃśa, which seemed to have been the language of choice for mystic poets such as Saraha, Tilopa, and Kaṇha, who wrote on the theme of innate wisdom, spontaneity, and bliss.
25. In Indo-Tibetan tradition, three different classifications of the Buddha's words (*buddhavacana*)—in sets of twelve, nine, and three—are recognized. The twelve are: (1) *sūtra* (discourses), (2) *geya* (mixed prose and verse discourses), (3) *vyākaraṇa* (declarations), (4) *gāthā* (verses), (5) *udāna* (aphorisms), (6) *itivṛttaka* (preambles), (7) *avadāna* (narratives), (8) *adbhutadharma* (miraculous events), (9) *jātaka* (former birth stories), (10) *vaipulya* (extensive discourses on the bodhisattvas), (11) *nidāna* (circumstantial teachings), and (12) *upadeśa* (defining instructions). When condensed into nine, numbers 7, 8, and 9 are subsumed in number 6 and not listed separately. Explanation of each of these twelve branches can be found in Asaṅga's *Abhidharmasamuccaya*. See Asaṅga 2001, 178–81.
26. Śāntideva himself mentions (in *Engaging in the Deeds of a Bodhisattva* 5.106) his *Compendium of Sūtras*, but this work seems to be lost.
27. Dignāga wrote several texts on logic and epistemology carrying the suffix "analysis" (*parikṣa*) in their titles. These include the *Sāmānyaparīkṣā*, *Nyāyaparīkṣā*, *Vaiśeṣikaparīkṣā*, and *Sāṃkhyaparīkṣā*, all of which are no longer extant.
28. The Seven Treatises of Abhidharma in the Sarvāstivāda system are

*Jñānaprasthāna* (Attainment of Knowledge) by Kātyāyanīputra, *Prakaraṇapāda* (Topic Divisions) by Vasumitra, *Vijñānakāya* (Compendium of Consciousness) by Devaśarmā, *Dharmaskandha* (Aggregate of Dharma) by Śāriputra, *Prajñaptiśāstra* (Treatise on Designation) by Maudgalyāyana, *Saṃgītiparyāya* (Statements on the Types of Being) by Mahākoṣṭha, and *Dhātukāya* (Compendium of Elements) by Pūrṇa.

29. The mental-object element (*chos kyi khams, dharmadhātu*) is also known as the phenomena element, and the mental-object base (*chos kyi skye mched, dharmāyatana*) is also known as the phenomena base.
30. The three-nature theory (*trisvabhāva*), a philosophical position of the Mind Only school of Buddhism, asserts that all phenomena are included in the three natures (*svabhāva*): (1) dependent nature (*paratantra*), which refers to all impermanent entities, (2) perfected nature (*pariniṣpanna*), which refers to emptiness or nonduality, and (3) imputed nature (*parikalpita*), which refers to all permanent phenomena other than emptiness. See Anacker 2005, 291. For a more detailed presentation see Garfield 1997.
31. Ultimate truth (*paramārthasatya*) refers to the ultimate nature of reality otherwise known as selflessness or emptiness, and conventional truth (*saṃvṛtisatya*) refers to all existent phenomena other than emptiness, such as ordinary sense objects.
32. Method (*upaya*) refers to both renunciation and compassion, which encompass the types of motivation that inspire a being to enter a spiritual path, and wisdom (*prajñā*) refers to the type of awareness that is capable of discerning the nature of reality, which is itself described in terms of emptiness or selflessness.
33. The form body (*rūpakāya*) refers to the physical bodies of a buddha, whereas the truth body (*dharmakāya*) refers to the mind of a buddha and the emptiness of that mind.
34. Tibetan scholars of both earlier and later periods composed numerous commentaries to both Asaṅga's *Compendium of Abhidharma* (*Abhidharmasamuccaya*), the basic text of the upper Abhidharma system, and Vasubandhu's *Treasury of Knowledge* (*Abhidharmakośa*), the basic text of the lower Abhidharma system, and the exposition and study of both textual systems spread widely in Tibet from the earliest period.
35. Nālandā University (Nālandā Mahāvihara) was founded near the ancient city of Rājagṛha in modern Bihar State, India, reputedly at the site of the birthplace of Śāriputra, the Buddha's foremost disciple in wisdom. Taranathā states that Aśoka constructed a great temple at this site, replacing an earlier *caitya* (shrine), but it was only later that it flourished as a university under Gupta and Pāla patronage from the sixth and twelfth centuries when it boasted a vast complex of monastic structures including a nine-story library. At its peak it housed up to fifteen hundred academic staff and ten thousand or more students, many from foreign lands. Its most famous scholars included Nāgārjuna, Āryadeva,

Buddhapālita, Bhāvaviveka, Asaṅga, Vasubandhu, Dignāga, Dharmakīrti, Śāntideva, Candrakīrti, and Kamalaśīla. See Buswell and Lopez 2014, 565–56.

36. This citation was not found within the sūtra division of the Kangyur. But one with a slightly different rendering occurs: "Just as gold is burned, cut, and polished; properly examine my speech, then practice with conviction. Do not accept it because of others who are learned." *Tantra of Great Power* (*Mahābala-tantra*), chap. 1, Toh 391, 216. Pd 79:627. A passage similar to this occurs in the *Kālāma Sutta* or *Kasamutti Sutta* in the Aṅguttara Nikāya of the Pali canon.
37. Buddhist texts often list three types of doubt: (1) doubt tending away from the fact (Tib. *don mi 'gyur gyi the tshom*), such as the doubt reflecting the statement "The mind is probably permanent"; (2) evenhanded doubt (*cha mnyams pa'i the tshom*), such as the doubt reflecting the statement "The mind may equally be either permanent or impermanent"; and (3) doubt tending to the fact (*don 'gyur gyi the tshom*), such as the doubt reflecting the statement "The mind is probably impermanent."
38. *Vyākhyāyukti*, chap. 1. Toh 4061, 30b, Pd 77:85.
39. *Ugraparipṛcchāsūtra*, chap. 19. Toh 63, 279b, Pd 42:831.
40. A bodhisattva—literally "a being" (*sattva*) who strives for "enlightenment" (*bodhi*)—refers to one who seeks to attain highest enlightenment for the sake of establishing all sentient beings in the state of enlightenment.
41. Defective instruction (*kālāpadeśa*) refers to instruction not found in the Buddhist canon and thus regarded as lacking canonical authority.
42. Tathāgata (*de bzhin gshegs pa*)—literally "one who proceeds (*gata*) in accordance with reality (*tathā*)"—refers to awakened or enlightened beings who continuously maintain within themselves the direct realization of the nature of selflessness or emptiness.
43. *Bodhisattvabhūmi*, chap. 17. Toh 4037, 136a, Pd 73:859. See Engle 2016, section 1.17.6.
44. *Bodhisattvabhūmi*, chap. 17. Toh 4037, 136a, Pd 73:859. See Engle 2016, section 1.17.6.
45. *Bodhisattvabhūmi*, chap. 17. Toh 4037, 136b, Pd 73:860. See Engle 2016, section 1.17.6.
46. *Bodhisattvabhūmi*, chap. 17. Toh 4037, 136b, Pd 73:860. See Engle 2016, section 1.17.6.
47. The Bodhisattva canon or *piṭaka* (*byang sems kyi sde snod*) is identified by Asaṅga in his *Compendium of Abhidharma* as the tenth of the twelve branches of Buddhist scripture, the category of extensive discourses (*vaipulya*). This consists of works teaching how one accomplishes highest perfect complete enlightenment, the ten powers, stainless transcendent wisdom, and the paths of bodhisattvas. See Asaṅga 2001, 180.
48. *Saṃsāra* is often used to refer to the world or universe in general, yet the essential

meaning of the term is the contaminated five aggregates or the contaminated mind-body continuum that is the basis of the imputation of the person.

49. *Karma* refers to the mental factor of intention or volition. When either virtuous or nonvirtuous intention arises in the mindstream of an ordinary being, such ethically potent intention leaves propensities or karmic seeds upon the mental consciousness. Such seeds in turn, once awakened, produce effects commensurate with their ethical value.
50. Affliction (*kleśa*) refers to the afflictive mental states or negative emotions that comprise six root afflictions and twenty branch afflictions.
51. Highest excellence (*niḥśreyasa*) means "having no higher virtue" and refers to liberation or enlightenment.
52. Turning "the wheel of Dharma" (*dharmacakra*) refers to the eleventh of the twelve principal deeds of a fully enlightened being (Buddha). It describes the process by which the Buddha taught Dharma (Buddhist doctrine), which is likened to setting a wheel in motion, from which the direct realization of the nature of reality or selflessness (Dharma) is established in the mindstreams of qualified students and constitutes their entry to the transcendent path.
53. The twelve links are: (1) ignorance, (2) karmic formation, (3) consciousness, (4) name and form, (5) the six sense bases, (6) contact, (7) feeling, (8) craving, (9) grasping, (10) existence, (11) birth, and (12) aging and death.
54. A citation with a different rendering occurs in the *Noble Dependent Origination Sūtra* (*Pratītyasamutpāda-sūtra*): "The Tathāgata proclaimed what phenomena are causally arisen, what are their causes, and what is their cessation. Just this was stated by the great Śrāmana." Toh 212, 125b, Pd 62:343.
55. *Śālistamba-sūtra*. Toh 210, 117b, Pd 62:318.
56. Mental formations or formative factors (*saṃskāra*) refer to the fourth of the five aggregates, the formation aggregate. This aggregate refers to all impermanent entities not already subsumed in the other four aggregates and includes the neutral, positive, and negative emotions that constitute forty-nine of the fifty-one mental factors in Asaṅga's list, plus time, person, impermanence, and so on.
57. *Śālistamba-sūtra*. Toh 210, 118b, Pd 62:321.
58. See chapter 21.
59. See chapter 11.
60. This topic will be explored in detail in volume 4 of the series.
61. For further discussion on this type of reason, see Klein 1991, 273.
62. See pp. 174–79.
63. Dharmakīrti's seven treatises include: *Exposition of Valid Cognition* (*Pramāṇavārttika*), *Ascertainment of Valid Cognition* (*Pramāṇaviniścaya*), *Drop of Reasoning* (*Nyāyabindu*), *Drop of Logic* (*Hetubindu*), *Inquiry into Relations* (*Saṃbandhaparīkṣā*), *Logic of Debate* (*Vādanyāya*), and *Proof of Other Minds* (*Saṃtānāntarasiddhi*).
64. Chapa Chökyi Sengé (Phya pa chos kyi seng ge, 1109–60), the sixth abbot of Sangphu Monastery, composed works on a variety of subjects, but these

three texts were specifically composed as primers for students of logic and epistemology (*pramāṇa*). His famous eighteen topics, listed in the *Root and Commentary of the Condensed Epistemology Eliminating the Darkness of the Mind* (*Tshad ma bsdus pa yid kyi mun sel rtsa 'grel*), marked the beginning of a Tibetan tradition of epistemological studies.

65. For a brief overview of these topics, see Onada, 1996, 187–291.
66. For a brief explanation of each of these topics, see appendix.
67. This consequence is typically presented to someone who accepts that sound is produced but asserts sound is permanent and accepts that if something is permanent it must not be produced.
68. *Abhidharmakośa*, 5.22. Toh 4089, 16b, Pd 79:38.
69. For a detailed presentation on the trilemma and tetralemma, see Tillemans 2016. See also Perdue 2014.
70. As H. H. the Dalai Lama has stated, "The path of reasoning that employs logical analysis will increasingly enhance the precision of one's insight and help one develop a quick mind. Whatever the category of knowledge, one will come to know it with certainty with an alert and agile mind; such are the great advantages of this method. Therefore, on the basis of this path of reason, through analyzing whether 'If it is this, it must be that' or 'if it is that, it is necessarily this,' and so on, in terms of the three possibilities, four possibilities, contradiction, and equivalence, one will gain access to all subjects without exception. And if this method is applied, the great advantage is that one will develop precise and detailed understanding of such topics."
71. The first volume covers the presentation of knowable objects, and the remaining four themes are covered in the second volume of *Science and Philosophy in the Indian Buddhist Classics*.
72. *Pramāṇavārttika* 4.263. Toh 4210, 149b, Pd 97:598. The verse numbering for Dharmakīrti's *Exposition of Valid Cognition* is based on the critical edition of the Tibetan text found in Jinpa 2016.
73. "Ascertainable objects" (*prameya*) refers literally to phenomena that are measurable in a broad sense, but it also refers to immeasurable phenomena such as time, which is beginningless and endless; or the four immeasurables: love, compassion, equanimity, and joy; or immeasurable permanent phenomena such as unconditioned space, which extends beyond measure in any direction.
74. Capable of materiality (*rūpaṇa*) signifies that which is capable of physical interaction.
75. *Madhyamakāvatāra*, 6.202. Toh 3861, 214a, Pd 60:541.
76. *Abhidharmakośabhāṣya*, 1.13. Toh 4090, 32b, Pd 79:79. See Pruden, 1988, 1:70.
77. *Abhidharmakośabhāṣya*, 1.13. Toh 4090, 32b, Pd 79:79. See Pruden, 1988, 1:70.
78. Nonindicative form, such as the form of a vow, is material form that does not reveal or indicate the mind that motivates it.
79. Vasubandhu, *Abhidharmakośabhāṣya*, 1.13. Toh 4090, 32b, Pd 79:79. See Pruden 1988, 1:70–71.
80. This refers to things that are apprehended by means of an abstract image (*bkra*

*bar 'dzin pa, cikrīkāra, nimittīkāra*), which indicates the way the conceptual mind apprehends its conceived object by accessing an object-universal or generic image.

81. *Abhidharmasamuccaya*, chap. 1. Toh 4049, 45a, Pd 76:118. See Asaṅga 2001, 3.
82. Meditative equipoise (*samāhita*) refers to mental equilibrium that has arisen through having cultivated meditative stability that single-pointedly focuses on its object.
83. *Abhidharmasamuccaya*, chap. 1. Toh 4049, 45a, Pd 76:118. See Asaṅga 2001, 3.
84. For a detailed explanation of meditation on repulsiveness (*aśubhabhāvana*), see Buddhaghosa 1991, 169–85.
85. *Abhidharmasamuccayavyākhyā*. Toh 4054, 2b, Pd 76:959.
86. Mental-object form refers to form of the phenomena base or phenomena-base form (*chos kyi skye mched kyi gzugs, dharmāyatana*).
87. The term "translucent form" (*rūpaprasāda*) is also translated as "sense organ."
88. Mental cognition (*manas*) is equivalent to mind (*citta*), primary mind (*mūlacitta*), and consciousness (*vijñāna*).
89. *Viniścayasaṃgraha*, chap. 22. Toh 4038, 203a, Pd 74:491.
90. Asaṅga, *Abhidharmasamuccaya*, chap. 1. Toh 4049, 46b, Pd 76:121. See Asaṅga 2001, 5–6.
91. *Abhidharmakośaṭīkā-lakṣaṇānusāriṇī*, chap. 1. Toh 4093, 27b, Pd 81:66.
92. *Abhidharmakośaṭīkā-sputārtha*, chap. 1. Toh 4092, 21b, Pd 80:51.
93. Vasubandhu, *Abhidharmakośabhāṣya*, 1.10. Toh 4090, 30a, Pd 79:74. See Pruden 1988, 1:64.
94. Vasubandhu, *Abhidharmakośabhāṣya*, 1.10. Toh 4090, 30a, Pd 79:74. See Pruden 1988, 1:64.
95. *Abhidharmasamuccayavyākhyā*, chap. 1. Toh 4054, 3b, Pd 76:961.
96. *Abhidharmasamuccayavyākhyā*, chap. 1. Toh 4054, 3b, Pd 76:961.
97. *Yogācārabhūmi*, chap. 1. Toh 4035, 2b, Pd 72:673.
98. Vasubandhu, *Abhidharmakośabhāṣya*, 1.10. Toh 4090, 30b, Pd 79:75. See Pruden 1988, 1:65.
99. This implies a pleasant cuckoo's song: sound made by the cuckoo out of joy and not for communication.
100. The sound of a lute is a pleasant sound that does not convey meaning to sentient beings in the sense that it does not convey linguistic meaning. In other words, to the uneducated listener such music conveys only pleasant sound, not complex meaning.
101. Vasubandhu, *Abhidharmakośabhāṣya*, 1.10. Toh 4090, 30b, Pd 79:75. See Pruden 1988, 1:65.
102. *Abhidharmakośabhāṣya*, 1.34. Toh 4090, 43a, Pd 79:105. See Pruden 1988, 1:99.
103. See Asaṅga 2001, 9.
104. Asaṅga, *Abhidharmasamuccaya*, chap. 1. Toh 4049, 46b, Pd 76:121. See Asaṅga 2001, 5.

105. Vasubandhu, *Abhidharmakośabhāṣya*, 4.74. Toh 4090, 204a, Pd 79:504. See Pruden 1988, 2:653.
106. A "phrase" refers to a linguistic or grammatical unit of two or more words that expresses a complete thought.
107. *Pañcaskandhavivaraṇa*. Toh 4067, 24a, Pd 77:725.
108. Dharmakīrti, *Pramāṇavārttika*, 3.36–37. Toh 4210, 119b, Pd 97:529.
109. Jinaputra/Yaśomitra, *Abhidharmasamuccayavyākhyā*, chap. 2. Toh 4054, 12b, Pd 76:983.
110. Vasubandhu, *Abhidharmakośa*, 1.36abc. Toh 4089, 3a, Pd 79:6. See Pruden 1988, 1:102.
111. Vasubandhu, *Abhidharmakośabhāṣya*, 1.36. Toh 4090, 44a, Pd 79:108. See Pruden 1988, 1:102.
112. Vasubandhu, *Abhidharmakośabhāṣya*, 1.10. Toh 4090, 30b, Pd 79:75. See Pruden 1988, 1:66.
113. *Abhidharmasamuccaya*, chap. 1. Toh 4049, 46b, Pd 76:122. See Asaṅga 2001, 6.
114. Jinaputra/Yaśomitra, *Abhidharmasamuccayavyākhyā*, chap. 1. Toh 4054, 4a, Pd 76:962.
115. *Pañcaskandhabhāṣya*. Toh 4065, 39b, Pd 77:769.
116. Vasubandhu, *Abhidharmakośabhāṣya*, 1.10. Toh 4090, 30b, Pd 79:75. See Pruden 1988, 1:66.
117. Tikta (*tig ta*) refers to a species of wild gentian (*radix gentianae*), a bitter herb used to treat stomach ailments.
118. Pṛthivībandhu, *Pañcaskandhabhāṣya*. Toh 4065, 40a, Pd 77:770.
119. Jinaputra/Yaśomitra, *Spuṭārthābhidharmakośavyākhyā*, chap. 1. Toh 5593, 23a, Pd 80:55.
120. *Abhidharmasamuccaya*, chap. 1. Toh 4049, 46b, Pd 76:122. See Asaṅga 2001, 6.
121. Arura or chebulic myrobalan is the common name for *Terminalia chebula*, the fruit of which is used extensively in Tibetan medicine as a general panacea.
122. Candranandana, *Candrikāprabhāsa-aṣṭāṅgahṛdayavivṛtti*, chap. 10. Toh 4312, 132a, Pd 113:316.
123. Candranandana, *Candrikāprabhāsa-aṣṭāṅgahṛdayavivṛtti*, chap. 1. Toh 4312, 10b, Pd 113:23.
124. *Pañcaskandhaprakaraṇavaibhāṣya*, chap. 1. Toh 4066, 202a, Pd 77:540.
125. Asaṅga, *Viniścayasaṃgraha*, chap. 6. Toh 4038, 43b, Pd 74:125.
126. *Abhidharmakośabhāṣya*, 1.10. Toh 4090, 30b, Pd 79:76. See Pruden 1988, 1:66.
127. *Pañcaskandhavivaraṇa*. Toh 4067, 6b, Pd 77:682.
128. Sthiramati, *Pañcaskandhaprakaraṇavaibhāṣya*, chap. 1. Toh 4066, 203a, Pd 77:543.
129. Asaṅga, *Abhidharmasamuccaya*, chap. 1. Toh 4049, 47a, Pd 76:122. See Asaṅga 2001, 6.

130. Sthiramati, *Pañcaskandhaprakaraṇavaibhāṣya*, chap. 2. Toh 4066, 203a, Pd 77:542.
131. "Translucent form" refers to the physical sense organ. Asaṅga, *Abhidharmasamuccaya*, chap. 1. Toh 4049, 46a, Pd 76:120. See Asaṅga 2001, 5.
132. *Abhidharmakośaṭīkā-lakṣaṇānusāriṇī*, chap. 1. Toh 4093, 95a, Pd 81:234.
133. In this section the subtle physical components of the sense faculties are described in terms of subtle particles (*rdul pha rab, paramāṇu*) and atoms (*rdul phran, aṇu*), but the actual description suggests a much coarser level of material form.
134. *Pañcaskandhabhāṣya*. Toh 4065, 37a, Pd 77:763.
135. The *Tibetan-Sanskrit Dictionary* of J. S. Negi gives Skt. *umakāpuṣpa*, Tib. *zar ma'i me tog*, which points to the flower of the flax plant (Skt. *ūma*) rather than to that of sesamum (Skt. *tila*). Common flax or linseed, *Linum usitatissimum*, is a member of the genus *Linum* in the family *Linaceae*, and possesses a vivid blue color like blue Varanasi silk, which corresponds to Negi's description, whereas the color of the sesamum flower appears more whitish. See Negi 1993, 5380.
136. Mental-object form is also known as "form of the phenomena element" or "phenomena-element form" (*chos kyi khams kyi gzugs*) or "form of the phenomena base" or "phenomena-base form" (*chos kyi skye mched gyi gzugs*).
137. *Vyākhyāykti*, chap. 1. Toh 4061, 36a, Pd 77:98.
138. *Abhidharmasamuccaya*, chap. 1. Toh 4049, 47a, Pd 76:122. Walpola Rahula notes that form arising from meditative expertise (*dbang 'byor ba, vaibhūtika*) is form produced by the supernatural powers. See Asaṅga 2001, 6.
139. "Pratimokṣa vows" refers to vows of individual liberation, which refers to the eight types of ethical precepts of ordained and lay Buddhist practitioners.
140. "Negative commitments" (*asaṃvara, sdom min*) implies both a commitment to engaging in nonvirtuous acts and a lack of restraint in nonvirtuous acts.
141. "Interim precepts" (*bar ma*) imply an intermediate or preliminary stage to generating vows and so on.
142. The point being made here is that such forms emerge primarily through the performance of a rite, but the rite does not exhaust all its conditions.
143. On the meditation on repulsiveness, see Buddhaghosa 1991, 169–85.
144. Jinaputra/Yaśomitra, *Abhidharmasamuccayavyākhyā*, chap. 1. Toh 4054, 123a, Pd 76:1278.
145. The topic of conceptual mind and its relation to its object—concepts—is explored in greater detail in volume 2.
146. Vasubandhu, *Abhidharmakośa*, 1.9ab. Toh 4089, 2a, Pd 79:4. See Pruden 1988, 1:63.
147. Pūrṇavardhana, *Abhidharmakośaṭīkā-lakṣaṇānusāriṇī*, chap. 1. Toh 4093, 26a, Pd 81:63.
148. *Abhidharmakośabhāṣya*, 4.3. Toh 4090, 169a, Pd 79:416. See Pruden 1988, 2:560.

149. *Saṃdhinirmocanasūtraṭīkā*, chap. 36. Toh 4016, 103b, Pd 68:979. For details on the life of Wonch'uk (613–96), see Powers 1992 and Hopkins 1999, 44.
150. *Karmasiddhiprakaraṇa*. Toh 4062, 144b, Pd 77:379. See Anacker 2005, 49.
151. *Madhyamakāvatārabhāṣya*, chap. 6. Toh 3862, 265a, Pd 60:708.
152. Candrakīrti, *Pañcaskandhaprakaraṇa*. Toh 3866, 242b, Pd 60:1543.
153. *Abhidharmasamuccaya*, chap. 1. Toh 4049, 46a, Pd 76:120. See Asaṅga 2001, 4.
154. *Pañcaskandhabhāṣya*. Toh 4065, 35a, Pd 77:758.
155. Pūrṇavardhana, *Abhidharmakośaṭīkā-lakṣaṇānusāriṇī*, chap. 1. Toh 4093, 34a, Pd 81:82.
156. Guṇaprabha, *Pañcaskandhavivaraṇa*. Toh 4067, 4a, Pd 77:675.
157. *Pañcaskandhaprakaraṇavaibhāṣya*. Toh 4066, 197b, Pd 77:528.
158. *Abhidharmakośaṭīkā-sphuṭārtha*, chap. 1. Toh 4092, 28b, Pd 80:68.
159. Vasubandhu, *Abhidharmakośabhāṣya*, 1.13abcd. Toh 4090, 32a, Pd 79:79. See Pruden 1988, 69–70.
160. Kātyāyanīputra, *Mahāvibhāṣā*, 127, 935.
161. Pūrṇavardhana, *Abhidharmakośaṭīkā-lakṣaṇānusāriṇī*, chap. 1. Toh 4093, 34b, Pd 81:83.
162. Sthiramati, *Pañcaskandhaprakaraṇavaibhāṣya*, chap. 1. Toh 4066, 197b, Pd 77:529.
163. Pṛthivībandhu, *Pañcaskandhabhāṣya*. Toh 4065, 35b, Pd 77:759. See Asaṅga 2001, 4.
164. *Viniścayasaṃgraha*, chap. 6. Toh 4038, 52b, Pd 74:124.
165. Pṛthivībandhu, *Pañcaskandhabhāṣya*. Toh 4065, 35b, Pd 77:760.
166. Kātyāyanīputra, *Mahāvibhāṣā*, 131, 196.
167. Jinaputra/Yaśomitra, *Abhidharmakośaṭīkā-sphuṭārtha*, chap. 2. Toh 4092, 113b, Pd 80:270.
168. *Bhūmivastu*, chap. 3. Toh 4035–37, 28b, Pd 72:736.
169. Pṛthivībandhu, *Pañcaskandhabhāṣya*. Toh 4065, 36a, Pd 77:760.
170. Pṛthivībandhu, *Pañcaskandhabhāṣya*. Toh 4065, 36a, Pd 77:761.
171. *Abhidharmasamuccaya*, chap. 2. Toh 4049, 77a, Pd 76:195. See Asaṅga 2001, 91.
172. Jinaputra/Yaśomitra, *Abhidharmakośaṭīkā-sphuṭārtha*, chap. 2. Toh 4092, 115a, Pd 80:274.
173. *Ratnāvalī*, 1.84–85. Toh 4158, 110a, Pd 96:295. Also see Hopkins 2007, 105.
174. *Catuḥśatakaṭīkā*, chap. 2. Toh 3865, 52a, Pd 60:1058.
175. *Prajñāpāramitāsañcayagāthā*. Toh 11b, Pd 34:26.
176. Jinaputra/Yaśomitra, *Abhidharmakośaṭīkā-sphuṭārtha*, chap. 1. Toh 4092, 12b, Pd 80:30.
177. Here the reference is to the ancient materialist schools such as Cārvāka and Lokāyata.
178. *Skhalitapramathanayuktihetusiddhi*. Toh 3847, 20a, 57:827.
179. This topic will be discussed in the final volume in the series.

180. Vasubandhu, *Abhidharmakośa*, 2.35–36a. Toh 4089, 5a, Pd 79:11. See Pruden 1988, 1:206.
181. Vasubandhu, *Abhidharmakośa*, 2.45ab. Toh 4089, 5b, Pd 79:12. See Pruden 1988, 1:233.
182. Vasubandhu, *Abhidharmakośabhāṣya*, 2.45. Toh 4090, 78b, Pd 79:195. See Pruden 1988, 1:233.
183. Kātyāyanīputra, *Mahāvibhāṣā*, 126, 874.
184. *Abhidharmakośabhāṣya*, 2.45. Toh 4090, 78b, Pd 79:195. See Pruden 1988, 1:233.
185. *Abhidharmakośabhāṣya*, 2.45. Toh 4090, 79a, Pd 79:196. See Pruden 1988, 1:234.
186. *Abhidharmakośabhāṣya*, 4.73. Toh 4090, 203b, Pd 79:500. See Pruden 1988, 2:650.
187. *Pramāṇavārttikaṭīkā*, chap. 1. Toh 4220, 18a, Pd 99:42.
188. *Viniścayasaṃgraha*, chap. 3. Toh 4038, 22b, Pd 74:52.
189. *Abhidharmakośabhāṣya*, 2.57. Toh 4090, 84b, Pd 79:210. See Pruden 1988, 1:251.
190. Asaṅga, *Abhidharmasamuccaya*, chap. 1. Toh 4049, 52a, Pd 76:134. See Asaṅga 2001, 19.
191. Asaṅga, *Abhidharmasamuccaya*, chap. 1. Toh 4049, 52a, Pd 76:135. See Asaṅga 2001, 19–21.
192. Vasubandhu, *Pañcaskandhaprakaraṇa*. Toh 4059, 14b, Pd 77:42. See Anacker 2005, 70.
193. *Pramāṇavārttika*, 2.182. Toh 4210, 114b, Pd 97:516. Also see Jackson 1993, 377.
194. *Śālistambakaṭīkā*. Toh 4001, 155a, Pd 67:403.
195. *Pramāṇavārttika*, 2.60. Toh 4210, 109b, Pd 97:505. Also see Jackson 1993: 251–52.
196. *Pramāṇavārttika*, 2.62. Toh 4210, 109b, Pd 97:505. Also see Jackson 1993, 252.
197. Dharmakīrti, *Pramāṇavārttika*, 2.25. Toh 4210, 108b, Pd 97:502. See Jackson 1993, 212.
198. *Aṣṭasāhasrikāprajñāpāramitā-sūtra*, chap. 31. Toh 12, 278b, Pd 33:667. See Conze 1975b, 292.
199. *Pramāṇavārttikapañjikā*, chap. 35. Toh 4217, 266b, Pd 98:647.
200. *Pramāṇavārttika*, 1.37. Toh 4210, 96a, Pd 97:472. See Dunne 2004, 337–38.
201. *Abhidharmakośa*, 1.5–6. Toh 4089, 2a, Pd 79:4. See Pruden 1988, 1:59–60 and 4:1275.
202. *Pañcaskandhaprakaraṇa*. Toh 4059, 16a, Pd 77:45. See Anacker 2005, 72–73.
203. *Abhidharmasamuccaya*, chap. 1. Toh 4049, 53b, Pd 76:138. See Asaṅga 2001, 23–24.
204. *Pramāṇavārttika*, 1.85–86. Toh 4210, 98a, Pd 97:477.
205. For further discussion on the definiens and definiendum, see Purdue 1992, 61–74.

206. *Mahāyānasūtrālaṃkāra*, 12.36. Toh 4020, 14b, Pd 70:834. See Thurman 2004, 130.
207. Here "basis" refers to "instance" (*mtshan gzhi*).
208. For example, Dignaga's definition of "perceptual cognition," as presented in his *Compendium on Epistemology* (*Pramāṇasamuccaya*), in terms of "being free of conception" is primarily aimed at dispelling what is seen as the misconception of the Nyāyāyikas who accept a determinate form of perception as well. For detailed discussion of Dignaga's definition of perception, see Hattori 1968.
209. For further explanation on these eight modes of pervasion plus examples, see Wilson 1992, 390–94.
210. For a more extensive presentation of one and many, see Dreyfus 1997, 173–79. Also see Perdue 1992, 346–50; 463–64.
211. Dharmakīrti, *Pramāṇavārttika*, 1.45. Toh 4210, 96b, Pd 97:473.
212. Dharmakīrti, *Pramāṇavārttika*, 3.47. Toh 4210, 125b, Pd 97:542.
213. Śākyamati, *Pramāṇavārttikaṭīkā*, chap. 3. Toh 4220, 125b, Pd 99:312.
214. *Pramāṇavārttika*, 1.40–42. Toh 4210, 96a, Pd 97:473.
215. For further discussion, see Perdue 1992, 481–98, 525–30.
216. Dharmakīrti, *Pramāṇavārttika*, 1.59. Toh 4210, 97a, Pd 97:475.
217. See Dreyfus 1997, chap. 9, "Universals in the Collected Topics," 171–73. For an introduction to particulars, see Dreyfus, chap. 6, "Introducing Universals," 127–42. Also see Perdue 1992, 621–93.
218. See Perdue 1992, 630.
219. Although a vase and a pillar by themselves are type-universals, the conjunction of the two can only be a particular since there is nothing in the world that can be considered to be an example of it.
220. See Perdue 1992, 695–771.
221. See Perdue 1992, 695–771, where the topic of Chapa's *rdzas ldog* is extensively presented.
222. For more detail on substantial phenomena and abstract phenomena, see Perdue 1992, 695–771.
223. *Madhyamakāloka*. Toh 3887, 135b, Pd 62:1119.
224. *Madhyamakāloka*. Toh 3887, 135b, Pd 62:1119.
225. For a different presentation of this point, see Yoysuya 1999, 24 n54.
226. *Madhyamakāloka*. Toh 3887, 135b, Pd 62:1119.
227. *Nyāyabinduṭīkā*. Toh 4230–31, 76b, Pd 105:204.
228. *Pramāṇaviniścayaṭīkā*, chap. 2. Toh 4227, 205a, Pd 104:1263.
229. *Pramāṇaviniścayaṭīkā*, chap. 2. Toh 4227, 205b, Pd 104:1264.
230. *Pañcaviṃśatisāhasrikāprajñāpāramitā-sūtra*, chap. 8. Toh 9, 184a, Pd 26:406. See Conze 1975b, 119.
231. Dharmakīrti, *Pramāṇavārttika*, 1.39. Toh 4210, 96a, Pd 97:473.
232. *Śālistamba-sūtra*. Toh 210, 116a, Pd 62:315.

233. Dharmakīrti, *Pramāṇavārttika*, 1.38. Toh 4210, 96a, Pd 97:473. See Dunne 2004, 337–38.
234. For an introduction to the theory of *apoha*, see Dreyfus 1997, 205–16.
235. On the nature of *apoha* theory, see Dreyfus 1997, 217–32. For difficult points related to *apoha* theory, see Tillemans 1999, 209–46.
236. *Prajñāpradīpa*, chap. 5. Toh 3853, 94a, Pd 57:1032.
237. *Tattvasaṃgraha*, 20.139. Toh 4226, 37b, Pd 107:92.
238. Bhāvaviveka, *Tarkajvālā*, 3.3. Toh 3856, 59b, Pd 58:150.
239. Bhāvaviveka, *Tarkajvālā*, chap 3. Toh 3856, 59b, Pd 58:149.
240. *Prajñāpradīpaṭīkā*. Toh 3859, 63b, Pd 58:1018.
241. *Piṇḍanivarta*, v. 8. Toh 4293, 252b, Pd 109:1750.
242. Examples of these four implicative negations are respectively: "A mountainless plain," "Fat Devadatta does not eat food at night," "Fat Devadatta who does not eat during the day is not thin," and "Śākyamuni is either a brahman or kṣatriya but not a brahman."
243. *Pratītyasamutpādādivibhaṅganirdeśa*. Toh 3995, 7b, Pd 66:730.
244. Nabidharma, *Piṇḍanivartana*, vv. 4–5. Toh 4293, 251a, Pd 109:1749.
245. Nabidharma, *Piṇḍanivartana*, vv. 6–7. Toh 4293, 251a, Pd 109:1750.
246. In other words, fifteen negations are subsumed in eight negations as follows: (1) nonexistence, (2) another, (3) similarity, (4) inferiority that includes three negations—inferiority, timidity, weakness, (5) subtlety that includes six negations—subtlety, small amount, swift passage, smallness, no extra, from some not all, (6) incompatibility, (7) antidote, and (8) absence.
247. Nabidharma, *Piṇḍanivartana*, v. 7. Toh 4293, 251a, Pd 109:1750.
248. Nabidharma, *Piṇḍanivartana*, v. 7. Toh 4293, 251a, Pd 109:1750.
249. Nabidharma, *Piṇḍanivartana*, v. 8. Toh 4293, 251a, Pd 109:1750.
250. Nabidharma, *Piṇḍanivartana*, v. 9. Toh 4293, 251a, Pd 109:1750.
251. Nabidharma, *Piṇḍanivartana*, vv. 10–11. Toh 4293, 251a, Pd 109:1750.
252. Nabidharma, *Piṇḍanivartana*, v. 11. Toh 4293, 251a, Pd 109:1750.
253. *Saṃdhinirmocana-sūtra*, chap. 10. Toh 106, 51b, Pd 49:120. In this passage a contrast is being drawn between the Buddha's omniscient mind, for whom these facts are evident, and sentient beings, for whom they remain hidden.
254. *Abhidharmakośabhāṣya*, chap. 9. Toh 4090, 82a, Pd 79:876.
255. *Pramāṇavārttika*, 3.63. Toh 4210, 121a, Pd 97:531.
256. *Pramāṇavārttika*, 4.48. Toh 4210, 141a, Pd 97:578. See Tillemans 2000, 78.
257. *Pramāṇavārttika*, 4.51cd. Toh 4210, 141a. Pd 97:578. See Tillemans 2000, 81.
258. Dharmakīrti, *Pramāṇavārttika*, 1.214. Toh 4210, 102b, Pd 97:488.
259. Dharmakīrti, *Pramāṇavārttika*, 1.215. Toh 4210, 102b, Pd 97:488.
260. Dharmakīrti, *Pramāṇavārttika*, 1.214ab. Although, as cited here, these lines are found in the Tibetan version of the *Pramāṇavārttika*, Dharmakīrti is here quoting Dignāga's *Pramāṇasamuccaya*. Toh 4210, 102b, Pd 97:488.
261. *Catuḥśataka*, 12.5. Toh 3846, 13a, Pd 57:809. See Rinchen and Sonam 2008, 241.

262. Dharmakīrti, *Pramāṇavārttika*, 2.145. Toh 4210, 113a, Pd 97:513. See Jackson 1993, 335.
263. For a discussion on traditional Buddhist atomic theory, see Dreyfus 1997, 83–94.
264. *Catuḥśatakaṭīkā*. Toh 3865, 153b, Pd 60:1296.
265. *Tattvasaṃgrahapañjikā*, chap 15. Toh 4267, 257b, Pd 107:669. See Jha 1986, 895.
266. *Tarkajvālā*, chap. 7. Toh 3856, 242b, Pd 58:590.
267. Bhāvaviveka, *Tarkajvālā*, chap. 7. Toh 3856, 243b, Pd 58:593.
268. Kātyāyanīputra, *Mahāvibhāṣā*, 136, 354.
269. Pūrṇavardhana, *Abhidharmakośaṭīkā-lakṣaṇānusāriṇī*, chap. 2. Toh 4093, 131a, Pd 81:329.
270. *Abhidharmasamuccaya*, chap. 2. Toh 4049, 77b, Pd 76:196. See Asaṅga 2001, 91.
271. Jinaputra/Yaśomitra, *Abhidharmasamuccayavyākhyā*, chap. 4. Toh 4054, 38b, Pd 76:1048.
272. *Alambanaparīkṣāvṛtti*. Toh 4206, 86b, Pd 97:433.
273. *Alambanaparīkṣāṭīkā*. Toh 4265, 181b, Pd 106:483.
274. *Viniścayasaṃgraha*, chap. 6. Toh 4038, 80a, Pd 74:117.
275. Bhāvaviveka, *Tarkajvālā*, chap. 5. Toh 3856, 209b, Pd 58:510.
276. *Prajñāpradīpaṭīkā*. Toh 3859, 92b, Pd 58:1089.
277. *Catuḥśataka*, 9.19. Toh 3846, 11a, Pd 57:803. See Rinchen and Sonam 2008, 211.
278. *Catuḥśatakaṭīkā*, chap. 9. Toh 3865, 154a, Pd 60:1298.
279. *Bāhyārthasiddhi*, v. 56. Toh 4244 ,191b, Pd 106:514.
280. *Abhidharmakośabhāṣya*, 3:85–86. Toh 4090, 154b, Pd 79:381. See Pruden 1988, 2:474.
281. *Lalitavistara-sūtra*. Toh 95, 77a, Pd 46:185.
282. A type of small particle (Skt. *avagaṇa*; Tib. *a ba ga na*) whose meaning is something isolated or separated from its companions.
283. Another term for a small particle (Skt. *vaṭika*; Tib. *ba ti ka*) that also refers to a pawn in chess.
284. *Lokaprajñapti*, chap. 4. Toh 4086, 9b, Pd 78:650.
285. *Vinayavibhaṅga*. Toh 5, 239a, Pd 8:562.
286. Guṇaprabha, *Vinayasūtra*. Toh 4117, 26a, Pd 88:898.
287. *Condensed Kālacakra Tantra*, *Kālacakra-laghutantra*, 1.13. Toh 362, 23b, Pd 77:60. *Abhidharmakośa* presents a slightly different explanation: one *krośa* (*rgyang grags*) equals 500 spans (*dhanu*, *'dom*) or 2,000 cubits (*hasta*, *khru*) or roughly one kilometer, while one yojana (*dpag tshad*) equals eight krośa or roughly eight kilometers.
288. Pūrṇavardhana, *Abhidharmakośaṭīkā-lakṣaṇānusāriṇī*, chap. 2. Toh 4093, 131b, Pd 81:329.
289. Vasubandhu, *Abhidharmakośa*, 2.22. Toh 4089, 4b, Pd 79:10. See Pruden 1988, 1:185.

290. Pūrṇavardhana, *Abhidharmakośaṭīkā-lakṣaṇānusāriṇī*, chap. 2. Toh 4093, 131b, Pd 81:330.
291. *Abhidharmakośabhāṣya*, 2.22. Toh 4090, 64a, Pd 79:159. See Pruden 1988, 1:187.
292. *Abhidharmasamuccaya*, chap. 2. Toh 4049, 77b, Pd 76:196. See Asaṅga 2001, 91.
293. Asaṅga, *Bhūmivastu*, chap. 3. Toh 4035–37, 27a, Pd 72:733.
294. Bhāvaviveka, *Tarkajvālā*, chap 5. Toh 3856, 210a, Pd 58:512.
295. Āryadeva, *Catuḥśataka*, 7.10. Toh 3846, 8b, Pd 58:798. See Rinchen and Sonam 2008, 174.
296. *Catuḥśatakaṭīkā*, chap. 9. Toh 3865, 152b, Pd 60:1294.
297. *Jñānasārasamuccayanibandhana*. Toh 3852, 38b, Pd 57:883.
298. Bhāvaviveka, *Tarkajvālā*, chap. 7. Toh 3856, 242b, Pd 58:590.
299. *Bāhyārthasiddhi*, v. 46. Toh 4244, 191a, Pd 106:513.
300. *Bāhyārthasiddhi*, vv. 47–48. Toh 4244, 191a, Pd 106:514.
301. *Abhidharmakośabhāṣya*, 1.43. Toh 4090, 49b, Pd 79:122. Also see Pruden 1988, 1:121.
302. *Abhidharmakośabhāṣya*, 1.43. Toh 4090, 49b, Pd 79:123. Also see Pruden 1988, 1:121.
303. Kamalaśīla, *Madhyamakālaṃkārapañjikā*, chap. 1. Toh 3886, 92b, Pd 62:1002.
304. Śubhagupta, *Bāhyārthasiddhi*, v. 52. Toh 4244, 191a, Pd 106:514.
305. Vasubandhu, *Viṃśatikā*, v. 12. Toh 4056, 3b, Pd 77:9. See Anacker 2005, 167–68.
306. Vasubandhu, *Viṃśatikāvṛtti*. Toh 4057, 7b, Pd 77:20. See Anacker 2005, 169.
307. *Bhūmivastu*, chap. 3. Toh 4035–37, 26b, Pd 72:732.
308. *Madhyamakālaṃkāravṛtti*. Toh 3885, 59a, Pd 62:912.
309. *Sarvadharmāsvabhāvasiddhi*. Toh 3889, 288b, Pd 62:1516.
310. Āryadeva, *Catuḥśataka*, 9.12–16. Toh 3846, 10b, Pd 57:803. See Rinchen and Sonam 2008, 208–9.
311. *Catuḥśatakaṭīkā*, chap. 9. Toh 3865, 152b, Pd 60:1294.
312. For a detailed explanation of these six categories of the Vaiśeṣika school in a form accessible to a contemporary reader, see Potter 1977, 47–146.
313. This passage is cited in Kamalaśīla's *Commentary on the Compendium on Reality* (*Tattvasaṃgrahapañjikā*). Toh 4267, 192a, Pd 107:501. See Jha 1986. Also see Tsongkhapa 2010, 398.
314. The label Vaibhāṣika (Tib. *bye brag smra ba*) refers in general to the early Buddhist schools whose tenet systems and belief structures are presented in Kātyāyanīputra's *Great Treatise on Differentiation* (*Mahāvibhāṣā*) and in particular to the Sarvāstivāda, a central school in the development of Abhidharma. For a more detailed presentation, see Hopkins 2003, 208–44.
315. *Abhidharmakośabhāṣya*, 5.23. Toh 4090, 239a, Pd 79:586. This passage of *Autocommentary* is a word commentary to chapter 5, verse 23, lines a–d of *Treasury of Knowledge*, and the relevant passages have been marked in the text. Also see Pruden 1988, 3:806.

316. Pūrṇavardhana, *Abhidharmakośaṭīkā-lakṣaṇānusāriṇī*, chap. 5. Toh 4093, 112b, Pd 81:1199.
317. *Abhidharmasamuccaya*, chap. 1. Toh 4049, 53a, Pd 76:136. See Asaṅga 2001, 21.
318. *Madhyantavibhaṅga*, chap. 3. Toh 4021, 42b, Pd 70:908. See Anacker 2005, 243.
319. *Madhyāntavibhaṅgaṭīkā*, v. 20. Toh 4027, 14b, Pd 71:35. See Anacker 2005, 243.
320. Nāgārjuna, *Mūlamadhyamakakārikā*, 19.1. Toh 3824, 11a, Pd 57:27. See Siderits and Katsura 2013, 208.
321. Nāgārjuna, *Mūlamadhyamakakārikā*, 19.3. Toh 3824, 11a, Pd 57:27. See Siderits and Katsura 2013, 209.
322. *Catuḥśataka*, 11.1. Toh 3846, 12a, Pd 57:806. See Rinchen and Sonam 2008, 227.
323. *Catuḥśatakaṭīkā*, chap. 11. Toh 3865, 171b, Pd 60:1339.
324. Vasubandhu, *Abhidharmakośa*, 5.26a–b. Toh 4089, 16b, Pd 79:38. See Pruden 1988, 3:808.
325. Vasubandhu, *Abhidharmakośabhāṣya*, 5.25. Toh 4090, 239b, Pd 79:588. See Pruden 1988, 3:808.
326. Vasubandhu, *Abhidharmakośabhāṣya*, 5.25. Toh 4090, 239b, Pd 79:588. See Pruden 1988, 3:809.
327. Vasubandhu, *Abhidharmakośabhāṣya*, 5.25. Toh 4090, 240a, Pd 79:588. See Pruden 1988, 3:809.
328. Vasubandhu, *Abhidharmakośabhāṣya*, 5.25. Toh 4090, 240a, Pd 79:589. See Pruden 1988, 3:809.
329. Vasubandhu, *Abhidharmakośabhāṣya*, 5.25. Toh 4090, 29a, Pd 79:71. Pruden 1988, 3:809.
330. *Pramāṇavārttika*, 1.281. Toh 4210, 105a, Pd 97:494.
331. See Tsongkhapa 2010, 400.
332. Asaṅga, *Abhidharmasamuccaya*, chap. 1. Toh 4049, 60b, Pd 76:155. See Asaṅga 2001, 44.
333. *Subāhuparipṛcchā-sūtra*, chap. 2. Toh 70, 175b, Pd 43:485.
334. Candrakīrti, *Catuḥśatakaṭīkā*, chap. 11. Toh 3865, 171b, Pd 60:1338.
335. A moment required to complete an action (*bya rdzogs kyi skad cig ma*) varies according to the time taken to accomplish a specific action, such as, for instance, the time taken for eye consciousness to perceive a visual object.
336. *Abhidharmakośa*, 3.88–89a–b. Toh 4089, 10a, Pd 79:22. See Pruden 1988, 2:475.
337. *Abhidharmakośabhāṣya*, 3.85. Toh 4090, 154b, Pd 79:381. See Pruden 1988, 2:474.
338. Dharmakīrti, *Pramāṇavārttika*, 3.495ab. Toh 4210, 137a, Pd 97:570.
339. Candrakīrti, *Catuḥśatakaṭīkā*, chap. 1. Toh 3865, 38a, Pd 60:1025.
340. *Prasphuṭapadā*, chap. 7. Toh 3796, 89a, Pd 52:916.
341. "Inconceivable states" refer to a buddha's body, speech, and mind. The twelve deeds of a buddha are: (1) the act of leaving Tuṣita, (2) the act of entering his

mother's womb, (3) the act of birth, (4) the act of engaging in sport, (5) the act of abiding with his female retinue, (6) the act of teaching the means of renunciation, (7) the act of practicing austerities, (8) the act of proceeding to the site where enlightenment is attained, (9) the act of taming the māras, (10) the act of attaining perfect enlightenment, (11) the act of turning the wheel of Dharma, and (12) the act of passing into nirvāṇa.

342. *Sūtrasamuccaya*, chap. 5. Toh 3934, 213a, Pd 64:595.
343. *Mañjuśrīmūla-tantra*, chap. 18. Toh 543, 201b, Pd 88:626.
344. *Vajraḍāka*, chap. 26. Toh 371, 64b, Pd 78:152.
345. *Pañcakrama*, v. 29. Toh 1802, 49a, Pd 18:139.
346. *Ratnāvalī*, 1.69. Toh 4158, 109b, Pd 96:294. See Hopkins 2007, 103.
347. For how momentariness is depicted in Dharmakīrti's system, see Dreyfus 1997, 60–65.
348. Vasubandhu, *Abhidharmakośabhāṣya*, 2.46. Toh 4090, 80, Pd 79:200. See Pruden 1988, 1:242.
349. Jinaputra/Yaśomitra, *Abhidharmakośaṭīkā-sputārtha*, chap. 2. Toh 4092, 158b, Pd 80:378.
350. Vasubandhu, *Abhidharmakośabhāṣya*, 2.46. Toh 4090, 81b, Pd 79:202. See Pruden 1988, 1:242.
351. Bhāvaviveka, *Tarkajvālā*, chap. 3. Toh 3856, 73a, Pd 58:183.
352. *Pramāṇaviniścaya*, chap. 2. Toh 4211, 183a, Pd 97:685.
353. Jinaputra/Yaśomitra, *Abhidharmakośaṭīkā-sputārtha*, chap. 2. Toh 4092, 163b, Pd 80:390.
354. Asaṅga, *Vastusaṃgrahaṇī*, chap. 5. Toh 4039, 179b, Pd 74:1189.
355. Āryadeva, *Catuḥśataka*, 14.22. Toh 3846, 16a, Pd 57:815. See Rinchen and Sonam 2008, 273.
356. *Udānavargavivaraṇa*, chap. 18. Toh 4100, 25a, Pd 83:653.
357. Pṛthivībandhu, *Pañcaskandhabhāṣya*. Toh 4065, 88b, Pd 77:891.
358. *Yogācārabhūmi*, chap. 3. Toh 4035, 24b, Pd 72:726.
359. The "hour" mentioned here is forty-eight minutes long.
360. *Vādavidhikhagajavinayakṣānti-sūtra*. Toh 263, 150a, Pd 67:368.
361. Dharmakīrti, *Pramāṇavārttika* 1.193cd. Toh 4210, 102a, Pd 97:486.
362. Dharmakīrti, *Pramāṇavārttika*, 4.281ab. Toh 4210, 150b, Pd 97:599.
363. Dharmakīrti, *Pramāṇavārttika*, 4.282–83b. Toh 4210, 150b, Pd 97:599.
364. Dharmakīrti, *Pramāṇaviniścaya*, chap. 1. Toh 4211, 179b, Pd 97:677.
365. The reasons here refer to the three proofs proposed earlier to demonstrate the existence of a prior mind, namely, the creator: (1) things operate in a temporally ordered way, (2) they possess specific shapes, and (3) they are capable of function.
366. Dharmakīrti, *Pramāṇavārttika*, 2.10. Toh 4210, 108a, Pd 97:501. See Jackson 1993, 195.
367. Dharmakīrti, *Pramāṇavārttika*, 2.23. Toh 4210, 108b, Pd 97:502. See Jackson 1993, 210.
368. *Madhyamakahṛdaya*, 9.134. Toh 3855, 36b, Pd 58:85.

369. *Tarkajvālā*, chap. 3. Toh 3856, 110b, Pd 58:269.
370. In Sāṃkhya philosophy "primal nature" (Skt. *prakṛti*; Tib. *spyi gtso bo*) refers to the first principle (*tattva*) or cause of the material universe. It is composed of the three qualities (*guṇa*): *sattva*, *rajas*, and *tamas*, and twenty-three components, such as intellect (*buddhi*), self (*āhaṃkāra*), and mind (*manas*). See Larson and Bhattacharya 1987.
371. This phrase, "merely this conditionedness" (Skt. *idam pratyaya*; Tib. *rkyen nyid 'di pa tsam*), is a crucial aspect of the Buddhist principle of dependent origination and its approach to understanding the origination of everything, including the entire cosmos, purely in terms of cause and effect dynamic. This crucial term is often translated as "conditionality" or "specific conditionality," such as in Ñaṇamoli's translation: "Because there is a condition, or because there is a total of conditions, for these states, beginning with ageing-and-death, . . . it is called *specific conditionality* [*idapaccayatā*]" (Buddhaghosa 1991, 526).
372. Asaṅga, *Abhidharmasamuccaya*, chap. 1. Toh 4049, 64b, Pd 76:167. See Asaṅga 2001, 56.
373. *Pratītyasamutpādādivibhaṅganirdeśa*, chap. 1. Toh 3995, 5a, Pd 66:725.
374. *Śālistamba-sūtra*. Toh 210, 116a, Pd 62:315. Almost identical passages on dependent origination are found in the Pali canon as well, e.g., "When this exists, that comes to be; with the arising of this, that arises. When this does not exist, that does not come to be; with the cessation of this, that ceases. That is, with ignorance as condition, formation comes to be" (Ñāṇamoli 2015, 927).
375. *Abhidharmasamuccaya*, chap. 1. Toh 4049, 66a, Pd 76:169. See Asaṅga 2001, 58.
376. Rain whose drops are like the shafts of a carriage. See Pruden 1988, 2:452.
377. *Abhidharmakośabhāṣya*, 3.46. Toh 4090, 144a, Pd 79:355. See Pruden 1988, 2:452.
378. Vasubandhu, *Abhidharmakośa*, 3.73–74. Toh 4089, 9b, Pd 79:21. See Pruden 1988, 2:468–69.
379. *Lalitavistara-sūtra*, chap. 12. Toh 95, 77a, Pd 46:186.
380. *Vinayavibhaṅga*, chap. 3. Toh 3, 49a, Pd 5:128.
381. *Vinayavibhaṅga*, chap. 3. Toh 3, 49b, Pd 5:130.
382. *Vinayavibhaṅga*, chap. 3. Toh 3, 49b, Pd 5:130.
383. *Vinayavibhaṅga*, chap. 3. Toh 3, 49b, Pd 5:130.
384. *Vinayavibhaṅga*, chap. 3. Toh 3, 50a, Pd 5:131.
385. *Vinayavibhaṅga*, chap. 3. Toh 3, 50b, Pd 5:133.
386. Maudgalyāyana, *Lokaprajñāpti*. Toh 4086, Tengyur, mngon pa, *i*. This is the first of three Abhidharma texts (the other two being *Kāraṇaprajñapti* and *Karmaprajñapti*) belonging to a corpus known collectively as the *Prajñapti Treatise* (*Gdags pa'i bstan bcos*), which is attributed to Maudgalyāyana, one of the two principal disciples of the Buddha. This corpus is part of the Seven Abhidharma Texts of the Sarvāstivāda school of Abhidharma.
387. The five are: (1) degeneration of time: the existence of dispute, (2) degeneration of lifespan: currently lifespan overall is becoming shorter, (3) degeneration of

view: the difficulty in transforming wrong view, (4) degeneration of affliction: an enduring strong sense of self, and (5) degeneration of beings: the difficulty of taming the mindstream and the senses.

388. The structure is cone-shaped with the narrower parts including the sharp tip submerged beneath the great ocean.
389. Puṇḍarīka, *Vimalaprabhā*. Toh 1347, 156a, Pd 6:367. See Khedrup Norsang Gyatso 2004, 79.
390. For a concise presentation of the Kālacakra system views on the formation of world systems as well as its cosmology, see Khedrup Norsang Gyatso 2004, part 1, chaps. 4, 6, 7.
391. Puṇḍarīka, *Vimalaprabhā*. Toh 1347, 256b, Pd 6:1278.
392. The general series editor Thupten Jinpa asserts this statement makes no sense and could be an error in the Tibetan text.
393. For a more detailed treatment of the differences between the Abhidharma and Kālacakra systems of cosmology, see Khedrup Norsang Gyatso 2004, 145–57.
394. *Avataṃsaka-sūtra*, chap. 22. Toh 44, 361a, Pd 38:795.
395. *Vajrapāṇi-abhiṣekamahā-tantra*, chap. 1. Toh 496, 3b, Pd 87:7. Also see Cleary 1993, 202–7.
396. *Avataṃsaka-sūtra*, chap. 4. Toh 44, 94a, Pd 38:208. Also see Cleary 1993, 189.
397. *Avataṃsaka-sūtra*, chap. 9. Toh 44, 144b, Pd 35:323. For an alternative translation based on the Chinese edition, see Cleary 1993, 250.
398. *Pitāputrasamāgama-sūtra*, chap. 26. Toh 60, 130b, Pd 42:312. The four great rivers and their flowing into the "lake that never heats up" (Anavatapta) is probably a reference to the ancient Indian myth of how four great rivers emerge from the four sides of Mount Kailash and flow into the sacred lake Manasarovar. In fact Mount Kailash does lie near the sources of Asia's greatest rivers—the Indus, the Sutlej, the Brahmaputra, and the Karnali, the latter being a tributary of the famed River Ganges.
399. Asaṅga, *Yogācārabhūmi*, chap. 2. Toh 4035, 17b, Pd 72:709.
400. Kātyāyanīputra, *Mahāvibhāṣā*, 133, 225.
401. Vasubandhu, *Abhidharmakośa*, 3.102. Toh 4089, Pd 79:24. See Pruden 1988, 2:495.
402. Puṇḍarīka, *Vimalaprabhā*, chap. 4. Toh 1347, 275b, Pd 6:1324.
403. *Catuḥśataka*, 8.25. Toh 3846, 10a, Pd 57:802. See Rinchen and Sonam 2008, 198.
404. *Mūlamadhyamakakārikā*, 11.8. See Siderits and Katsura 2013, 126; and Tsongkhapa 2010, 275.
405. *Prajñāpradīpa*, chap. 11. Toh 3853, 139b, Pd 57:1145.
406. *Vimalaprabhā*, chap. 7. Toh 1347, 155b, Pd 6:367.
407. *Vimalaprabhā*, chap. 7. Toh 1347, 156a, Pd 6:367.
408. The seven planets refer to Sun, Moon, Mars, Mercury, Jupiter, Venus, and Saturn, and to this are added three more, namely, Kālāgni, Rāhu, and Ketu. The planet Rāhu (Tib. *sgra can*) is associated with eclipse, the planet Kālāgni with destruction of the universe at the end of the world, and the planet Ketu with

the comets. The Kālacakra tantra presents a complex and detailed explanation of how these planets come to be formed in a sequential order and their association with the emergence of calenderial time in terms of days, weeks, months, and years. For a detailed explanation of these points in English by the Tibetan master Khedrup Norsang Gyatso, see Khedrup Norsang Gyatso 2004, chap. 6.

409. For a detailed presentation of this material, see Khedrup Norsang Gyatso 2004, 105–44.
410. Puṇḍarīka, *Vimalaprabhā*, chap. 9. Toh 1347, 180a, Pd 6:425.
411. Puṇḍarīka, *Vimalaprabhā*, chap. 9. Toh 1347, 195a, Pd 6:460. See Khedrup Norsang Gyatso 2004, 120.
412. Puṇḍarīka, *Vimalaprabhā*, chap. 9. Toh 1347, 193a, Pd 6:456. See Khedrup Norsang Gyatso 2004, 143.
413. Although there is consensus that Āryabhaṭa was active in Nālandā, there are differing opinions on his sectarian affiliation, that is, whether he was a Buddhist or a non-Buddhist scholar. He was born in 476 CE at Pāṭaliputra and later studied at Nālandā, which housed an astronomical observatory, an institution he reputedly later came to head. His contribution to astronomy and mathematics was unprecedented. In mathematics he expanded the concept of positional notation [decimal system], the notion of zero, the approximation of pi, trigonometry, indeterminate equations, and algebra. In astronomy he clarified the motion of planets, the mechanism of eclipses, sidereal periods, and heliocentrism, and he attained fame for his two works on astronomy, the *Āryabhaṭayaṃ* and *Āryabhaṭasiddhanta*.
414. Āryabhaṭa, *Āryabhaṭayaṃ*, 4.6. For an alternative translation of this and subsequent verses cited from the text, see Clark 2006, 64–66.
415. Āryabhaṭa, *Āryabhaṭayaṃ*, 4.9.
416. Āryabhaṭa, *Āryabhaṭayaṃ*, 4.10.
417. Āryabhaṭa, *Āryabhaṭayaṃ*, 4.37.
418. One "finger span" (*mtho*) refers to the length covered by extending the thumb and the ring finger, which is said to be equivalent to twelve finger widths. Two such finger spans equal a cubit, or twenty-four finger widths. Interestingly, as can be seen in the subsequent citation, in the *Treasury of Knowledge* as well as its *Autocommentary*, Vasubandhu does not list the finger span among his units of measurement.
419. *Abhidharmakośabhāṣya*, 3.85–88. Toh 4090, 155a, Pd 79:381. See Pruden 1988, 2:474.
420. *Kālacakra-laghutantra*, 1.13. Toh 362, 23b, Pd 77:60.
421. Kātyāyanīputra, *Mahāvibhāṣā*, 135, 350. By "generic name," the text makes the point that the Sanskrit word for eon, *kalpa*, has the connotation of differentiation of temporal segmentation, and the eon being the "ultimate" temporal length in the measurement, it is given the name that connotes this general notion of segmentation of time.
422. Vasubandhu, *Abhidharmakośa*, 3.88–89. Toh 4089, Pd 79:22.
423. Vasubandhu, *Abhidharmakośabhāṣya*, 3.88–89. Toh 4090, 155b, Pd 79:382.

The final sentence in quotation marks indicates a section in verse from the *Abhidharmakośa*. Also see Pruden 1988, 2:475.

424. The Tibetan calendar, which is based on the Kālacakra tantra, employs a complex system of calculation designed to harmonize the three different ways of measuring the lunar, solar, and zodiac year. Also it employs various measures to resolve the disparity between the length of lunar and solar months by adding an extra month every four years. For a detailed analysis of the Tibetan calendar system and its Kālacakra roots, see Henning 2007.
425. Vasubandhu, *Abhidharmakośa*, 3.90–93. Toh 4089, 10a, Pd 79:23. See Pruden 1988, 2:476–79.
426. Pūrṇavardhana, *Abhidharmakośaṭīkā-lakṣaṇānusāriṇī*, chap. 3. Toh 4093, 339a, Pd 81:840.
427. This number far exceeds the duration of the universe calculated by science, which is currently estimated as 13.4 billion years.
428. Daśabalaśrīmitra, *Saṃskṛtāsaṃskṛtaviniścaya*, chap. 8. Toh 3897, 122b, Pd 63:327.
429. Puṇḍarīka, *Vimalaprabhā*, chap. 8. Toh 1347, 169a, Pd 6:399. See Khedrup Norsang Gyatso 2004, 95.
430. On the status of the *Miscellaneous Sūtra* (*mdo sil bu*, *muktakasūtra*) cited by Vasubandhu below, see the discussion at Pruden 1988, 2:543n506.
431. Unlike European languages, Tibetan (as well as Sanskrit) has separate terms for ten thousand (*khri*), hundred thousand (*'bum*), ten million (*bye ba*), hundred million (*dung phyur*), and so on.
432. Here Vasubandhu is commenting on the text *Miscellaneous Discourses* (*Muktakāgama*), which is his source, and notes that the names of eight numbers are missing in that text. In any case, the final number in the list of sixty numbers, "countless," is understood as the sixtieth power of ten.
433. *Abhidharmakośabhāṣya*, 3.93. Toh 4090, 157b, Pd 79:388. See Pruden 1988, 2:480.
434. *Lalitavistara-sūtra*, chap. 12. Toh 95, 76a, Pd 46:183.
435. *Avataṃsaka-sūtra*, chap. 36. Toh 44, 385a, Pd 36:808.
436. Nāga (Skt. *nāga*; Tib. *klu*) is a term applied equally to ordinary snakes as to mythical or legendary creatures that resemble snakes, such as powerful serpent-demons who intervene in human affairs or beings dwelling beneath the earth with human faces and serpent-like lower extremities. Often they possess the ability to change their form magically into a human form. In Tibet nāgas were held to dwell in lakes, streams, trees, and so on, where they guard hidden treasures and so on, whereas in China nāgas were often equated with dragons.
437. Garuḍas (Skt. *garuḍa*; Tib. *nam mkha' lding*) are mythical birds that feed on snakes or nāgas. Also called "fine-winged" (Skt. *suparṇa*), they combined the features of an animal and a divine being, like nāgas. They possess magical powers, are appointed as guardians by higher gods, or act as a mount of the gods and the means of their conveyance.
438. *Kāraṇaprajñapti*, 3.10. Toh 4087, 159b, Pd 79:1030.

439. *Nandagarbhakrāntinirdeśa-sūtra*, chap. 1. Toh 58, 238a, Pd 41:740.
440. *Vinayavibhaṅga*, chap. 28. Toh 3, 83b, Pd 6:195.
441. Vasubandhu, *Abhidharmakośabhāṣya*, 4.103. Toh 4090, 212a, Pd 41:651. See Pruden 1988, 2:686.
442. *Nandagarbhakrāntinirdeśa-sūtra*, chap. 1. Toh 57, 212a, Pd 41:651.
443. Asaṅga, *Yogācārabhūmi*. Toh 4035, 11b, Pd 72:695.
444. *Nandagarbhakrāntinirdeśa-sūtra*, chap. 1. Toh 58, 243a, Pd 41:753.
445. Vasubandhu, *Abhidharmakośabhāṣya*, 3.15. Toh 4090, 121a, Pd 79:299. See Pruden 1988, 2:395.
446. "Vairocana" (Tib. *rnam snang*) represents the purified aspect of the form aggregate, and the head of one of the five Buddha families in Vajrayāna Buddhism. Within the human body, the crown of the head is associated with the seat of Vairocana, hence the crown cakra (channel complex) is also referred to as the "Vairocana cakra."
447. The mind vajra (Skt. *cittavajra*; Tib. *thugs rdo rje*) in general refers to the mind of a buddha, but here it refers to the mental consciousness of the bardo being.
448. Nāgabodhi, *Samājasādhanavyavasthāli*. Toh 1809, 123b, Pd 18:343.
449. *Saṃvarodaya-tantra*, 2.14–15. Toh 373, 266a, Pd 78:777.
450. *Vajramālā-tantra*, chap. 17. Toh 445, 230a, Pd 81:805.
451. Pūrṇavardhana, *Abhidharmakośaṭīkā-lakṣaṇānusāriṇī*, chap. 3. Toh 4093, 279a, Pd 81:698.
452. This is a traditional argument for intermediate beings not being physically obstructed.
453. *Aṣṭāṅgahṛdayasaṃhitā*, 1.5–6. Toh 4310, 106b, Pd 111:294.
454. *Saṃvarodaya-tantra*, 2.27–28. Toh 373, 266b, Pd 78:778.
455. *Nandagarbhāvakrāntinirdeśa-sūtra*, chap. 1. Toh 57, 214a, Pd 41:659.
456. *Nandagarbhāvakrāntinirdeśa-sūtra*, chap. 1. Toh 57, 214b, Pd 41:659.
457. *Nandagarbhāvakrāntinirdeśa-sūtra*, chap. 1. Toh 57, 214b, Pd 41:660.
458. *Nandagarbhāvakrāntinirdeśa-sūtra*, chap. 1. Toh 57, 214b, Pd 41:660.
459. *Nandagarbhāvakrāntinirdeśa-sūtra*, chap. 1. Toh 57, 214b, Pd 41:660.
460. *Nandagarbhāvakrāntinirdeśa-sūtra*, chap. 1. Toh 57, 215a, Pd 41:660.
461. *Nandagarbhāvakrāntinirdeśa-sūtra*, chap. 1. Toh 57, 215a, Pd 41:661.
462. *Nandagarbhāvakrāntinirdeśa-sūtra*, chap. 1. Toh 57, 215a, Pd 41:661.
463. *Nandagarbhāvakrāntinirdeśa-sūtra*, chap. 1. Toh 57, 215a, Pd 41:661.
464. *Nandagarbhāvakrāntinirdeśa-sūtra*, chap. 1. Toh 57, 215a, Pd 41:661.
465. *Nandagarbhāvakrāntinirdeśa-sūtra*, chap. 1. Toh 57, 215b, Pd 41:662.
466. *Nandagarbhāvakrāntinirdeśa-sūtra*, chap. 1. Toh 57, 215b, Pd 41:662.
467. *Nandagarbhāvakrāntinirdeśa-sūtra*, chap. 1. Toh 57, 215b, Pd 41:662.
468. *Nandagarbhāvakrāntinirdeśa-sūtra*, chap. 1. Toh 57, 215b, Pd 41:662.
469. *Nandagarbhāvakrāntinirdeśa-sūtra*, chap. 1. Toh 57, 216a, Pd 41:663.
470. *Nandagarbhāvakrāntinirdeśa-sūtra*, chap. 1. Toh 57, 216a, Pd 41:664.
471. *Nandagarbhāvakrāntinirdeśa-sūtra*, chap. 1. Toh 57, 216b, Pd 41:664.
472. *Nandagarbhāvakrāntinirdeśa-sūtra*, chap. 1. Toh 57, 216b, Pd 41:665.
473. *Nandagarbhāvakrāntinirdeśa-sūtra*, chap. 1. Toh 57, 216b, Pd 41:665.

474. *Nandagarbhāvakrāntinirdeśa-sūtra*, chap. 1. Toh 57, 216b, Pd 41:665.
475. *Nandagarbhāvakrāntinirdeśa-sūtra*, chap. 1. Toh 57, 217a, Pd 41:666.
476. *Nandagarbhāvakrāntinirdeśa-sūtra*, chap. 1. Toh 57, 217a, Pd 41:666.
477. *Nandagarbhāvakrāntinirdeśa-sūtra*, chap. 1. Toh 57, 217a, Pd 41:667.
478. *Nandagarbhāvakrāntinirdeśa-sūtra*, chap. 1. Toh 57, 217a, Pd 41:667.
479. *Nandagarbhāvakrāntinirdeśa-sūtra*, chap. 1. Toh 57, 217a, Pd 41:667.
480. *Nandagarbhāvakrāntinirdeśa-sūtra*, chap. 1. Toh 57, 217b, Pd 41:667.
481. *Nandagarbhāvakrāntinirdeśa-sūtra*, chap. 1. Toh 57, 217b, Pd 41:667.
482. *Nandagarbhāvakrāntinirdeśa-sūtra*, chap. 1. Toh 57, 219a, Pd 41:670.
483. *Nandagarbhāvakrāntinirdeśa-sūtra*, chap. 1. Toh 57, 219a, Pd 41:671.
484. *Nandagarbhāvakrāntinirdeśa-sūtra*, chap. 1. Toh 57, 219a, Pd 41:671.
485. *Nandagarbhāvakrāntinirdeśa-sūtra*, chap. 1. Toh 57, 219b, Pd 41:671.
486. *Nandagarbhāvakrāntinirdeśa-sūtra*, chap. 1. Toh 57, 219b, Pd 41:671.
487. *Nandagarbhāvakrāntinirdeśa-sūtra*, chap. 1. Toh 57, 219b, Pd 41:672.
488. *Nandagarbhāvakrāntinirdeśa-sūtra*, chap. 1. Toh 57, 219b, Pd 41:672.
489. *Nandagarbhāvakrāntinirdeśa-sūtra*, chap. 1. Toh 57, 219b, Pd 41:672.
490. *Kālacakra-laghutantra*, v. 4. Toh 362, 40b, Pd 77:99.
491. The phases of "fish" and so forth refer to the ten *avatars* (incarnations) of Viṣṇu according to the Hindu mythology: the fish, the tortoise, the boar, the man-lion (*narasiṃha*), the dwarf, Ramacandra, Paraśurama, Kṛṣṇa, the Buddha, and the Kalki king Kīrti. The first four are cited as metaphors for the four stages of fetal development leading up to birth. See also Khedrup Norsang Gyatso 2004, 167–70.
492. In Kālacakra literature the number 360 is represented by the phrase "space taste fire," where *space* represents zero, *taste* represents six, *fire* represents three and this number is then reversed.
493. *Vimalaprabhā*, chap. 1. Toh 1347, 225a, Pd 6:530.
494. *Kālacakra-laghutantra*, v. 9. Toh 362, 41a, Pd 77:100.
495. *Kālacakra-laghutantra*, vv. 9–10. Toh 362, 41a, Pd 77:100.
496. Khedrup Norsang Gyatso 2004 makes this same point (p. 171): "From the time of conception to the end of the first month no channels have developed."
497. In Kālacakra literature the number 200 is represented by the phrase "empty empty eyes," where *empty* represents zero, *eyes* represents two, and the number is read in reverse.
498. Literally, "space taste fire."
499. Literally, "empty empty eyes."
500. *Kālacakra-laghutantra*, vv. 10–11. Toh 362, 41b, Pd 77:100.
501. *Kālacakra-laghutantra*, v. 11. Toh 362, 41a, Pd 77:101.
502. *Aṣṭāṅgahṛdayasaṃhitā*, 1.2. Toh 4310, 106b, Pd 111:294.
503. *Aṣṭāṅgahṛdayasaṃhitā*, 1.37. Toh 4310, 108a, Pd 111:298.
504. *Aṣṭāṅgahṛdayasaṃhitā*, 1.48. Toh 4310, 108b, Pd 111:299.
505. *Aṣṭāṅgahṛdayasaṃhitā*, 1.43–44. Toh 4310, 108b, Pd 111:299.
506. *Aṣṭāṅgahṛdayasaṃhitā*, 1.54. Toh 4310, 109a, Pd 111:299.

507. *Aṣṭāṅgahṛdayasaṃhitā*, 1.55. Toh 4310, 109a, Pd 111:299.
508. *Aṣṭāṅgahṛdayasaṃhitā*, 1.55. Toh 4310, 109a, Pd 111:299.
509. *Aṣṭāṅgahṛdayasaṃhitā*, 1.56. Toh 4310, 109a, Pd 111:299.
510. *Aṣṭāṅgahṛdayasaṃhitā*, 1.56. Toh 4310, 109a, Pd 111:299.
511. *Aṣṭāṅgahṛdayasaṃhitā*, 1.61–62. Toh 4310, 109a, Pd 111:300.
512. *Aṣṭāṅgahṛdayasaṃhitā*, 1.71–72. Toh 4310, 109a, Pd 111:301.
513. The Tibetan editors of the *Compendium* have read Naropa's passage as referring to "coarse common mode of being" and "subtle common mode of being" as the twofold mode of being of entities. It seems better, however, to read the text as stating simply that there are two modes of entities, that of the body and that of the mind (*dngos po'i gnas lugs*), in fact a standard classification found at the beginning of numerous completion stage manuals of the highest yoga tantra. Tsongkhapa, in his *Annotations on the Summary of Fives Stages* (vol. *tsha*), reads the passage as referring to the specific modes of being of the body and mind and interprets the term "common" as referring to the indivisibility of the body and mind at the subtlest level, which is their common or shared mode of being.
514. *Pañcakramasaṅgrahaprakāśa*, v. 2. Toh 2333, 276b, Pd 26:1747.
515. Tibetan *rtsa*. This etymological connotation of being fundamental or a root seems to apply to the Tibetan word but not to the Sanskrit term *nāḍī*. The Sanskrit term refers to channels and veins and connotes something that is tubelike and facilitates movement of the air or fluids within.
516. *Nandagarbhāvakrāntinirdeśa-sūtra*, chap. 1. Toh 58, 241a, Pd 41:741.
517. *Saṃvarodaya-tantra*, v. 20. Toh 373, 273a, Pd 78:793.
518. *Vajramālā-tantra*, 30.9. Toh 445, 242a, Pd 81:833.
519. *Vajramālā-tantra*, 30.10. Toh 445, 242a, Pd 81:833.
520. *Saṃpūṭi-tantra*, v. 14–15. Toh 381, 112b, Pd 79:305.
521. *Hevajra-tantra*, chap. 1. Toh 417, 2b, Pd 80:4.
522. *Vajramālā-tantra*, 30.11. Toh 445, 242a, Pd 81:834.
523. *Vajramālā-tantra*, 30.11–12. Toh 445, 242a, Pd 81:834.
524. *Vajramālā-tantra*, 30.2–4. Toh 445, 242a, Pd 81:833.
525. *Saṃvarodaya-tantra*, vv. 1–2. Toh 373, 272a, Pd 78:792.
526. *Ḍākārṇava-tantra*, 5.2. Toh 372, 153b, Pd 78:443.
527. *Vajramālā-tantra*, 6.10. Toh 445, 216b, Pd 81:772.
528. Interestingly, the channels of body, speech, and mind are also referred to by their alternative names, which echo Sāṃkhya terminology of the three qualities of the so-called primal substance or *prakṛti*. Known as *sattva* (Tibetan *snying stobs*, courage), *rajas* (rendered as *rdul*, atoms), and *tamas* (Tibetan *mun pa*, darkness), according to the Saṃkhya view, these three qualities characterize primal nature, and it is the combinations of these three that give rise to the manifest world of everyday experience.
529. *Vimalaprabhā*, chap. 2. Toh 1347, 239b, Pd 6:565.

530. On the explanation of the nature and functions of these five root winds, according to the Guhyasamāja system, see Tsongkhapa 2013, 222–28.
531. On these five branch winds, especially from the point of view of their specific functions, see Tsongkhapa 2013, 228–33.
532. *Sandhivyākaraṇa-tantra*, 1.27–28. Toh 444, 158b, Pd 81:614. The names nāga, turtle, and so forth are alternative names of the five branch winds, known also as "moving," "roving," and so on, as listed earlier.
533. *Pañcakrama*, 1.12. Toh 1802, 45a, Pd 18:129.
534. *Vajramālā-tantra*, 3.8. Toh 445, 213b, Pd 81:765.
535. Āryadeva, *Caryāmelāpakapradīpa*, chap. 2. Toh 1803, 64b, Pd 18:180.
536. Āryadeva, *Caryāmelāpakapradīpa*, chap. 2. Toh 1803, 64b, Pd 18:180.
537. In the tantras, the most extensive presentations of channels, winds, and drops are found in the Guhyasamāja cycle of texts and their associated commentarial treatises. There are also detailed presentations in other tantric texts, such as *Saṃvarodaya Tantra* of the Cakrasaṃvara cycle of texts, the *Hevajra Tantra*, as well as the Kālacakra tantra and its commentary, *Stainless Light*.
538. *Explanatory Tantra*, 2.23–24, 27. There are four medical tantras—the *Root Tantra*, the *Explanatory Tantra*, the *Instruction Tantra*, and the *Subsequent Tantra*. There is a divergence of opinion among Tibetan scholars, especially in medical science, on whether these "four medical tantras" (*rgyud bzhi*) are scriptures attributable to the Buddha or whether they were treatises composed by subsequent scholars, possibly the Tibetan medical scholar Yuthok Yönten Gömpo. For a brief discussion of this question by the influential Tibetan medical scholar and historian Desi Sangyé Gyatso, see his *Mirror of Beryl* (2010), especially part 3, 13.
539. *Explanatory Tantra*, 4.8, 31.
540. Together with "wind" and "phlegm," bile is one of the three humors that are thought to sustain the body and mind and its various states, such as health and illness. The three humors therefore constitute the basic principles of Tibetan medicine for understanding all characteristics of pathology and treatment. For an accessible introduction to the principles of Tibetan medicine, see Dhonden 2000.
541. *Explanatory Tantra*, 4.9, 32.
542. *Explanatory Tantra*, 4.10, 32.
543. *Explanatory Tantra*, 4.11–13, 32.
544. *Explanatory Tantra*, 4.13, 32.
545. Organs or viscera (*don snod*). The five organs (*don snod lnga*) are usually the heart, liver, kidney, spleen, lungs.
546. *Explanatory Tantra*, 4.13–16, 32.
547. *Explanatory Tantra*, 4.18, 32.
548. The editors are referring to traditional Tibetan medical lore, according to which a distinction is made between three important aspects of a person's life: "vital essence" (*bla*), "lifespan" (*tshe*), and "life-force" (*srog*).

549. *Four Medical Tantras of Dratang*, 617.
550. *Explanatory Tantra*, 4.19, 32.
551. Candranandana, *Candrikāprabhāsa-aṣṭāṅgahṛdayavivṛtti*, chap. 1. Toh 4312, 7b, Pd 113:16.
552. *Somarāja: A Medical Treatise*, 25.5, 86.
553. *Somarāja: A Medical Treatise*, 5.15–18, 27.
554. *Explanatory Tantra*, 5.18, 36.
555. *Explanatory Tantra*, 5.25–28, 36.
556. *Explanatory Tantra*, 2.12, 26.
557. *Somarāja: A Medical Treatise*, 1.7–11, 10.
558. *Somarāja: A Medical Treatise*, 1.12, 10.
559. *Explanatory Tantra*, 4.8, 31.
560. This text speaks of the brain in terms of male and female typologies, implying the existence of both masculine and feminine qualities in a single brain—where the harder and more rigid segments of the brain are masculine and the softer and more pliant segments are feminine. But it also implies that the entire brain may possess qualities indicative of a male brain or female brain.
561. *Somarāja: A Medical Treatise*, 48.26–32, 120.
562. *Instruction Tantra*, 83.3, 471.
563. *Explanatory Tantra*, 4.4, 31.
564. In general, especially in the context of highest yoga tantra, the Tibetan word *rtsa* is translated as "channel," but here in the context of traditional Tibetan medicine the same word may also be translated as "nerve" in accordance with conventional physiology.
565. *Explanatory Tantra*, 4.11, 32.
566. *Somarāja: A Medical Treatise*, 49.36, 124.
567. *Somarāja: A Medical Treatise*, 48.6–8, 119.
568. *Somarāja: A Medical Treatise*, 3.31–33, 19.
569. *Explanatory Tantra*, 4.16–18, 32.
570. *Somarāja: A Medical Treatise*, 2.11, 15.
571. *Somarāja: A Medical Treatise*, 4.30, 21.
572. *Instruction Tantra*, 83.4, 471.
573. *Somarāja: A Medical Treatise*, 4.30–33, 21.
574. *Somarāja: A Medical Treatise*, 49.33, 123.
575. Tib. *don snying lnga*. The five internal organs are the heart, lungs, liver, spleen, and kidneys.
576. *Somarāja: A Medical Treatise*, 97.29, 270.
577. This is most probably a reference to the brain stem. In contemporary science too, the brain stem, which links the brain with the spinal cord, is a crucial part of the brain associated with the motor and sensory systems.
578. *Somarāja: A Medical Treatise*, 49.27, 123.
579. *Somarāja: A Medical Treatise*, 4.40, 22.
580. *Explanatory Tantra*, 5.25, 36.

581. Tib. *snod (drug)*. In the Tibetan medical system, the body's internal organs are grouped into two separate categories of "five internal organs" (*don snying lga*) and "six vessels" (*snod drug*), the latter including the digestive tract, intestines, entrails (*long ga*), gallbladder, urinary bladder (*lgang pa*), and seminal vesicle (*bsam se'u*).
582. *Śālistambakasūtraṭīkā*, chap. 3. Toh 3986, 39a, Pd 65:829.
583. *Bhūmivastu*, chap. 9. Toh 4035, 101b, Pd 72:914.
584. *Pratītyasamutpādādivibhaṅganirdeśa-sūtra*. Toh 211, 124b, Pd 62:340.
585. This is a reference to the phenomenon known as *thukdam*, where individuals, generally monks with advanced meditative experience, although clinically declared dead, exist in a state in which their body remains without decomposition for days, in some cases even weeks. According to the Tibetan tradition, these individuals are understood to remain in the clear light state of death, the subtlest level of consciousness. One of the most well-known cases in recent memory is that of the Dalai Lama's senior tutor, Kyabjé Ling Rinpoche. See Dalai Lama 2017.
586. *Pramāṇavārttika*, 2.40. Toh 4210, 109a, Pd 97:503. See Jackson 1993, 226.
587. *Saddharmānusmṛtyupasthāna-sūtra*, chap. 22. Toh 287, 73a, Pd 71:168.
588. *Pramāṇavārttika*, 2.62bc. Toh 4210, 109b, Pd 97:505.
589. Dharmakīrti, *Pramāṇavārttika*, 2.165ab. Toh 4210, 113b, Pd 97:515. According to Dharmakīrti, "From this too it is established" means: from this argument too—that which is not itself consciousness [or its potentiality] cannot be a substantial cause of consciousness—it is established that there is no absolute beginning to the cycle of existence.
590. *Pramāṇavārttikapañjikā*, chap. 2. Toh 4217, 71b, Pd 98:171.
591. Dharmakīrti, *Pramāṇavārttika* 2.50. Toh 4210, 109b, Pd 97:504.
592. Dharmakīrti, *Pramāṇavārttika* 2.75–77. Toh 4210, 110a, Pd 97:507.
593. *Tattvasaṃgrahapañjikā*, chap. 27. Toh 4267, 96a, Pd 107:1222. For an alternative translation see Jha 1986, 900.
594. Dharmakīrti, *Pramāṇavārttika*, 2.73–74b. Toh 4210, 110a, Pd 97:506.
595. Kamalaśīla, *Tattvasaṃgrahapañjikā*, chap. 27. Toh 4267, 96a, Pd 107:1222. For an alternative translation see Jha 1986, 900.
596. Dharmakīrti, *Pramāṇavārttika*, 2.49ab. Toh 4210, 109a, Pd 97:504.
597. Vasubandhu, *Abhidharmakośabhāṣya*, 3.41. Toh 4090, 141a, Pd 79:347. See Pruden 1988, 3:443.
598. Vasubandhu, *Abhidharmakośabhāṣya*, 3:41. Toh 4090, 141a, Pd 79:347. See Pruden 1988, 3:443.
599. *Pañcakramaṭīkākramārthaprakāśikā*, chap. 1. Toh 1842, 199a, Pd 19:530.
600. *Vajramālā-tantra*, 68.71–73. Toh 445, 276b, Pd 81:915.
601. *Pañcaviṃśatisāhasrikāprajñāpāramitā-sūtra*, chap. 39. Toh 9, 356b. The Tibetan word for these microorganisms is *srin 'bu*, which is also the term used in the Tibetan medical system for "germs," or "parasites." In contemporary scientific and medical terminology, these microorganisms are bacteria whose

symbiotic relationship with their host body is critical to the overall health of the host body.

602. *Saddharmānusmṛtyupasthāna-sūtra*. Toh 287, 118a, Pd 71:275.
603. *Saddharmānusmṛtyupasthāna-sūtra*. Toh 287, 118b, Pd 71:276.
604. The term *tsu ra tha* has not been identified.
605. *Saddharmānusmṛtyupasthāna-sūtra*. Toh 287, 147a, Pd 71:344.
606. *Saddharmānusmṛtyupasthāna-sūtra*. Toh 287, 126a, Pd 71:295.
607. *Saddharmānusmṛtyupasthāna-sūtra*. Toh 287, 130a, Pd 71:304.
608. *Saddharmānusmṛtyupasthāna-sūtra*. Toh 287, 131b, Pd 71:307.
609. *Thod pa rko rlog du gyur pa*, likely a reference to a migraine.
610. *Saddharmānusmṛtyupasthāna-sūtra*. Toh 287, 131b, Pd 71:308.

# Glossary

Abhidharma (*chos mngon pa*). The higher (*abhi*) doctrine (*dharma*) that constitutes Buddhist philosophy. It forms the curriculum studied and contemplated in the higher training of Buddhist wisdom.

absorption (*bsam gtan, dhyāna*). *Also* meditative absorption. Four levels of meditative concentration that correspond to the four levels of the form realm.

affliction (*nyon mongs, kleśa*). Afflictive mental states or negative emotions. There are six root afflictions and twenty branch afflictions.

aggregated atom (*bsags rdul phra rab, saṃgrahaparamāṇu*). The smallest composite unit of matter.

aggregates (*phung po, skandha*). The five perishable heaps or composites that act as the basis of designation of "I" or the person: form (*gzugs, rūpa*), feeling (*tshor ba, vedanā*), discernment (*'du shes, saṃjñā*), [karmic] formation (*'du byed, saṃskāra*), and consciousness (*rnam shes, vijñāna*).

ascertainable object (*gzhal bya, prameya*). That which is comprehended by valid cognition. Equivalent to existent, established basis, object, knowable object, and phenomenon.

assembly (*tshogs pa, sāmagrī*). One of the nine factors of the upper Abhidharma system that indicates the assembly or convergence of conditions in relation to cause and effect.

atom (*rdul phran, aṇu*). The smallest viable aggregation of subtle particles. In the desire realm an atom is said to comprise eight substantial particles. In the form realm an atom comprises just six substantial particles.

base (*skye mched, āyatana*). *Also* sense base, sense basis. The sense objects and the sense organs, including the mental sense organ, that act to multiply or proliferate minds and mental factors. There are twelve bases: the six objects of form, sound, smell, taste, tactility, and phenomena bases, plus the six sense bases of eye sense base, ear sense base, nose sense base, tongue sense base, body sense base, and mental sense base.

being and not being (*yin gyur min gyur*). Chapa Chökyi Sengé's fifteenth topic. See appendix.

bodhisattva (*byang chub sems dpa'*). A "being" or "courageous one" (*sattva*) who strives for "enlightenment" (*bodhi*) for the sake of establishing all sentient beings in the state of enlightenment.

branch winds (*yan lag gi rlung*). Secondary wind-energies that move in the channels that constitute the subtle physiology of a being. There are five: moving wind (*rgyu ba*), roving wind (*rnam par rgyu ba*), perfectly flowing wind (*yang dag par rgyu ba*), intensely flowing wind (*rab tu rgyu ba*), and definitely flowing wind (*nges par rgyu ba*).

cakra (*'khor lo, cakra*). *Also* channel center (*rtsa 'khor, mūlacakra*). Branch or radial channels that extend horizontally from the vertical central channel like the spokes of a wheel. There are five major cakras: the crown great-bliss cakra (*spyi bo'i rtsa 'khor bde chen gyi 'khor lo*) with thirty-two spokes, the throat perfect-enjoyment cakra (*mgrin pa'i rtsa 'khor longs spyod pa'i 'khor lo*) with sixteen spokes, the heart reality cakra (*snying ga'i rtsa 'khor chos kyi 'khor lo*) with eight spokes, the navel emanation cakra (*lte ba'i rtsa 'khor sprul pa'i 'khor lo*) with sixty-four spokes, the secret bliss-sustaining cakra (*gsang ba'i rtsa 'khor bde skyong gi 'khor lo*) with eight spokes. Also six main cakras are asserted: (1) the crown cakra, (2) the throat cakra, (3) the heart cakra, (4) the navel cakra, (5) the forehead cakra, and (6) the secret cakra.

causal relation (*de byung 'brel, tadutpattisaṃbandha*). A type of relationship between two or more phenomena, where the related entities do not share the same intrinsic or essential nature. For example, fire that is causally related to its smoke or the seed that is causally related to the tree it produces.

cause (*rgyu, hetu*). That which produces or generates an effect. Differentiated as either direct cause or indirect cause; or substantial cause and cooperative condition.

cause and effect (*rgyu 'bras*). Chapa Chökyi Sengé's eighth topic. See appendix.

channel (*rtsa, nāḍī*). In the three categories of "channels," "winds," and "drops," the channels form an intricate network that constitute the body's subtle physiology.

coexisting entities (*grub bde gcig, ekayogakṣema*). Impermanent phenomena that possess three temporal characteristics: they arise simultaneously, endure simultaneously, and cease simultaneously.

collection of letters (*yi ge'i tshogs, vyañyjanakāya*). One of the fourteen factors of the lower Abhidharma system. It refers to the body of letters that give rise to nouns and phrases.

collection of nouns (*ming gi tshogs, nāmakāya*). One of the fourteen factors of the lower Abhidharma system. The body of appellations that indicate the essential nature of phenomena.

collection of phrases (*tshigs kyi tshogs, padakāya*). One of the fourteen factors of the lower Abhidharma system. The body of phrases that indicate the attributes of nouns.

color, white and red (*kha dog dkar mar*). The first of Chapa Chökyi Sengé's eighteen topics. See appendix.

conceived object (*zhen yul, adhyavasāyaviṣaya*). Object apprehended by conception, or conceptual mental consciousness.

condition (*rkyen, pratyaya*). Conditions are impermanent entities that assist causes in producing their effects.

conditioned phenomenon (*'dus byas, saṃskṛta*). Phenomenon qualified by the three characteristics of arising, enduring, and ceasing. Equivalent to thing, product, and impermanent.

consciousness (*rnam shes, vijñāna*). Mind which is aware of the mere nature of its object. The fifth of the five aggregates.

consequence (*thal 'gyur, prasaṅga*). *Also* trimodal consequence. Formulation of the three modes (*tshul gsum*) to express a consequence worthy of repudiation.

contingent relation (*med na mi 'byung gi 'brel ba, avinabhāvinasaṃbandha*). *Also* invariable relation. One phenomenon bears a contingent relation with a second if they are distinct from each other, and if the second does not exist the first will necessarily not exist. It has two types: causal relation and intrinsic relation.

continuity (*'jug pa, pravṛtti*). One of the nine factors of the upper Abhidharma system that indicates that the continuum of cause and effect has not being severed.

contradiction and noncontradiction (*'gal ba dang mi 'gal ba*). Chapa Chökyi Sengé's third topic. See appendix.

contradictory (*'gal ba, virodha*). *Also* mutual exclusion, contradiction. Distinct phenomena that share no common locus. There are two types: contradictory through incompatibility and contradictory through mutual exclusion; or direct contradiction and indirect contradiction.

contradictory through incompatibility (*lhan cig mi gnas 'gal, sahānavastāvirodha*). *Also* contradictory through not abiding together. Distinct phenomena that cannot assist each other in a supportive capacity. It has two types: contradictory material forms that are incompatible and contradictory cognitions that are incompatible.

contradictory through mutual exclusion (*phan tshun spang 'gal, parasparaparihāri-virodha*). Distinct phenomena that mutually exclude or eliminate each other.

cooperative condition (*lhan cig byed rkyen, sahakāripratyaya*). That which produces an effect that exists outside its own substantial continuum. That which produces the attributes of the basis constituting the effect, instead of the basis of those attributes.

definiendum (*mtshon bya, lakṣya*). *Also* object defined. That which possesses all three properties of an imputed existent (*prajñāptisat*): it is an object defined, it is established with respect to its basis, and it does not serve as the definiendum of any other definition.

definiens (*mtshan nyid, lakṣaṇa*). *Also* definition, characteristics. That which possesses the three properties of a referent existent (*dravyasat*): it is a definition, it is established with respect to its basis, and it does not serve as a definiens of any other object defined.

definition (*mtshan nyid, lakṣaṇa*). *See* definiens.

definitions and definiendum (*mtshan nyid dang mtshon bya*). Chapa Chökyi Sengé's tenth topic. See appendix.

dependent origination (*rten cing 'brel bar 'byung ba, pratītyasamutpāda*). *Also* dependent arising, interdependence. The fact of phenomena arising in dependence on other phenomena.

derivative form (*'byung gyur, bhautika*). Material form derived from the primary elements (*bhūta*). It is equivalent to derived form.

derived form (*rgyur byas pa'i gzugs, upādāyarūpa*). Material form derived from or caused by the primary elements (*bhūta*). It is equivalent to derivative form.

Dharma (*chos*). With a capital D, refers to the teachings of the Buddha. With a lowercase d, refers to any existent entity.

direct cause (*dngos rgyu, sākṣātkāraṇa*). Direct producer of an effect.

direct contradiction (*brgyud 'gal, sākṣādvirodha*). *Also* directly contradictory. Mutually exclusive phenomena that directly harm each other, such as hot and cold contact.

direct contradiction and indirect contradiction (*dngos 'gal dang brgyud 'gal*). Chapa Chökyi Sengé's thirteenth topic. See appendix.

discernment (*'du shes, saṃjñā*). *Also* discrimination. A mental factor that apprehends the unique mark or characteristics of its object. The third of the five aggregates.

discrete physical unit (*rdul phran, āṇu*). A term applied to minuscule material states or molecular structures such as the cells of the sense organs and so on.

distinct conditioned entities (*'dus byas su gyur pa'i ngo bo tha dad*). Impermanent entities that do not possess the same substance.

distinct entities (*ngo bo tha dad*). *Also* distinct essential natures. Different phenomena that do not possess the same essential nature.

distinct substance (*rdzas tha dad, dravyabheda*). Impermanent entities that do not possess the same substantial reality or energy, such as a diamond and a feather.

distinction (*so sor nges pa, pratiniyama*). One of the nine factors of the upper Abhidharma system that indicates that cause and effect are distinct states.

distinction and nondistinction (*tha dad dang tha mi dad*). Chapa Chökyi Sengé's sixth topic. See appendix.

drops (*thig le, bindu*). Drops comprise a composite of four things: wind, consciousness, and the vital essence of semen and blood that exist within the channels that pervade the body. In general there are two types of drops: white drop (*thig le dkar po*) and red drop (*thig le dmar po*). They are further enumerated in four types: drop generating sleep (*gnyis 'thug po skyed pa'i thig le*), drop generating dreams (*rmi lam skyed pa'i thig le*), drop generating the waking state (*sad pa skyed pa thig le*), and drop generating the fourth state (*bzhi pa'i gnas skabs skyed pa'i thig le*).

eight basic substances (*rdul rdzas brgyas*). *Also* eight basic particles. These constitute the four primary elements of the particles of earth, water, fire, and wind, plus the four particles of form, smell, taste, and tactility.

eight modes of pervasion (*khyab pa sgo brgyad*). These are the two modes of entailment in terms of being (*yin khyab mnyams gnyis*), the two modes of entailment in terms of not being (*min khyab mnyams gnyis*), the two modes of entailment in terms of existing (*yod khyab mnyams gnyis*), and the two modes of entailment in terms of not existing (*med khyab mnyams gnyis*).

eighteen topics (*gnas rnams bsdus tshan bco brgyad*). The eighteen topics formulated by Chapa Chökyi Sengé as epistemological tools for analyzing the nature of reality: (1) color, white and red, (2) substantial phenomena and abstract phenomena, (3) contradiction and noncontradiction, (4) universals and particulars, (5) relation and absence of relation, (6) distinction and nondistinction, (7) presence and absence, (8) cause and effect, (9) the first stage, intermediate stage, and later stage, (10) definitions and definiendum, (11) multiple reasons and multiple predicates, (12) forward and reverse negation, (13) direct contradiction and indirect contradiction, (14) two types of logical entailment, (15) being and not being, (16) the negation of being and not being, (17) understanding existence and understanding nonexistence, and (18) understanding permanent phenomena and understanding things. See appendix.

element (*khams, dhātu*). *See also* realms. The essential constituents of existent phenomena. There are eighteen types: the six objects, such as the elements of visible form, sound, smell, taste, tactility, and phenomena otherwise known as "mental objects"; the six sense bases—the elements of the eye, ear, nose, tongue, body, and mental sense base; and the six types of consciousness—the elements of eye, ear, nose, tongue, body, and mental consciousness.

enumeration (*grangs, saṃkhyā*). One of the nine factors of the upper Abhidharma system that indicates distinct separate formative factors.

established basis (*gzhi grub, āśrayasiddhi*). That which is established by valid cognition. Equivalent to existent, object, knowable object, and phenomenon.

exclusion (*sel ba, apoha*). *Also* conceptual exclusion. That comprehended through the application of words to exclude an object of negation. For example, the words "he is male" excludes the possibility that he is female. It is equivalent to exclusion of other and negation.

exclusion of other (*gzhan sel, anyāpoha*). That comprehended by means of the mind that directly comprehends it, directly excluding its object of negation. For example, the object-universal or generic image of a vase. It is equivalent to exclusion and negation.

existent (*yod pa, bhāva*). That which is observed by valid cognition. Equivalent to existent, established basis, object, knowable object, and phenomenon.

facilitate (*phan 'dogs pa*) *Also* benefit. In the process of causation, the cause facilitates or acts to benefit the generation of its effect.

factor (*ldan min 'du byed, viprayuktasaṃskāra*). An abbreviated form of "nonassociated formative factor."

figurative noun (*btags ming*). The noun or name subsequently applied to something because of its resemblance or its relationship to another object. Thus it has two types: "figurative name based on relationship" (*'brel pa rgyu mtshan du byas pa'i btags ming*) and "figurative name based on resemblance" (*'dra ba rgyu mtshan du byas pa'i btags ming*).

first stage, intermediate stage, later stage (*snga btsan bar btsan phyi btsan*). The ninth of Chapa Chökyi Sengé's eighteen topics. See appendix.

five aggregates (*phung po lnga, skandha*). *See* aggregates.

five channel centers (*rtsa 'khor lnga*). *See* cakra.

five elements of exposition (*bshad thabs yan lag lnga*). The basic features that aid the reader in comprehending a treatise. Thus any composition should have a purpose (*dgos pa*), a summary presentation (*bsdus don*), a word meaning (*tshig don*), objections and rebuttals (*brgal len*), and transition markers (*mtshams sbyar*).

five embryonic stages: arbuda embryo (*mer mer po*), kalala embryo (*nur nur po*), peśin embryo (*ltar ltar po*), ghana embryo (*mkhrung 'gyur*), and praśākhā embryo (*rkang lag 'gyus pa*). The order in which these are presented differs in different texts.

five ways that primary elements cause derivative forms (*'byung bas 'byung gyur gyi rgyu byed tshul*). These include the generating cause (*skyed rgyu*), the supporting cause (*brten pa'i rgyu*), the stabilizing cause (*gnas pa'i rgyu*), the dependent cause (*rton pa'i rgyu*), and the enhancing cause (*'phel ba'i rgyu*).

flower bank arrays (*me tog gi gzhi'i snying po'i rgyan bkod pa dang ldan pa*). Vast cosmic structures containing myriads of stars.

form (*gzugs, rūpa*). Any material or physical entity that is capable of physical interaction whether on a subtle or coarse level.

form arising from meditative expertise (*dbang 'byor ba'i gzugs*). One of the five types of mental-object form. It refers to forms such as fire, water, and so on that are clearly perceived in meditative concentration, and the process by which that concentration exercises control over fire, water, and so on.

form derived through a rite (*yang dag par blangs pa las byung ba'i gzugs*). One of the five types of mental-object form. This refers to the vows generated in the mindstream of a being owing to enacting a rite. In this context it has three types: pratimokṣa vows (*so thar gyi sdom pa*), negative commitments (*sdom min*), and interim precepts (*bar mas bsdus pa'i rnam rig min*).

form emerging from a process of deconstruction (*bsdus pa las gyur pa'i gzugs*). One of the five types of mental-object form. It refers to the subtle particles that emerge from the process of mentally deconstructing coarse material form into its component parts. It is apprehended by mental cognition alone, such as the eight substantial particles.

form of open space (*mngon par skabs yod kyi gzugs*). One of the five types of mental-object form. The vacant space that appears as a whitish-vacuity to mental consciousness alone.

form that is imagined (*kun btags pa'i gzugs*). *Also* imaginary form. One of the five types of mental-object form. Imaginary form that appears to meditative states, such as the meditation of repulsiveness, or the form that appears in dreams.

formal reasoning (*gtan tshigs, hetu*). A statement designed to settle an issue through revealing a specific reason.

formation (*chags pa*). One of the four cyclic stages in the evolution of the cosmos: the eon of voidness, the eon of formation, the eon of endurance, and the eon of destruction.

formation (*'du byed, saṃskāra*). *Also* formative factor, factor, conditioning factor. The fourth of the five aggregates.

forward negation and reverse negation (*dgag pa phar tshur gnyis*). Chapa Chökyi Sengé's twelfth topic. See appendix.

four channels taught in medical texts (*gso rig pa'i gzhung du rtsa'i rnam gzhag*). These include the developmental channel (*chags pa'i rtsa*), existence channel (*srid pa'i rtsa*), related channels (*'brel pa'i rtsa*), and life-sustaining channel (*tshe gnas pa'i rtsa*).

four characteristics of conditioned things (*'dus byas kyi mtshan nyid*). Four factors of the lower Abhidharma system: arising (*skye ba, jāti*), enduring (*gnas pa, sthiti*), aging (*rga ba, jāra*), and disintegrating (*'jig pa, anityatā*).

four kinds of permutation (*mu gsum mu bzhir grags pa'i tshul*). *Also* four possibilities. These include the trilemma or four possibilities (*mu gsum*), tetralemma or three possibilities (*mu bzhi*), contradiction or mutual exclusion (*'gal ba*), equivalence or mutual inclusion (*don gcig*).

four reliances, principles of the (*rton pa'i bzhi rnam gzhag*). The four are: (1) rely on the teachings, not the person, (2) rely on the meaning, not the words, (3) rely on the definitive meaning, not the provisional meaning, and (4) rely on transcendent wisdom, not ordinary consciousness.

four stages of the embryo in Kālacakra (*dus 'khor las mngal du lus chags rim bzhi*): the fish phase (*nya*) from the first month to the end of the second, the tortoise phase (*rus sbal*) from the third month to the end of the fourth, the boar phase (*phag rgod*) from the fifth month until just prior to birth, and the man-lion phase (*mi'i seng ge*), which is the final stage of birth itself.

four types of birth (*skye ba'i gnas bzhi*): birth from an egg (*sgo nga las skye ba*), birth from the womb (*mngal las skye ba*), birth from heat and moisture (*drod gsher las skye ba*), and spontaneous supernatural birth (*rdzus te skye ba*).

four types of response (*dri ba'i len bzhi*): response as a categorical statement (*mgo gcig tu lung bstan pa*), response as a qualified statement (*rnam par phye ste lung bstan pa*), response in the form of a question (*dris nas lung bstan pa*), and response in the form of silence (*gzhag par lung bstan pa*).

garuḍa (*khyung, nam mkha' lding, garuḍa*). A class of mythical bird, also called "those with beautiful wings" (*suparṇa*). They are often portrayed as enemies of nāgas, yet like nāgas they combine features of an animal and divine being. They possess magical powers, they are appointed as guardians by higher gods, and they often act as a mount of specific gods and thus are seen as the means of their conveyance.

great elements (*'byung ba chen po, mahābhūta*). Smallest discrete units of matter. Equivalent to the primary elements (*'byung ba, bhūta*).

Guhyasamāja (*gsangs ba 'dus pa*). One of the two major Buddhist tantric systems, the other being the Kālacakra system.

highest excellence (*nges legs, niḥśreyasa*). A term meaning "definite good" or "having no higher virtue"; refers to liberation or enlightenment.

highest yoga tantra (*bla na med pa'i rnal 'byor rgyud, yogānuttaratantra*). The highest of the four classes of Buddhist tantra within the Guhyasamāja system.

homogeneity (*skal mnyam, nikāyasabhāga*). One of the fourteen factors of the lower Abhidharma system. A type of substance that gives rise to the similarity of conduct, thought, and characteristics of sentient beings.

impermanent (*mi rtag pa, anitya*). Phenomena that are momentary. Equivalent to thing, product, and conditioned phenomenon.

implicative negation (*ma yin dgag, paryudāsapratiṣedha*). *See* negation. A negation where another positive phenomena is implied in the wake of negating the negandum.

imputed existent (*btags yod, prajñāptisat*). From the perspective of the lower schools, in general any phenomena whose comprehension depends on understanding something other than it. For example, "a person," which is understood through comprehending the basis of designation of that person, that is, the five aggregates, which are not the person.

indicative form (*rnam par rig byed kyi gzugs, vijñaptirūpa*). Form of the body that reveals or indicates the underlying motivation of an action.

indifferentiable coexisting substance (*grub bde dbyer med kyi rdzas gcig*). Distinct impermanent entities that simultaneously arise, endure, and cease, and whose substance is not divisible.

indirect cause (*brgyud rgyu, paraṃparyakāraṇa*). Indirect producer of an effect.

indirect contradiction (*dngos 'gal, pāraṃparyavirodha*) *Also* indirectly contradictory. Mutually exclusive phenomena that do not directly harm each other, such as strongly billowing smoke and cold contact.

inherent existence (*rang bzhin gyi yod pa, svabhāvasat*). Existence by way of an essential nature that appears to inhere in an object.

instance (*mtshan gzhi*). An example that establishes the relation between the definiens and the definiendum.

instance-isolate (*gzhi ldog, vyatirekādhikaraṇa*). Instance or basis that illustrates the specific defining characteristics of something. For example, a pot illustrates the meaning of "momentary," which is the definition of impermanence.

interim precepts (*bar ma*). *Also* intermediate precepts. This implies the stage when vows have been taken but have not yet arisen in the mindstream of the person.

intrinsic relation (*bdag cig 'brel, tādātmyasaṃbandha*). *Also* same-nature relation. A type of relationship between two or more phenomena where the related entities share the same intrinsic or essential nature. For example, the impermanence of a pot that is intrinsically related to the pot.

isolate (*ldog pa, vyatireka*). *Also* conceptual isolate. That which appears to conception through negating its opposite. *See* self-isolate *and* instance-isolate.

Kālacakra (*dus 'khor*). "Wheel of Time." One of the two major Buddhist tantric systems, the other being the Guhyasamāja system.

karma (*las*). The mental factor of intention or volition that encompasses the free will of an individual. When virtuous or nonvirtuous intention arises in the mindstream of an ordinary being, such intention lays down imprints or karmic seeds upon the mental consciousness. Such seeds in turn, once awakened, produce effects commensurate with the ethical value of the intention that laid down those karmic seeds.

Ketu (*dkar ma du ba mjug ring*). A planet identified in the Kālacakra tantra called the "dragon's tail." Also identified as the descending node or the ninth planet.

knowable object (*shes bya, jñeya*). *Also* object of knowledge. That which is fit to be an object of awareness. Equivalent to existent, established basis, object, and phenomenon.

life (*tshe*). *Also* lifespan. The period of union of body and mind.

life-force (*srog, jīvitendriya*). One of the fourteen factors of the lower Abhidharma system. It refers to the faculty that constitutes the basis of a being's consciousness and bodily heat.

linguistic sounds (*rjod byed gyi sgra*). A specific type of sound that reveals its content or subject matter. There are three types: nouns or names (*ming*), predicated phrases (*tshig*), and letters (*yi ge*).

location (*yul, deśa*). One of the nine factors of the upper Abhidharma system that indicates that cause and effect exist in all ten directions.

logical entailment (*khyab, vyāpti*). *Also* logical pervasion. *See* eight modes of pervasion.

lower Abhidharma system (*mngon pa 'og ma*). The system of Abhidharma set out in Vasubandhu's *Treasury of Knowledge* (*Abhidharmakośa*) and its commentarial tradition.

Madhyamaka (*dbu ma*). A school of Buddhism also known as the "Middle Way," which avoids falling into the extreme of eternalism or nihilism.

Madhyamaka philosophy (*dbu ma lta ba*). The view that avoids all extremes. Though all Buddhist schools assert a path that avoids extremes, here it specifically refers to the Mahāyāna view that asserts all entities lack true existence.

maṇḍala (*dkyil 'khor*). The abode of the deity.

many (*tha dad; nānā, bhinna*). *Also* distinct, separate. Phenomena that are diverse or multiple, such as a horse and a goat.

material form (*gzugs, rūpa*). *See* form.

meditative attainment of cessation (*'gog pa'i snyoms 'jug, nirodhasamāpatti*). One of the fourteen factors of the lower Abhidharma system. The substance that temporarily stops minds and mental factors in the mindstreams of ārya beings in dependence on the mind of the summit of existence, which represents the subtlest mind within the three realms of existence.

meditative attainment of nondiscernment (*'du shes med pa'i snyoms 'jug, asaṃjñāsamāpatti*). One of the fourteen factors of the lower Abhidharma system. The substance that acts as the cause of nondiscernment as its result. It brings about the cessation of mind and mental factors until one rises from that meditative attainment.

meditative equipoise (*mnyam gzhag, samāhita*). The mind in single-pointed equilibrium that engages emptiness or selflessness.

mental faculty (*yid dbang, mana-indriya*). *Also* mental sense power. The dominant condition of mental consciousness. One of the twenty-two faculties listed in Abhidharma.

mental-object base (*chos kyi skye mched, dharmāyatana*). One of the twelve bases. The object of mental consciousness that is otherwise known as the "phenomena base."

mental-object element (*chos kyi khams, dharmadhātu*). One of the eighteen elements. The object of mental consciousness that is otherwise known as the "phenomena element."

mental-object form (*chos kyi skye mched pa'i gzugs, dharmāyatanikarūpa*). *Also* form of the phenomena base, or phenomena-base form. Subtle material form that is accessed by mental conception alone, not by the five senses. According to Asaṅga it has five types: form emerging from a process of deconstruction (*bsdus pa las gyur pa*), form of open space (*mngon par skabs yod pa*), form of vows derived through a rite (*yang dag par blangs pa las byung ba*), form that is imagined (*kun btags pa*), and form arising from meditative expertise (*dbang 'byor ba, vaibhūtika*).

multiple reasons and multiple predicates (*rtags mang bsal mang*). The eleventh of Chapa Chökyi Sengé's eighteen topics. See appendix.

nāga (*klu*). This term applies equally to ordinary snakes such as cobras as to mythical or legendary creatures that resemble snakes, such as powerful serpent-demons who intervene in human affairs or beings dwelling beneath the earth with human faces and serpent-like lower extremities, often with the capacity to change their form magically into that of a human. In Tibet nāgas were believed to dwell in lakes, streams, trees, and so on, where they often guarded hidden treasures. In China nāgas were often equated with dragons.

name (*ming, nāma*). *See* noun.

negation (*dgag pa, pratiṣedha*). An object that is understood through the mind directly comprehending it directly negating its object of negation (negandum). For example, selflessness. Equivalent to exclusion and exclusion of other. There are two main types: implicative negation and nonimplicative negation.

negation of being and negation of not being (*yin log min log*). The sixteenth of Chapa Chökyi Sengé's eighteen topics. See appendix.

negative commitment (*sdom min, asaṃvara*). A commitment to engage in nonvirtuous acts or a lack of restraint in performing nonvirtuous acts.

nonassociated formative factors (*ldan min 'du byed, viprayuktasaṃskāra*). *Also* nonassociated conditioning factors. Abbreviated as "factors." They encompass all impermanent phenomena not included in the class of material form or the class of mental phenomena. There are fourteen classified in the lower Abhidharma system: obtainment, nonobtainment, homogeneity, state of nondiscernment, meditative attainment of nondiscernment, meditative attainment of cessation, faculty of life-force, the four characteristics of conditioned things, the collection of nouns, the collection of phrases, the collection of letters. There are a further nine classified in the upper Abhidharma system: continuity, distinction, union, rapidity, sequence, time, location, enumeration, and assembly. *See the separate entries for each of these in this glossary.*

nonimplicative negation (*med dgag, prasajyapratiṣedha*). A negation where another positive phenomena is not implied in the wake of negating the negandum.

nonindicative form (*rnam par rig byed min pa, avijñaptirūpa*). Mental-object form that does not reveal or indicate to others the motivation that gave rise to it.

nonobtainment (*ma thob pa, aprāpta*). One of the fourteen factors of the lower Abhidharma system. The substance that constitutes the nonpossession of something that is not obtained.

nonthing (*dngos med, abhāva*). Any existent phenomena incapable of producing an effect. Equivalent to unproduced phenomena, unconditioned phenomena, and permanent phenomena.

noun (*ming, nāma*). Refers to a noun or name, and applies to that which expresses or indicates just the thing itself. There are two types: proper noun (*dngos ming*) and figurative noun (*btags ming*). Alternatively, there are arbitrary nouns initially formulated (*'dod rgyal gyi ming*) and subsequent derived nouns (*rjes su grub pa'i ming*).

object (*yul, viṣaya*). That which is cognized by awareness. Equivalent to existent, established basis, knowable object, and phenomenon.

obtainment (*thob pa, prāpta*). One of the fourteen factors of the lower Abhidharma system indicating a substance enabling one to possess that which is obtained within one's mindstream.

one (*gcig, ekatva*). *Also* same, identical. A phenomenon that is not diverse or multiple. One itself may have various types.

one and many (*gcig dang tha dad*). Identity and distinction. A dichotomy that contrasts singular entities with multiple entities.

permanent (*rtag pa, nitya, nityatva*). Phenomena that are not momentary. For example, unconditioned space that exists as the negation of physical obstruction. Equivalent to nonthing, unproduced phenomena, and unconditioned phenomena.

phenomena (*chos, dharma*). That which retains or holds its essential nature. Equivalent to existent, established basis, object, knowable object.

physical form (*gzugs, rūpa*). *See* form, material form.

predicated phrase (*tshig, sandhi*). Sound that expresses or indicates a thing that is qualified by an attribute; for example, "conditioned phenomena are perishable."

presence and absence (*rjes su 'gro ldog*). Chapa Chökyi Sengé's seventh topic. See appendix.

primary element (*'byung ba, bhūta*). The smallest discrete unit of material form. There are four: the primary elements of earth, water, fire, and wind.

principles of reasoning (*rigs pa'i bzhi*). There are four types: the principle of nature (*chos nyid kyi rigs*), the principle of function (*byed pa'i rigs*), the principle of dependence (*ltos pa'i rigs*), and the principle of evidence (*'thad pa'i rigs*).

pristine wisdom (*ye shes, jñāna*). *See* transcendent wisdom. The components of clarity and cognizance defining awareness that endures from beginningless time.

product (*byas pa, kṛta*). Phenomena produced through causes and conditions. Equivalent to thing, conditioned phenomenon, and impermanent.

productive cause (*skyed byed kyi rgyu*). *Also* efficient cause. The principal cause of its effect.

proof statement (*sgrub ngag, sādhanavākya*). A statement formulated as a syllogism to prove a thesis.

propensity (*bag chags, vāsanā*). *Also* latent imprint. Latent predispositions imprinted on consciousness that are the dormant potency derived from the internal mind's familiarity with external positive, negative, and neutral objects.

proper noun (*dngos ming*). This is the noun or name initially applied to signify an object, such as "fire." It is equivalent to "primary name" (*ming gsto bo*) and "arbitrary name" (*'dod rgyal gyi ming*).

rapidity (*mgyogs pa, java*). One of the nine factors of the upper Abhidharma system that indicates that cause and effect arise swiftly.

realms (*khams, dhātu*). *See also* element. The main abodes of sentient beings. There are three: the desire realm, form realm, and formless realm.

referent existence (*rdzas yod, dravyasat*). Equivalent to "substantial existent."

relation (*'brel ba, saṃbandha*). *See* contingent relation. It has two types: causal relation and intrinsic relation.

relation and absence of relation (*'brel ba dang ma 'brel ba*). Chapa Chökyi Sengé's fifth topic. See appendix.

repudiation (*sun 'byin, dūṣaṇa*). Criticism of another's standpoint that is seen to be erroneous.

root winds (*rtsa ba'i rlung lnga*). Primary wind-energies that move in the channels (*nāḍī*) that form the subtle physiology within the human body. These are five: life-sustaining wind (*srog 'dzin*), downward-voiding wind (*thur sel*), upward-flowing wind (*gyen rgyu*), pervading wind (*khyab byed*), and accompanying wind (*mnyam gnas*).

same (*gcig, ekatva*). *See* one.

same coexisting entity (*grub bde gcig, ekayogakṣema*). Phenomena that possess the same arising, endurance, and cessation.

same coexisting substance (*grub bde rdzas gcig*). Distinct impermanent substances that arise simultaneously, endure simultaneously, and cease simultaneously.

same conditioned entity (*'dus byas su gyur pa'i ngo bo gcig*). Impermanent entities that possess the same substance.

same entity (*ngo bo gcig, ekavastu, ekarūpata*). *Also* same nature, same essential nature. Different phenomena that possess the same entity or nature when their individual natures cannot exist separated from each other.

same intrinsic nature (*bdag nyid gcig*, *ekātmatā*). Equivalent to same nature, same entity.

same nature (*rang bzhin gcig*, *ekasvabhāva*). *Also* same essence. *See* same entity.

same substance (*rdzas gcig*, *ekadravya*). Different impermanent entities that possess the same substantial reality or energy, such as a pot and the impermanence of that pot.

same substantial type (*rdzas rigs gcig*). Different impermanent entities that arise from the same substantial cause, such as two leaves that grow on the same tree.

saṃsāra (*'khor ba*). The term is often applied to the world or universe in general, yet its essential meaning is the contaminated five aggregates, or the contaminated mind-body continuum that forms the basis of imputation of the self. It is the process by which beings cycle within the six realms of mundane existence, taking rebirth in one life after another by the power of karma and the afflictive emotions.

self-isolate (*rang ldog*, *svavyatireka*). The unique identity of something expressed as a double negative, or the opposite of that which is not that entity. For example, the self-isolate of a pot is the very pot itself and described as that which is not not that pot.

sense faculty (*dbang po*, *indriya*). There are six sense faculties: eye sense faculty, ear sense faculty, nose sense faculty, tongue sense faculty, body sense faculty, and mental sense faculty.

sequence (*go rim*, *anukrama*). One of the nine factors of the upper Abhidharma system indicating that cause and effect each arise discretely or separately.

six features of form (*gzugs rnam pa drug*). These are characteristics (*mtshan nyid*), location (*gnas*), augmented or diminished (*phan gnod*), basis of activity (*byed pa'i rten*), characteristic of activity (*byed pa'i mtshan nyid*), and ornamentation (*rgyan*).

six vessels (*snod drug*). These are the digestive tract (*pho ba*), intestines (*rgyu ma*), entrails (*long ga*), gallbladder (*mkhris thum*), urinary bladder (*lgang pa*), and seminal vesicle (*bsam se'u*).

śrāvaka (*nyan thos*). "Hearer" or "listener," those who first listen to instruction, then critically reflect on it, and finally meditate on it with a mind imbued with mental stability. Students of Buddhism who enter a systematic method (often described as a vehicle) in order to attain personal or individual enlightenment.

state of nondiscernment (*'du shes med pa*, *āsaṃjñika*). *Also* nondiscerning agent. One of the fourteen factors of the lower Abhidharma system, it refers to a type of substance that temporarily stops minds and mental factors in the mindstreams of gods born in [the realm of] nondiscernment.

substantial cause (*nyer len gyi rgyu*, *upādānakāraṇa*). That which produces an effect that exists within its own substantial continuum. That which produces, as its effect, the basis of the attributes instead of the attributes of the basis.

substantial existent (*rdzas yod*, *dravyasat*). From the perspective of the lower schools, in general any phenomena whose comprehension does not depend on understanding something other than it. For example, a pot is understood by the senses without need for understanding something else.

substantial phenomena and abstract phenomena (*rdzas chos ldog chos*). The second of Chapa Chökyi Sengé's eighteen topics. See appendix.

subtle particle (*rdul phra rab*, *paramāṇu*). The smallest discrete unit of matter.

suchness (*de bzhin nyid*, *tathātā*). The nature of reality. Suchness is synonymous with emptiness or selflessness and refers to the ultimate level of reality of any phenomenon.

syllogistic consequence (*thal 'gyur*, *prasaṅga*). *See* consequence.

tathāgata (*de bzhin gshegs pa*). Literally "one who proceeds (*gata*) in accordance with reality (*tathā*)." Applies to enlightened beings who continuously maintain within themselves the direct realization of the nature of selflessness or emptiness.

terrestrial divisions (*sa dum bu*, *bhūmikhaṇḍa*). The twelve sections of the earth corresponding to the twelve zodiac houses.

thing (*dngos po*, *bhava*). *Also* functional thing. That which is capable of producing an effect. Equivalent to product, conditioned phenomenon, and impermanent.

three conditions (*rkyen gsum*). These refer to the condition of the absence of the prior design of a creator (*g.yo ba med pa'i rkyen*), the condition of impermanence (*mi rtag pa'i rkyen*), and the condition of potentiality (*nus pa'i rkyen*).

three elements of the definition (*mtshan mtshon gzhi gsum*). These are: definiendum (*mtshon bya*), definiens (*mtshan nyid*), and the basis or instance (*mtshan gzhi*).

three general characteristics of form (*gzugs kyi spyi'i mtshan nyid*). These are: all form impedes objects, all form exhibits increase and decrease, and all form is capable of tactility.

three specific characteristics of form (*gzugs kyi rang gi mtshan nyid*). These are: the translucent sense organs (*dang ba*), the perceived objects of the sense organs (*dang ba'i gzung ba*), the perceived objects of mental cognition (*yid kyi gzung ba*).

three types of ascertainable object (*gzhal bya rnam gsum*). These are: (1) evident phenomena (*mthong ba mngon gyur*, *abhimukhi*), (2) slightly hidden phenomena (*cung zad lkog gyur*, *king zad lkog*), and (2) extremely hidden phenomena (*shin tu lkog gyur*, *atyarthaparokmena*).

three types of unconditioned phenomena (*'dus ma byas kyi chos gsum*). These are: space (*nam mkha'*, *ākāśa*), analytical cessation (*so sor brtags 'gog*, *pratisaṃkhyānirodha*), and nonanalytical cessation (*so sor brtags min gyis 'gog pa*, *apratisaṃkhyānirodha*).

three-nature theory (*mtshan nyid gsum*, *trisvabhāva*). A philosophical position of the Buddhist Mind Only school (Cittamātra), which asserts that all phenomena are included in the three natures (*svabhāva*): (1) dependent nature (*gzhan dbang*, *paratantra*), which refers to all impermanent entities, (2) perfected nature (*yongs sgrub, pariniṣarini*), which refers to emptiness or nonduality, and (3) imputed nature (*kun btags, parikalpita*), which refers to all permanent phenomena other than emptiness.

time (*dus, kāla*). One of the nine factors of the upper Abhidharma system that indicates "cause and effect occurring in a continuous stream."

transcendent wisdom (*ye shes, jñāna*). *See* pristine wisdom. The components of clarity and cognizance defining awareness that attain the uncontaminated state.

trimodal consequence (*thal gyur, prasaṅga*). *See* consequence.

truth body (*chos sku, dharmakāya*). One of the two bodies of a buddha, manifesting for the purpose of oneself. It has two types: the nature body and the pristine wisdom dharma body.

turning the wheel of Dharma (*chos kyi 'khor lo bskor ba, dharmacakrapravartana*) The eleventh of the twelve deeds of a fully enlightened being (Buddha). Describes the process of teaching Buddhist doctrine (Dharma), which is likened to setting a wheel in motion, by which the direct realization of the nature of reality or selflessness (Dharma) is established in the mindstreams of qualified students and constitutes their entry to the transcendent path.

twelve bases (*skye mched bcu gnyis, dvādaśāyatana*). Consist of the six objects: the bases of visible form, sound, smell, taste, tactility, and phenomena otherwise known as "mental objects"; and the six sense bases: of the eye, ear, nose, tongue, body, and mental sense base (which includes all primary minds and mental factors).

twelve links of dependent origination (*rten 'brel bcu gnyis*). These are: (1) ignorance, (2) karmic formation, (3) consciousness, (4) name and form, (5) the six sense bases, (6) contact, (7) feeling, (8) craving, (9) grasping, (10) existence, (11) birth, and (12) aging and death.

two types of logical entailment (*khyab mnyam rnam pa gnyis*). The fourteenth of Chapa Chökyi Sengé's eighteen topics. See appendix.

unconditioned phenomena (*'dus ma byas kyi chos, asaṃskṛtatva*). Phenomena devoid of the three characteristics of arising, ceasing, and enduring; phenomena not conditioned by time. There are eight types: suchness of auspicious things, suchness of inauspicious things, suchness of neutral things, space, analytical cessation, nonanalytical cessation, the immovable, cessation of discernment and feeling.

understanding existence and understanding nonexistence (*yod rtogs med rtogs*). The seventeenth of Chapa Chökyi Sengé's eighteen topics. See appendix.

understanding permanent phenomena and understanding things (*rtag rtogs dngos rtogs*). The eighteenth of Chapa Chökyi Sengé's eighteen topics. See appendix.

union (*'byor ba, yoga*). One of the nine factors of the upper Abhidharma system that indicates that the effect is in harmony with its cause.

unitary substantial particle (*rdzas rdul phra rab, dravyaparamāṇu*). The smallest discrete unit of matter, and eight such particles comprise an atom in the desire realm.

universal (*spyi*, *sāmānya*). A phenomenon that encompasses its specific particulars. For example, the universal or generic set "thing" encompasses or includes the particular or subset "water." There are three types: type universal (*rigs spyi*), composite universal (*tshogs spyi*), and object universal (*don spyi*).

universals and particulars (*spyi dang bye brag*). Chapa Chökyi Sengé's fourth topic. See appendix.

unproduced phenomena (*ma byas pa'i chos*, *akṛta*). Phenomena that are unarisen, or have not been generated by causes and conditions. Equivalent to nonthing, unconditioned phenomena, and permanent.

upper Abhidharma system (*mngon pa gong ma*). The system of Abhidharma set out in Asaṅga's *Compendium of Abhidharma* (*Abhidharmasamuccaya*) and its commentarial tradition.

Vaibhāṣika (*bye brag smra ba*). A follower of the *Great Treatise of Differentiation* (*Mahāvibhāṣā*), one who adheres to a tenet system taught in that work.

vital essence (*bla*). The vitality of the body.

wheel of Dharma (*chos kyi 'khor lo*, *dharmacakra*). The essence of Buddhist doctrine that refers to the cause, result, and identity of the state beyond sorrow (*nirvāṇa*). This doctrine, which encompasses the three higher trainings that lead to nirvāṇa, is said to resemble a wheel: the hub represents the higher training in ethics, the spokes represent the higher training in wisdom, and the outer rim the higher training in concentration. *See* turning the wheel of Dharma.

winds (*rlung*, *prāṇa*). In the three tantric categories of channels, winds, and drops, winds are the energy-winds that move in the network of channels that pervade the body's subtle physiology.

zodiac band (*khyim gyi go la*). The celestial band that marks the paths of the sun, moon, and planets along the ecliptic; the ecliptic.

zodiac constellations (*khyim rgyu skar*). The constellations that mark the twelve houses of the zodiac.

zodiac houses (*khyim gyi gzhi*). The twelve lunar mansions.

# Bibliography

## Sūtras and Tantras


*Condensed Kālacakra Tantra. Kālacakra-laghutantra. Dus 'khor bsdus rgyud.* Toh 362, Tantra, ka.

*Condensed Perfection of Wisdom. Prajñāpāramitāsañcayagāthā. Shes rab gyi pha rol du phyin pa sdud pa tshigs su bcad pa.* Toh 13, Prajñāpāramitā, ka.

*Ḍākā Ocean Tantra. Ḍākārṇava-tantra. Mkha' 'gro rgya mtsho'i rgyud.* Toh 372, Tantra, kha.

*Descent into Laṅka Sūtra. Laṅkāvatāra-sūtra. Lang kar gshegs pa'i mdo.* Toh 107, Sūtra, ca.

*Detailed Explanations of Discipline. Vinayavibhaṅga. 'Dul ba rnam par 'byed pa.* Toh 3, Vinaya, ca.

*Eight Thousand Verse Perfection of Wisdom Sūtra. Aṣṭasāhasrikāprajñāpāramitā-sūtra. Sher phyin brgyad stong pa.* Toh 12, Prajñāpāramitā, ka.

*Extensive Display Sūtra. Lalitavistara-sūtra. Rgya cher rol pa'i mdo.* Toh 95, Sūtra, kha.

*Flower Ornament Sūtra. Avataṃsaka-sūtra. Mdo sde phal po che.* Toh 44, Avataṃsaka, ka, ah.

*Hevajra Tantra. Hevajra-tantra. Skye'i rdo rje'i rgyud.* Toh 417, Tantra, nga.

*Nanda's Sūtra on Entering the Womb. Nandagarbhāvakrāntinirdeśa-sūtra. Dga' bo la mngal na gnas pa bstan pa'i mdo / Mngal gnas kyi mdo.* Toh 57, Ratnakūṭa, ga, 205–36.

*Nanda's Sūtra on Entering the Womb. Nandagarbhāvakrāntinirdeśa-sūtra. Dga' bo mngal du 'jug pa bstan pa'i mdo / Mngal du 'jug pa'i mdo.* Toh 58, Ratnakūṭa, ga, 237–48.

*Noble Dependent Origination Sūtra. Pratītyasamutpāda-sūtra. Rten cing 'brel bar 'byung ba'i mdo.* Toh 212, Sūtra, tsha.

*Questions of the Householder Ugra Sūtra. Ugraparipṛcchā-sūtra. Khyim bdag drag shul can gyis zhus pa'i mdo.* Toh 63, Ratnakūṭa, nga.

*Questions of Subāhu Sūtra. Subāhuparipṛcchā-sūtra. Lag bzang gis zhus pa'i mdo.* Toh 70, Ratnakūṭa, ca.

*Rice Seedling Sūtra. Śālistamba-sūtra. Sa lu ljang pa'i mdo.* Toh 210, Sūtra, tsha.

*Root Tantra of Mañjuśrī. Mañjuśrīmūla-tantra. 'Jam dpal rtsa rgyud.* Toh 543, Tantra, na.

*Saṃpuṭa Tantra. Saṃpūṭi-tantra. Yang dag pa sbyor ba zhes bya ba'i rgyud.* Toh 381, Tantra, kha.

*Saṃvarodaya Tantra. Saṃvarodaya-tantra. Bde mchog 'byung ba'i rgyud [Rgyud sdom 'byung].* Toh 373, Tantra, kha.

*Sūtra on the First Dependent Origination and Its Divisions. Pratītyasamutpādādināvibhaṅganirdeśa-sūtra. Rten cing ra 'byung ba dang po dang rnam par dbye ba bstan pa'i mdo.* Toh 211, Sūtra, tsha.

*Sūtra on the Foundation of Mindfulness. Saddharmānusmṛtyupasthāna-sūtra. Dam pa'i chos dran pa nyer gzhag.* Toh 287, Sūtra, sha.

*Sūtra on Gaganavarṇa's Patient Training. Gaganavarṇavinayakṣānti-sūtra. Nam mkha'i mdog gis 'dul ba'i bzod pa'i mdo.* Toh 263, Sūtra, 'a.

*Sūtra on the Meeting of the Father and Son. Pitāputrasamāgama-sūtra. Yab sras mjal ba'i mdo.* Toh 60, Ratnakūṭa, nga.

*Tantra of Great Power. Mahābala-tantra. Stobs po che'i rgyud.* Toh 391, Tantra, ga.

*Tantra Predicting Realization. Sandhivyākaraṇa-tantra. Gsang ba 'dus pa'i bshad rgyud dgongs pa lung ston.* Toh 444, Tantra, ca.

*Twenty-Five Thousand Verse Perfection of Wisdom Sūtra. Pañcaviṃśatisāhasrikāprajñāpāramitā-sūtra. Sher phyin nyi khri.* Toh 9, Prajñāpāramitā, ka.

*Unraveling the Intention Sūtra. Saṃdhinirmocana-sūtra. Dgongs pa nges par 'grel pa'i mdo.* Toh 106, Sūtra, ca.

*Vajra Garland Tantra. Vajramālā-tantra. Gsang ba 'dus pa'i bshad rgyud rdo rje phreng ba.* Toh 445, Tantra, ca.

*Vajraḍāka Tantra. Vajraḍāka-tantra. Rdo rje mkha' 'gro'i rgyud.* Toh 371, Tantra, *kha.*

*Vajrapāṇi Initiation Tantra. Vajrapāṇi-abhiṣekamahā-tantra. Lag na rdo rje dbang bskur ba'i rgyud chen po.* Toh 496, Tantra, da.

## Canonical Treatises

Āryadeva. *Establishing the Proofs Refuting Distorted Views. Skhalitapramathanayuktihetusiddhi. 'Khrul pa bzlog pa'i rigs pa gtan tshigs grub pa.* Toh 3847, Madhyamaka, tsha.

———. *Four Hundred Stanzas. Catuḥśataka. Dbu ma bzhi brgya pa.* Toh 3846, Madhyamaka, tsha.

———. *Lamp on the Compendium of Practice. Caryāmelāpakapradīpa. Spyod pa bsdus pa'i sgron ma.* Toh 1803, Tantra, ngi.

Asaṅga. *Bodhisattva Ground. Bodhisattvabhūmi. Byang chub sems dpa'i sa.* Toh 4037, Cittamātra, wi.

———. *Compendium of Abhidharma. Abhidharmasamuccaya. Chos mngon pa kun las btus pa.* Toh 4049, Cittamātra, ri.

———. *Compendium of Ascertainment. Viniścayasaṃgraha. Rnam par gtan la dbab pa bsdu ba.* Toh 4038, Cittamātra, zhi–zi.

———. *Compendium of Bases. Vastusaṃgraha. Gzhi bsdu ba.* Toh 4039, Cittamātra, wi.

———. *Facts of the Grounds. Bhūmivastu. Sa'i dngos gzhi.* Toh 4035–37, Cittamātra, tshi, dzi, wi. The *Bhūmivastu* includes three sections: *Bāhubhūmika* (Toh 4035), *Śrāvakabhūmi* (Toh 4036), *Bodhisattvabhūmi* (Toh 4037).

———. *Śrāvaka Ground. Śrāvakabhūmi. Nyan thos kyi sa.* Toh 4036, Cittamātra, dzi.

———. *Yogācāra Ground. Yogācārabhūmi. Rnal 'byor spyod pa'i sa.* Toh 4035, Cittamātra, *tshi.*

Avalokitavrata (Spyan ras gzigs brtul zhugs). *Explanation of the Lamp of Wisdom. Prajñāpradīpaṭīkā. Shes rab sgron ma'i 'grel bshad.* Toh 3859, Madhyamaka, wa.

Bhāvaviveka. *Blaze of Reasoning. Tarkajvālā. 'Grel pa rtog ge 'bar ba.* Toh 3856, Madhyamaka, dza.

———. *Heart of the Middle Way. Madhyamakahṛdaya. Dbu ma'i snying po'i tshig le'ur byas pa.* Toh 3855, Madhyamaka, dza.

———. *Lamp of Wisdom. Prajñāpradīpa. Rtsa she'i 'grel pa shes rab sgron ma.* Toh. 3853, Madhyamaka, tsha.

Bodhibhadra (Byang chub bzang po). *Discourse on the Compendium of the Essence of Wisdom. Jñānasārasamuccayanibandhana. Ye shes snying po kun las btus pa bshad sbyar.* Toh 3852, Madhyamaka, tsha.

Candragomin. *Praise of All Tathāgatas. Sarvatathāgatastotra. Bshags bstod.* Toh 1151, Stotra, ka.

Candrakīrti. *Autocommentary on Entering the Middle Way. Madhyamakāvatārabhāṣya. Dbu ma la 'jug pa'i rang 'grel.* Toh 3862, Madhyamaka, 'a.

———. *Clear Words. Prasannapadā. Dbu ma rtsa ba'i 'grel pa tshig gsal ba zhes bya ba.* Toh 3860, Madhyamaka, 'a.

———. *Commentary on the Four Hundred Stanzas. Catuḥśatakaṭīkā. Byang chub sems pa'i rnal 'byor spyod pa bzhi brgya pa'i rgya cher 'grel pa.* Toh 3865, Madhyamaka, ya.

———. *Entering the Middle Way. Madhyamakāvatāra. Dbu ma la 'jug pa.* Toh 3861, Madhyamaka, 'a.

———. *Treatise on the Five Aggregates. Pañcaskandhaprakaraṇa. Phung po lnga'i rab tu byed pa.* Toh 3866, Madhyamaka, ya.

Candranandana (Zla ba la dga' ba). *Moonlight Commentary on the Essence of the Eight Branches. Candrikāprabhāsa-aṣṭāṅgahṛdayavivṛtti. Yan lag brgyad pa'i snying po'i rnam 'grel zla zer.* Toh 4312, Cikitsāvidyā, ko.

Daśabalaśrīmitra (Stobs bcu dpal bshes gnyen). *Ascertaining Conditioned and Unconditioned Phenomena. Saṃskṛtāsaṃskṛtaviniścaya. 'Dus byas dang 'dus ma byas rnam par nges pa.* Toh 3897, Madhyamaka, ha.

Devendrabuddhi (Lha dbang blo). *Commentary on the Exposition of Valid Cognition. Pramāṇavārttikapañjikā. Tshad ma rnam 'grel gyi dka' 'grel.* Toh 4217, Pramāṇa, che.

Dharmakīrti. *Ascertainment of Valid Cognition. Pramāṇaviniścaya. Tshad ma rnam nges.* Toh 4211, Pramāṇa, ce.

———. *Autocommentary on the Exposition of Valid Cognition. Pramāṇavārttikavṛtti. Tshad ma rnam 'grel rang 'grel.* Toh 4216, Pramāṇa, ce.

———. *Drop of Reasoning. Nyāyabinduprakaraṇa. Rigs pa'i thigs pa rab tu byed pa.* Toh 4212, Pramāṇa, ce.

———. *Exposition of Valid Cognition. Pramāṇavārttika. Tshad ma rnam 'grel gyi tshig le'ur byas pa.* Toh 4210, Pramāṇa, ce.

———. *Logic of Debate. Vādanyāya. Rtsod pa'i rigs pa.* Toh 4218, Pramāṇa, che.

Dharmamitra. *Clarifying Words. Prasphuṭapadā. Phar phyin gyi 'grel bshad tshig gsal.* Toh 3796, Prajñāpāramitā, nya.

Dharmottara. *Explanation of Ascertainment of Valid Cognition. Pramāṇaviniś-cayaṭīkā. Tshad ma rnam nges kyi 'grel bshad 'thad ldan.* Toh 4227, Cittamātra, dze.

———. *Extensive Commentary on the Drop of Reasoning. Nyāyabinduṭīkā. Rigs pa'i thigs pa'i rgya cher 'grel pa.* Toh 4230–31, Cittamātra, we.

———. *Proof of Momentariness. Kṣaṇabhaṅgasiddhi. Skad cig ma 'jig pa grub pa.* Toh 4253, Cittamātra, zhe.

Dignāga. *Autocommentary on the Analysis of Objective Conditions. Alambana-parīkṣāvṛtti. Dmigs pa brtag rang 'grel.* Toh 4206, Cittamātra, ce.

Guṇaprabha. *Exposition of the Five Aggregates. Pañcaskandhavivaraṇa. Phung po lnga'i rnam par 'grel pa.* Toh 4067, Cittamātra, si.

———. *Vinaya Sūtra. Vinayasūtra. Dul ba'i mdo.* Toh 4117, Vinaya, wu.

Jinaputra/Yaśomitra. *Elucidation of the Compendium of Abhidharma. Abhidharma-samuccayavyākhyā. Mngon pa kun las btus pa'i rnam par bshad pa.* Toh 4054, Cittamātra, li.

———. *Explanation of the Treasury of Knowledge. Abhidharmakośaṭīkā-spuṭārtha. Chos mngon pa mdzod kyi 'grel bshad (Chos mngon pa mdzod kyi 'grel bshad don gsal).* Toh 4092, Abhidharma, gu–ngu.

Kamalaśīla. *Commentary on the Compendium on Reality. Tattvasaṃgrahapañjikā. Tshad ma'i de kho na nyid bsdus pa'i dka' 'grel.* Toh 4267, Pramāṇa, je.

———. *Commentary on the Ornament of Madhyamaka. Madhyamakālaṃkāra-pañjikā. Dbu ma rgyan gyi dka' 'grel.* Toh 3886, Madhyamaka, sa.

———. *Extensive Commentary on the Rice Seedling Sūtra. Śālistambakaṭīkā. Sa lu'i ljang pa'i mdo'i rgya cher 'grel.* Toh 4001, Sūtra, ji.

———. *Light of the Middle Way. Madhyamakāloka. Dbu ma snang ba.* Toh 3887, Madhyamaka, sa.

———. *Proof of the Absence of Inherent Existence. Sarvadharmāsvabhāvasiddhi. Rang bzhin med pa nyid du grub pa.* Toh 3889, Madhyamaka, sa.

———. *Stages of Meditation. Bhāvanākrama. Sgom pa'i rimpa.* Toh 3908, Madhya-maka, ki.

Kātyāyanīputra. *Great Treatise on Differentiation. Mahāvibhāṣā. Chos mngon pa bye brag tu bshad pa chen po.* Translated from the Chinese by Lozang Chopak [Hwatsun] (1945–49). Book (*Bris ma*) of the Gaden Potang Library. Published by Krung go'i bod rig pa'i dpe skrun khang, 2011. Chinese edition translated by Xuanzang, Taishō 27, no. 1545.

Lakṣmi. *Clarifying the Meaning of the Five Stages. Pañcakramaṭīkākramārtha-prakāśikā. Rim lnga'i 'don gsal bar byed pa.* Toh 1842, Tantra, chi.

Mahākoṣṭha (Mahākauṣṭhila). *Statements on the Types of Being. Saṃgītiparyāya. 'Gro ba'i rnam grangs.* One of the seven treatises of the Sarvāstivāda Abhidharma Piṭaka. Chinese edition composed by Śāriputra, translated by Xuanzang, T26, no. 1536, in 20 fascicles.

Maitreya. *Distinguishing the Middle from the Extremes. Madhyantavibhaṅga. Dbus dang mtha' rnam par 'byed pa'i tshig le'ur byas pa.* Toh 4021, Cittamātra, phi.

———. *Ornament of the Mahāyāna Sūtras. Mahāyānasūtrālaṃkāra. Theg pa chen po mdo sde'i rgyan.* Toh 4020, Cittamātra, phi.

Matṛceṭa (Slob dpon Pha khol/dpa' bo). *Core Summary of the Eight Branches. Aṣṭāṅgahṛdayasaṃhitā. Yan lag brgyad pa'i snying po bsdus pa.* Toh 4310, Cikitsāvidyā, he.

Maudgalyāyana. *Presentation on Causation. Kāraṇaprajñapti. Rgyu gdags pa.* Toh 4087, Abhidharma, i.

———. *Presentation on the World. Lokaprajñapti. 'Jig rten gzhags pa.* Toh 4086, Abhidharma, i.

Muktikalaśa. *Commentary on Proof of Momentariness. Kṣaṇabhaṅgasiddhivivaraṇa. Skad cig ma 'jig pa grub pa'i rnam par 'grel pa.* Toh 4254, Pramāṇa, zhe.

Nabidharma. *Summary of Negations. Piṇḍanivartana. Ldog pa bsdus pa bstan pa'i tshig le'ur byas pa.* Toh 4293, Śabdavidyā, she.

Nāgabodhi (Klu'i byang chub). *Presentation of the Guhyasamāja Sādhana. Samājasādhanavyavasthāli. Rnam gzhag rim pa: 'Dus pa'i sgrub thabs rnam par gzhag pa'i rim pa.* Toh 1809, Tantra, ngi.

Nāgārjuna. *Compendium of Sūtras. Sūtrasamuccaya. Mdo kun las btus pa.* Toh 3934, dbu ma, ki.

———. *Extensive Explanation on the Rice Seedling Sūtra. Śālistambakasūtraṭīkā. Sa lu lhang pa'i rgya cher bshad pa.* Toh 3986, Sūtra, ngi.

———. *Five Stages. Pañcakrama. Rim pa lnga pa.* Toh 1802, Tantra, ngi.

———. *Fundamental Treatise on the Middle Way. Mūlamadhyamakakārikā. Dbu ma rtsa ba'i tshig le'ur byas pa shes rab ces bya ba.* Toh 3824, Madhyamaka, tsa.

———. *Precious Garland of Advice to the King. Ratnāvalī. Rgyal po la gtam bya ba rin po che'i phreng ba.* Toh 4158, Jātaka, ge.

Nāropa. *Clear Compilation of the Five Stages. Pañcakramasaṅgrahaprakāśa. Rim pa lnga bsdus pa gsal ba.* Toh 2333, Tantra, zhi.

Prajñāvarman. *Exposition of the Collection of Aphorisms. Udānavargavivaraṇa. Ched du brjod pa'i tshoms kyi rnam par 'grel pa.* Toh 4100, Abhidharma, thu.

Pṛthivībandhu (Sa'i rtsa lag). *Explication of the Five Aggregates. Pañcaskandhabhāṣya. Phung po lnga'i bshad pa.* Toh 4065, Cittamātra, si.

Puṇḍarīka. *Stainless Light: Commentary on Kālacakra. Vimalaprabhākālacakratantraṭīkā. Rtsa ba'i rgyud kyi rjes su 'jug pa stong phrag bcu gnyis pa dri ma med pa'i 'od ces bya ba / bsdus pa'i rgyud kyi rgyal po dus kyi 'khor lo'i 'grel bshad.* Toh 1347, Tantra, tha–da.

Pūrṇavardhana. *Investigating Characteristics: Explanatory Commentary on the Treasury of Knowledge. Abhidharmakośaṭīkā-lakṣaṇānusāriṇī. Mdzod kyi 'grel bshad mtshan nyid rjes 'brang.* Toh 4093, Abhidharma, cu.

Śākyamati. *Commentary on the Exposition of Valid Cognition. Pramāṇavārttikaṭīkā. Tshad ma rnam 'grel gyi 'grel bshad.* Toh 4220, Pramāṇa, nye.

Śāntarakṣita. *Autocommentary on the Ornament of Madhyamaka. Madhyamakālaṃkāravṛtti. Dbu ma rgyan gyi rang 'grel.* Toh 3885, Madhyamaka, sa.

———. *Compendium on Reality. Tattvasaṃgraha. De kho na nyid bsdus pa.* Toh 4226, Pramāṇa, ze.

———. *Establishment of Reality. Tattvasiddhiprakaraṇa. De kho na nyid grub pa zhes bya ba'i rab tu byed pa.* Toh 3708, Tantra, tsu.

———. *Ornament of Madhyamaka. Madhyamakālaṃkāra. Dbu ma rgyan.* Toh 3884, Madhyamaka, sa.

Śāntideva. *Compendium of Training. Śikṣāsamuccaya. Blsab pa kun las btus pa.* Toh 3939, Madhyamaka, khi.

———. *Engaging in the Bodhisattva's Deeds. Bodhisattvacaryāvatāra. Byang chub sems dpa'i spyod pa la 'jug pa.* Toh 3871, Madhyamaka, la.

Sthiramati. *Specific Explanation of the Five Aggregates. Pañcaskandhaprakaraṇavaibhāṣya. Phung po lnga'i rab tu byed pa bye brag tu bshad pa.* Toh 4066, Cittamātra, shi.

Śubhagupta (Dge srungs). *Proof of External Objects. Bāhyārthasiddhi. Phyi rol gyi don grub pa.* Toh 4244, Pramāṇa, zhe.

Vasubandhu. *Commentary on the Distinguishing the Middle from the Extremes. Madhyāntavibhāgabhāṣya. Dbu dang mtha' rnam par 'byed pa'i 'grel ba.* Toh 4027, Cittamātra, bi.

———. *Explanation on the First Dependent Origination and Its Divisions. Pratītyasamutpādādivibhaṅganirdeśa. Rten cing 'brel par 'byung ba dang po'i rnam par dbye ba bshad pa.* Toh 3995, Sūtra, chi.

———. *Rational System of Exposition. Vyākhyāyukti. Rnam bshad rigs pa.* Toh 4061, Cittamātra, shi.

———. *Treasury of Knowledge. Abhidharmakośa. Chos mngon pa'i mdzod kyi tshig le'ur byas pa.* Toh 4089, Abhidharma, ku.

———. *Treasury of Knowledge Autocommentary. Abhidharmakośabhāṣya. Chos mngon pa'i mdzod kyi bshad pa.* Toh 4090, Abhidharma, ku–khu.

———. *Treatise on the Establishment of Karma. Karmasiddhiprakaraṇa. Las grub rab tu byed pa.* Toh 4062, Cittamātra, shi.

———. *Treatise on the Five Aggregates. Pañcaskandhaprakaraṇa. Phung po lnga rab byed.* Toh 4059, Cittamātra, shi.

———. *Twenty Verses. Viṃśatika. Nyi shu pa.* Toh 4056, Cittamātra, shi.

———. *Twenty Verses Autocommentary. Viṃśatikāvṛtti. Nyi shu pa'i rang 'grel.* Toh 4057, Cittamātra, shi.

Vinītadeva ('Dul ba lha). *Commentary on the Analysis of Objective Conditions. Alambanaparīkṣāṭīkā. Dmigs brtag 'grel bshad.* Toh 4265, Pramāṇa, zhe.

Wonch'uk (Wen tshegs). *Commentary to the Unraveling the Intention Sūtra. Saṃdhinirmocanasūtraṭīkā. Dgongs 'grel gyi mdo'i rgya cher 'grel.* Toh 4016, Sūtra, thi.

## Noncanonical Treatises

*Explanatory Tantra*. *Gso rig rgyud bzhi bshad rgyud*. One of the Four Medical Tantras.

*Four Medical Tantras*. *Gso rigs pa'i rgyud bzhi*. Chengdu: Mi rigs dpe skrun khang, 2007.

*Four Medical Tantras of Dratang*. *Grwa thang rgyud bzhi*. Peking: Mi rigs dpe skrun khang, 2005 (TBRC: W29627).

*Instruction Tantra*. *Gso rig rgyud bzhi man ngag rgyud*. One of the Four Medical Tantras.

*Somarāja: A Medical Treatise*. *Sman dpyad zla ba'i rgyal po*. Chengdu: Mi rigs dpe skrun khang, 2007.

## Other Sources Cited

Anacker, Stephen. 2005. *Seven Works of Vasubandhu: The Buddhist Psychological Doctor*. Delhi: Motilal Banarsidass Publishers.

Āryabhaṭa. *Āryabhaṭayaṃ*. *Skar rtsis kyi bstan bcos*. See Clark 2006.

Asaṅga. 2001. *Abhidharmasamuccaya*: *The Compendium of the Higher Teaching (Philosophy)*. Originally translated into French by Walpola Rahula. English translation from the French by Sara Boin-Webb. Fremont, CA: Asian Humanities Press.

Blumenthal, James. 2004. *The Ornament of the Middle Way: A Study of the Madhyamaka Thought of Śāntarakṣita*. Ithaca, NY: Snow Lion Publications.

Buddhaghosa. 1991. *The Path of Purification (Visuddhimagga)*. Translated by Bhikkhu Ñāṇamoli. Kandy, Sri Lanka: Buddhist Publication Society.

Buswell, Robert, E. Jr., and Donald S. Lopez, Jr. 2014. *The Princeton Dictionary of Buddhism*. Princeton, NJ: Princeton University Press.

Cabezón, José Ignacio. 1992. *A Dose of Emptiness: An Annotated Translation of the sTong thun chen mo of mKhas-grub dGe-legs-dpal-bzang*. Albany: State University of New York Press.

Chapa Chökyi Sengé (Phywa pa chos kyi seng ge) (1109–69). 2002. *Root and Commentary of the Condensed Epistemology Eliminating the Darkness of the Mind*. *Tshad ma bsdus pa yid kyi mun sel rtsa 'grel*. TBRC vol. 2077, W12171.

———. 2006. *Epistemology Eliminating the Darkness of the Mind* (*Tshad ma yid kyi mun sel*). First arrangement (*Phyogs bsgrigs thengs dang po*) of the Collected Works (*gsungs 'bums*) of the Kadampas, vol. 9. Chengdu: Si khron mi rigs dpe skrun khang.

Clark, Walter Eugene. 2006. *Āryabhaṭīya of Āryabhaṭa: An Ancient Indian Work on Mathematics and Astronomy*. Chicago: University of Chicago Press, 1930; reprint: Kessinger Publishing.

Cleary, Thomas, trans. 1993. *The Flower Ornament Scripture: A Translation of the Avatamsaka Sutra*. Boston: Shambhala Publications.

Conze, Edward, trans. 1975a. *The Large Sutra on Perfect Wisdom with the Divisions of the Abhisamayālaṅkāra*. Berkeley: University of California Press.

———. 1975b. *The Perfection of Wisdom in Eight Thousand Lines and Its Verse Summary*. Bolinas, CA: Four Seasons Foundation.

Cox, Collette. 1995. *Disputed Dharmas: Early Buddhist Theories on Existence*. Tokyo: International Institute of Buddhist Studies.

Dalai Lama Tenzin Gyatso. 1999. *Ethics for the New Millennium*. New York: Riverhead Books.

———. 2002. *Essence of the Heart Sutra*. Boston: Wisdom Publications.

———. 2005. *The Universe in a Single Atom: The Convergence of Science and Spirituality*. New York: Harmony Books.

———. 2011. *Beyond Religion: Ethics for a Whole World*. Boston: Houghton Mifflin Harcourt.

———. 2017. *The Life of My Teacher: A Biography of Ling Rinpoché*. Boston: Wisdom Publications.

Desi Sangyé Gyatso. 2010. *Mirror of Beryl: A Historical Introduction to Tibetan Medical Science*. Translated by Gavin Kilty. Boston: Wisdom Publications.

Dhonden, Yeshe. 2000. *Healing from the Source*. Ithaca, NY: Snow Lion Publications.

Dreyfus, Georges, B. J. 1997. *Recognizing Reality: Dharmakirti's Philosophy and Its Tibetan Interpretations*. New York: State University of New York Press.

Dunne, John. 2004. *Foundations of Dharmakīrti's Philosophy*. Boston: Wisdom Publications.

Engle, Artemus. 2009. *Inner Science of Buddhist Practice: Vasubandhu's Summary of the Five Heaps with Commentary by Sthiramati*. Ithaca, NY: Snow Lion Publications.

———. 2016. *The Bodhisattva Path to Unsurpassed Enlightenment: A Complete Translation of the Bodhisattvabhūmi*. Boulder, CO: Snow Lion Publications.

Garfield, Jay. 1997. "Vasubandhu's Treatise on the Three Natures: A Translation and Commentary." *Asian Philosophy* 7.2: 133–54.

Hattori, Masaaki. 1968. *Dignāga, on Perception: Being the Pratyakṣapariccheda of Dignāga's Pramāṇasamuccaya from the Sanskrit Fragments and the Tibetan Version*. Cambridge, MA: Harvard University Press.

Henning, Edward. 2007. *Kālacakra and the Tibetan Calendar*. New York: American Institute of Buddhist Studies.

Hopkins, Jeffrey. 1999. *Emptiness in the Mind Only School*. Berkeley: University of California Press.

———. 2003. *Maps of the Profound: Jam-yang-shay-ba's Great Exposition of Buddhist and Non-Buddhist Views on the Nature of Reality*. Ithaca, NY: Snow Lion Publications.

———. 2007. *Nāgārjuna's Precious Garland*. Ithaca, NY: Snow Lion Publications.

———. 2011. *Tibetan-Sanskrit-English Dictionary*. Taipei: Digital Archives Section, Library and Information Center of Dharma Drum Buddhist College.

Jackson, Roger. 1993. *Is Enlightenment Possible?: Rgyal Tshab Rje on Knowledge, Rebirth, No-Self, and Liberation*. Ithaca, NY: Snow Lion Publications.

Jha, Ganganatha, trans. 1986. *The Tattvasaṁgraha of Shāntarakṣita: With the Commentary of Kamalashīla*, vol. 2. Delhi: Motilal Banarsidass.

Jinpa, Thupten. 2002. *Self, Reality and Reason in Tibetan Philosophy: Tsongkhapa's Quest for the Middle Way*. New York: Routledge Curzon.

———, ed. 2016. *Seven Pramāṇa Works of Dharmakīrti. Dpal chos kyi grags pa'i tshad ma sde bdun*. Delhi: Institute of Tibetan Classics.

Khedrup Norsang Gyatso. 2004. *Ornament of Stainless Light: An Explanation of the Kālacakra Tantra*. Translated by Gavin Kilty. Boston: Wisdom Publications.

Klein, Anne. 1987. *Knowledge and Liberation: Tibetan Buddhist Epistemology in Support of Transformative Religious Experience*. Ithaca, NY: Snow Lion Publications.

———. 1991. *Knowing, Naming, and Negation: A Sourcebook of Tibetan Sautrātika*. Ithaca, NY: Snow Lion Publications.

Larson, Gerald James, and Ram Shankar Bhattacharya, eds. 1987. *Encyclopedia of Indian Philosophies*, vol. 4, *Sāṃkhya: A Dualist Tradition in Indian Philosophy*. Princeton, NJ: Princeton University Press.

McGovern, William Montgomery. 1923. *A Manual of Buddhist Philosophy*. London: Kegan Paul.

Ñāṇamoli, Bhikkhu, trans. 2015. *The Middle-Length Discourses of the Buddha: A Translation of the Majjhima Nikāya*, 4th ed. Edited and revised by Bhikkhu Bodhi. Somerville, MA: Wisdom Publications.

Negi, J. S. 1993. *Tibetan-Sanskrit Dictionary*. Sarnath, India: Central Institute of Higher Tibetan Studies.

Ngawang Tashi (Ngag dbang bkra shis). 2001. *Sé Collected Topics. Bse bsdus grwa: Tshad ma'i dgongs don rtsa 'grel mkhas pa'i mgul rgyan*. Beijing: Mi rigs dpe skrun khang.

Nyanaponika Thera. 1998. *Abhidhamma Studies: Buddhist Explorations of Consciousness and Time*. Boston: Wisdom Publications.

Onada, Shunzo. 1996. "*bsDus grwa* Literature." In *Tibetan Literature: Studies in Genre,* edited by J. I. Cabezon and R. R. Jackson, 187–201. Ithaca, NY: Snow Lion Publications.

Perdue, Daniel, E. 1992. *Debate in Tibetan Buddhism*. Ithaca, NY: Snow Lion Publications.

———. 2014. *The Course in Buddhist Reasoning and Debate: An Approach to Analytical Thinking Drawn from Indian & Tibetan Sources*. Boston: Snow Lion Publications.

Popper, Karl. 1982. *Unended Quest: An Intellectual Biography*. London: Routledge.

Potter, Karl H., ed. 1977. *Encyclopedia of Indian Philosophies*, vol. 2, *Indian Metaphysics and Epistemology: The Tradition of Nyāya-Vaiśeṣika Up to Gaṅgeśa*. Princeton, NJ: Princeton University Press.

Powers, John. 1992. "Lost in China, Found in Tibet: How Wonch'uk Became the Author of the Great Chinese Commentary." *Journal of the International Association of Buddhist Studies* 15, no. 1: 95–103.

Pruden, Leo M., trans. 1988. *Abhidharmakośa Bhāṣyam of Vasubandhu.* Translated into French by Louis de la Vallée Poussin with English version by Leo M. Pruden. 4 vols. Freemont, CA: Asian Humanities Press.

Rinchen, Geshe Sonam, with Ruth Sonam. 2008. *Āryadeva's Four Hundred Stanzas on the Middle Way with Commentary by Gyeltsap.* Ithaca, NY: Snow Lion Publications.

Rogers, Katherine, M. 2014. *Tibetan Logic.* Boulder, CO: Shambhala Publications.

Siderits, Mark, and Shōryū Katsura. 2013. *Nāgārjuna's Middle Way: The Mūlamadhyamakakārikā.* Somerville, MA: Wisdom Publications.

Thurman, Robert, A. F., ed., trans. 2004. *The Universal Vehicle Discourse Literature (Mahāyānasūtrālamkāra) by Maitreyanātha/Āryāsanga Together with Its Commentary (Bhāsya) by Vasubandhu.* Translated from the Sanskrit, Tibetan, and Chinese by L. Jamspal, R. Clark, J. Wilson, L. Zwilling, M. Sweet, and R. Thurman. New York: American Institute of Buddhist Studies at Columbia University.

Tillemans, Tom, J. F. 1997. "On a Recent Translation of the *Saṃdhinirmocanasūtra.*" *Journal of the International Association of Buddhist Studies* 20.1: 153–64. https://journals.ub.uni-heidelberg.de/index.php/jiabs/article/viewFile/8858/2765.

———. 1999. *Scripture, Logic, Language: Essays on Dharmakīrti and His Tibetan Successors.* Boston: Wisdom Publications.

———. 2000. *Dharmakīrti's Pramāṇavārttika: An Annotated Translation of the Fourth Chapter (parārthānumāna),* vol. 1 (k. 1–148). Vienna: Verlag der Österreichischen Akademie der Wissenschaften.

———. 2016. "What Happened to the Third and Fourth Lemmas in Tibet?" In *How Do Mādhyamikas Think?: And Other Essays on the Buddhist Philosophy of the Middle.* Somerville, MA: Wisdom Publications.

Tsongkhapa. 1978–79. *Annotations on the Summary of Five Stages. Rim lnga bsdus pa'i zin bris.* In Collected Works (Shöl edition), 6:455–500. New Delhi, India: Mongolian Lama Guru Deva (TBRC W635).

———. 2010. *Ocean of Reasoning: A Great Commentary on Nāgārjuna's Mūlamadhyamakakārikā.* Translated by Geshe Ngawang Samten and Jay L. Garfield. Oxford: Oxford University Press.

———. 2013. *A Lamp to Illuminate the Five Stages: Teachings on Guhyasamāja Tantra.* Translated by Gavin Kilty. Boston: Wisdom Publications.

Wilson, Joe Bransford. 1992. *Translating Buddhism from Tibetan.* Ithaca, NY: Snow Lion Publications.

Yoysuya, Kodo. 1999. *A Critique of Svatantra Reasoning by Candrakirti and Tsongkha-pa.* Stuttgart: Franz Steiner Verlag.

# Index

## C

## O



## P

## Q

## Y



## Z

# About the Authors

HIS HOLINESS THE DALAI LAMA is the spiritual leader of the Tibetan people, a Nobel Peace Prize recipient, and a beacon of inspiration for Buddhists and non-Buddhists alike. He is admired also for his more than four decades of systematic dialogues with scientists exploring ways to developing new evidence-based approaches to alleviation of suffering and promoting human flourishing. He is the co-founder of the Mind and Life Institute and has helped to revolutionize traditional Tibetan monastic curriculum by incorporating the teaching of modern science. He is a great champion of the great Indian Nalanda tradition of science, philosophy, and wisdom practices.

THUPTEN JINPA is a well-known Buddhist scholar and has been the principal English-language translator for His Holiness the Dalai Lama for more than three decades. A former monk and a Geshe Lharampa, he also holds a BA in philosophy and a PhD in religious studies, both from Cambridge University. He is the author and translator of many books and teaches at McGill University in Montreal.

IAN JAMES COGHLAN (Jampa Ignyen) trained as a monk at Sera Jé Monastery, completing his geshé studies in 1995, and holds a PhD in Asian Studies from La Trobe University. Currently he is a translator for the Institute of Tibetan Classics and an adjunct research fellow at SOPHIS, Monash University. He lives with his partner, Voula, and dog, Pila, in Churchill, Victoria, Australia.

# About the Authors

# What to Read Next from the Dalai Lama

**Approaching the Buddhist Path**
**Library of Wisdom and Compassion series, volume 1**

**Buddhism**
*One Teacher, Many Traditions*

**The Compassionate Life**

**Essence of the Heart Sutra**
*The Dalai Lama's Heart of Wisdom Teachings*

**The Good Heart**
*A Buddhist Perspective on the Teachings of Jesus*

**Imagine All the People**
*A Conversation with the Dalai Lama on Money, Politics, and Life as It Could Be*

**Kalachakra Tantra**
*Rite of Initiation*

**The Life of My Teacher**
*A Biography of Kyabjé Ling Rinpoché*

**Meditation on the Nature of Mind**

**The Middle Way**
*Faith Grounded in Reason*

**Mind in Comfort and Ease**
*The Vision of Enlightenment in the Great Perfection*

**MindScience**
*An East-West Dialogue*

**Practicing Wisdom**
*The Perfection of Shantideva's Bodhisattva Way*

**Sleeping, Dreaming and Dying**
*An Exploration of Consciousness*

**The Wheel of Life**
*Buddhist Perspectives on Cause and Effect*

**The World of Tibetan Buddhism**
*An Overview of Its Philosophy and Practice*

## About Wisdom Publications

Wisdom Publications is the leading publisher of classic and contemporary Buddhist books and practical works on mindfulness. To learn more about us or to explore our other books, please visit our website at wisdompubs.org or contact us at the address below.

Wisdom Publications
199 Elm Street
Somerville, MA 02144 USA

We are a 501(c)(3) organization, and donations in support of our mission are tax deductible.

Wisdom Publications is affiliated with the Foundation for the Preservation of the Mahayana Tradition (FPMT).